PLANT PHYSIOLOGY

PLANT PHYSIOLOGY

With Reference to the Green Plant

2nd Edition

– Volume 3 –

EDWIN CYRUS MILLER

Professor, of Plant Physiology,
Kansas State College of Agriculture and Applied Science, and Plant Physiological, Kansas,
Agricultural Experimental Station

BIOTECH

2005

Biotech Books

Delhi - 110 035

ISBN 81-7622-154-6 (SET)
ISBN 81-7622-155-4 (Volume 1)
ISBN 81-7622-156-2 (Volume 2)
ISBN 81-7622-157-0 (Volume 3)

Published by : **BIOTECH BOOKS**
1123/74, Tri Nagar,
DELHI – 110 035
Phone: 27382765
e-mail: biotechbooks@yahoo.co.in

Showroom : 4762-63/23, Ansari Road, Darya Ganj,
NEW DELHI - 110 002
Phone: 23245578, 23244987

Printed at : **Chawla Offset Printers**
New Delhi – 110 052

PRINTED IN INDIA

To my wife

DELLA SLYE MILLER

PREFACE TO THE SECOND EDITION

In the preparation of this edition, various topics have been enlarged or revised to include the investigations and findings that have been made since the publication of the first edition. Certain topics that were overlooked in the preparation of that edition have been added. New material has been presented under appropriate headings.

Although no claim to completeness is made, it has been the effort of the author to present the more important citations relating to each subject. He has tried to present impartially the data that have been reported. He does not feel qualified to speak with authority on a majority of the topics, and his own ideas, if presented, thus would be of little consequence.

In order to conserve space, the questions that concluded each chapter in the first edition have been omitted.

The author is indebted in various ways to several of his associates at the Kansas State College of Agriculture and Applied Science, and he takes this occasion to thank them. He wishes to thank especially Hugh G. Gauch, for his invaluable aid in assembling and typing the bibliographies and other portions of the manuscript, and for aid in editing it; John C. Frazier, for aid in editing the manuscript, for helpful advice, and for unselfish service in many ways; Dr. Fritz Moore, head of the Department of Modern Languages, and Mrs. Irene Moore, for checking, respectively, the German and the French references; and Miss Esther Lewis, for clerical help in preparing the last half of the manuscript.

The author is especially indebted to Professor L. E. Melchers, head of the Department of Botany and Plant Pathology, Kansas State College of Agriculture and Applied Science, for his kindness in allotting clerical and stenographic help for the revision of this text.

EDWIN C. MILLER.

KANSAS STATE COLLEGE OF AGRICULTURE AND APPLIED SCIENCE,

PREFACE TO THE FIRST EDITION

There has existed for some time a need for an advanced text in plant physiology. The various texts by European investigators and teachers, although summarizing the work that has been done on the Continent, have failed to cover adequately the contributions of American and English plant physiologists. These contributions during the past two decades have been outstanding and dominate in many cases the work along certain lines. A summary of these contributions should be available to students, teachers, and investigators in the field of plant physiology.

It has been the intention of the author to bridge this gap in the literature and to summarize in this book the more important findings of English, American, and Continental plant physiologists. The material has been assembled in such form as to be available as a text for upperclassmen and graduate students and at the same time to be sufficiently comprehensive as a reference book for investigators in plant physiology.

The work has been confined entirely to the physiology of the green plant, as it has been thought that the processes and phenomena of the fungi belong, at present at least, to the province of plant pathology and allied fields. The discussions and summaries of the various investigators have been used freely, and in many cases their phraseology has been altered but little. The author has tried, however, in each instance to indicate the source of the information presented. Some topics are discussed more thoroughly than others. This, however, is to be expected as an author naturally is more familiar with those subjects which are in the field of his research.

With but few exceptions the references cited at the end of each chapter have been mentioned or discussed in the text. It is considered that these references will be of value to those who desire to familiarize themselves more thoroughly with any of the topics than they are herein discussed. The questions following each chapter cover all the topics discussed in that portion of the text. It has been the experience of the author that these are valuable to the student as an aid in guiding him to a mastery of the subject matter.

This text is the outgrowth of twenty years' experience as a teacher of plant physiology in the Kansas State Agricultural College and as plant physiologist of the Kansas Agricultural Experiment Station. The material included herein has been presented to classes composed for the most part of juniors, seniors, and graduate students. The favorable manner in which the subject matter and its method of presentation have been received in the classroom has prompted the author to present this volume to the public.

I am indebted to many of my associates for assistance in the preparation of this text. Most of the drawings were made by Mr. S. Fred Prince, formerly biological artist at the Kansas State Agricultural College. I am indebted to Miss Nora E. Dalbey, associate professor of botany in the Kansas State Agricultural College, for certain drawings and for helpful suggestions in the arrangement of the illustrations. I wish to thank Dr. Frank M. Schertz of the U. S. Bureau of Chemistry and Soils, Dr. John W. Shive of the New Jersey Agricultural Experiment Station, Dr. W. E. Tottingham of the University of Wisconsin, Dr. F. F. Blackman of Botany School, Cambridge, England, and Professor W. E. Davis of the Kansas State Agricultural College for permission to use certain figures as indicated in the text.

I am grateful to Mr. J. C. Frazier, formerly of the Department of Botany and Plant Pathology, Kansas State Agricultural College, for helpful advice in the preparation of the manuscript and to Dr. R. P. White, formerly of the same department, for criticism and advice in the preparation of Chap. II. I wish to thank Dr. J. V. Cortelyou, professor of modern languages at Kansas State Agricultural College, and Miss Madge Wardell, instructor in French at the same institution, for checking respectively the German and French references. I wish also to acknowledge the most valuable aid and advice of Miss Nellie Jacobs, clerk in the Department of Botany and Plant Pathology, Kansas State Agricultural College, in the preparation of the manuscript.

I am especially indebted to the following of my colleagues in the Kansas State Agricultural College for their constructive criticism in the preparation of the text: S. C. Salmon, professor of farm crops; Dr. P. L. Gainey, professor of bacteriology; R. J. Barnett, professor of horticulture; Dr. H. H. Haymaker, professor of plant pathology; W. E. Davis, professor of plant physiology; Dr. F. C. Gates, professor of taxonomy and ecology; Dr. F. L. Duley, professor of soils; Dr. C. W. Colver, professor of organic chemistry; and Dr. E. L. Tague, associate professor of chemistry. The author alone, however, is responsible for the statements in the text.

EDWIN C. MILLER.

KANSAS STATE AGRICULTURAL COLLEGE,

CONTENTS

CHAPTER I

CHAPTER II

CHAPTER III

CHAPTER IV

CHAPTER V

CHAPTER VI

CHAPTER VII

CHAPTER VIII

CHAPTER IX

CHAPTER X

CHAPTER XI

CHAPTER XII

PAGE

CHAPTER XIII

CHAPTER XIV

CHAPTER XII

THE TRANSLOCATION OF MATERIALS IN PLANTS

I. NEED OF TRANSLOCATION

Most of the green plants are highly specialized organisms. They are thus composed of numerous interdependent parts or organs that perform special functions. Each organ must be supplied by other organs with the materials necessary for its functions, while it in turn furnishes other organs with certain materials needed by them. Thus the epidermal cells of the roots absorb the compounds that enter the plant from the soil. These substances are for the most part not used entirely or directly by these cells that absorb them but are moved to other parts of the plant where they are utilized in the various life functions. For example, only a very small portion of the water that is absorbed by the epidermal cells of the roots is utilized by these cells. The greater portion of it is moved through the roots and stem to supply the needs of their cells and then on to the leaves to replace the water lost by transpiration. The movement of sugar is another striking example of translocation. This carbohydrate is manufactured only in the green cells of the leaf but is needed by every living cell of the plant for the manufacture of other foods, for respiration, for osmotic effects, and for other purposes. Sugar thus must be moved from the leaves to every living cell in the plant. The reserve foods of plants are practically all stored in places that are some distance removed from the regions in which they will eventually be utilized. Thus when the demand for them arises, they must be translocated to those parts of the plant in which they are needed.

II. TISSUES CONCERNED

In the broadest sense, all the living cells and certain of the nonliving cells play a role in the movement of materials in plants. In the lower plants where materials need to be moved for only short distances, no specialization of cells for translocation occurs and all cells are equally concerned in the process. Also, if, in the higher plants, materials are moved for relatively short distances only, the cells show no special adaptations for translocation. Thus the materials taken from the soil by the epidermal cells of the root are moved to the endodermis across the parenchyma cells of the cortex. Likewise, the raw materials needed by the chlorenchyma cells of the leaf, as well as their manufactured products, are moved across them to and from the fibrovascular bundles. These parenchyma cells, as well as those of the pith and cortex of the stem, show no special adaptations for translocation other than that they are possessed of thin cell walls that are readily permeable to solvent and solutes.

The tissues of the green plant that are specialized for the translocation of materials are the xylem, the phloem and possibly the latex system.

A. The Xylem

The cells of the xylem especially concerned in translocation are the tracheids, the vessels or tracheae, the wood parenchyma, or xylem parenchyma, and the xylem medullary rays. The tracheids are elongated cells with tapering ends. They may have relatively thin walls that are strengthened by spiral or annular thickenings, or the walls may be thick, lignified, and possessed of numerous pits, generally of the bordered type. The tracheids are living cells for varying periods after their formation but eventually lose their protoplasmic contents so that they consist only of their cell walls. The vessels or tracheae, according to Eames and MacDaniels (1925), include those xylem elements which have a large lumen and marked perforations, especially in the end cell walls. The development of these perforations in the end walls brings a series of cells into a definite tubelike system. The conduction in the vessels is thus direct and is in contrast with the indirect lines of conduction in a group of tracheids. The vessels frequently have thin walls that are strengthened by annular or spiral thickenings. In other cases, the walls are thickened but generally not to the degree of the walls of the tracheids, and the vessels then are possessed of pits in their lateral walls. These pits are usually more numerous and smaller than those of the tracheids. The vessels also lose their protoplasmic contents at an early period after their development. Malhotra (1931) found the average length in centimeters of tracheae was 50 for grape; 49 for peach; 46 for apricot; 42 for plum; 34 for prune, lemon, and Delicious apple; 24 for cherry; 20 for walnut; and 17 for Bartlett pear. The "wood" or "xylem parenchyma" is the name given to those living cells which are associated with the tracheids and vessels. These cells may be thin walled or thick walled and supplied with simple pits. They retain their protoplasmic contents and remain alive for a long period of time.

The xylem rays, or wood rays, are ribbonlike strips of tissue that extend radially into the xylem. These strips are composed of parenchyma cells that have their long axes at right angles to the long axes of the tracheids, vessels, and wood parenchyma. These cells are living cells and contain their protoplasmic contents for a long period of time, and, as mentioned in Chap. I, they remain alive longer than any other known plant cells. The ray cells have cell walls of varying degrees of thickness and have numerous simple pits on all their walls. According to Eames and MacDaniels (1925), the xylem rays may be one, two, or several cells in width, and in height they may range from one cell to many and thus may vary in height from a fraction of a millimeter to 8 to 10 mm. in extreme cases. In a given stem some of the xylem rays may extend from the cambium to the pith, while others extend from the cambium to varying distances in the xylem.

In this general description of the conducting elements of the xylem, it should be noted that those tracheids and vessels which have lost their protoplasmic contents and are yet concerned in conduction are in contact with some parts of their wall surface with one or more living cells. Abundant pits are also present in these contact areas. The living cells with which these nonliving elements are in contact may be either the xylem parenchyma or xylem rays, or both these elements.

B. The Phloem

The phloem is composed of the sieve tubes, companion cells, phloem parenchyma, and phloem rays. A sieve tube is an elongated living cell with a thin cellulose wall. It has a large vacuole, and the cytoplasm forms a thin peripheral layer. No nucleus is present in a sieve tube. The sieve tubes are joined end to end and are arranged more or less definitely in longitudinal rows. Their protoplasmic contents are connected through pores or perforations in their end and lateral walls. These pores are located in more or less definite areas which are termed "sieve plates." Although these sieve plates may occur over the end and side walls, they are limited to a considerable extent to the end walls (Fig. 34). The cell sap of many of the sieve tubes contains a relatively large amount of protein compounds of a slimy nature (Hill, 1901, 1908). The outstanding characteristics of the sieve tubes are thus the pores in their walls, which are perforated with conspicuous cytoplasmic strands, the absence of a nucleus, and the relatively large amount of protein compounds in their vacuoles.

Crafts (1931, 1933) found that young sieve tubes are nucleated and show the characteristics of ordinary living cells. As they mature the nuclei disintegrate, and the protoplasm apparently changes in its organization and becomes more permeable.

The companion cells are elongated, slender, parenchymatous cells that have a small vacuole and dense protoplasmic contents with a prominent nucleus. One companion cell is associated with each sieve tube, but it is often difficult to determine the sieve tube with which a given companion cell is associated (Fig. 34). A companion cell has numerous simple pits connecting with the sieve tube with which it is associated.

The phloem parenchyma cells have dense protoplasmic contents and possess simple pits in both their longitudinal and end walls. The relative amount of phloem parenchyma varies greatly with different types of phloem and with different plants. In some cases it makes up the greater portion of the phloem, while in others it is entirely absent. The phloem rays have the same general structure and arrangement as the xylem rays. It should be noted in this connection that the elements which compose the phloem are all living cells, which is in marked contrast with the elements of the xylem tissue.

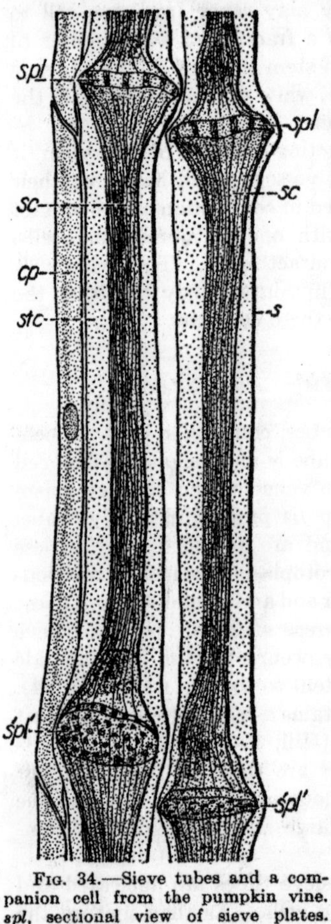

Fig. 34.—Sieve tubes and a companion cell from the pumpkin vine. *spl*, sectional view of sieve plates. *s'pl'*, surface view of sieve plates. *sc*, protoplasmic contents of sieve tube, here contracted to the center of the cell. *stc*, cell cavity caused by the contraction of the protoplasmic contents. *s*, cell wall of sieve tube. *cp*, companion cell.

C. The Latex System

Certain families of the Angiosperms including the *Papaveraceae, Lobeliaceae, Campanulaceae, Asclepiadaceae, Apocynaceae, Euphorbiaceae, Moraceae, Araceae,* and *Musaceae,* are characterized by the possession of an apparent additional conducting system in the form of laticiferous elements or latex tubes, which owe their name to the milky appearance of their contents. There are two distinct types of laticiferous organs: the articulated latex tubes or vessels and the nonarticulated latex tubes or latex cells. According to Eames and MacDaniels (1925), the former originate from rows of meristematic cells by either the complete or partial absorption of the separating walls early in the development of the vessel. In the mature plant a much branched anastomosing system is formed by the joining of more or less parallel ducts through connecting living cells and by the penetration of lateral outgrowths into the surrounding tissue. The latex cell on the other hand is structurally a single cell which as the plant grows develops into a branching system ramifying throughout the entire plant body. In this method of growth the tips of the growing cell penetrate between the cells of the tissue after the manner of a parasitic fungus. In this type of laticiferous duct, there is no anastomosing as in the latex vessel.

The laticiferous structures are living and are coenocytic in nature. The liquid that fills these structures is called "latex," is under pressure, and is apparently the cell sap of these cells. In color it is white, yellow, orange, or blood red depending on the plant in which it occurs. The latex is composed of water in which are dissolved or suspended mineral salts, sugars, proteins, tannins, gums, resins, alkaloids, starch, caoutchouc and proteolytic enzymes. It is commercially important as the source of opium, papain, India rubber, gutta-percha, and chicle (Haberlandt, 1914).

According to Barnes (1930) the function of conduction has been ascribed to the laticiferous system because of its relation to the nutritive cells of the leaves and the abundance of food in the latex. The carbohydrate and nitrogenous foods sometimes compose 30 per cent of the dry matter therein; they are most abundant at the beginning of active growth and development and are least in amount when growth is checked and a resting period is at hand. It may be, however, that the latex is concerned in the storage of materials in the plant or serves as a protection against predatory insects or animals. A thorough review of the physiology of latex is given by Moyer (1937).

III. MOVEMENT OF MATERIALS

A. Translocation of Water

The movement of water from the roots to the leaves, especially in the case of trees, has been of much general interest for several centuries. One of the most common questions asked by the general public of those individuals whom they know to be interested in the functions of plants concerns the manner in which the sap is moved through the stems of plants. It should be stated here that at present no answer that is entirely satisfactory can be given to this question. Although a large amount of experimental work has been undertaken to demonstrate the manner of water movement in plants, the evidence that has been obtained is in no way conclusive in many respects. In the following topics it is the intention to relate not only those facts which have been definitely established in regard to the translocation of water through the plant but also some of the theories concerning certain phases of this process that as yet have not been satisfactorily explained by the experimental evidence at hand.

1. Pathway of Water Movement.—The pathway through which most of the water travels from its entrance into the root until it is lost from the leaves into the atmosphere has been definitely established. The water enters the epidermal cells of the root, crosses the cells of the cortex, and enters the endodermis whence it moves into the lumina of the tracheids and vessels of the xylem of the root. It moves upward in these elements in the roots and thence through the same unit structures in the stem to the fibrovascular bundles of the leaf. From the xylem elements of the fibrovascular bundles of the leaf, it moves through the various chlorenchyma cells to those bordering on the intercellular air spaces or on the substomatal air cavities. Here it is lost from the cell walls in the form of vapor and escapes through the stomata to the external atmosphere. Some water may travel upward through the pith and cortex of the root and stem of certain plants, but apparently the amount moved in these regions is relatively small, except in young plants, compared with that moved through the channels that have just been mentioned. The lateral movement of water apparently takes place for the most part through the xylem rays and phloem rays in the case of young dicotyledons and through the general parenchymatous tissue of the stem in the case of the monocotyledons.

The pathway of water in the xylem has been determined directly by Peirce (1929), Harvey (1930, 1931), and Baker and James (1933) with the use of colored liquids and has been proved in an indirect manner by noting the lack of sufficient water to supply the leaves when the xylem elements are blocked or severed. The portion of the xylem concerned in water conduction has, however, in the case of woody plants been in considerable dispute. The work of Overton (1925), Overton and Smith (1926), and MacDougal, Overton, and Smith (1929) has thrown much light upon this controversy. Their work indicates that the portion of the xylem concerned in water translocation varies markedly in different plants. Thus by using acid fuchsin it was found for the willow (*Salix lasiolepis*) that the colored liquid rose in every annual layer in stems that were 11 or more years of age. It was also found that the liquid passed only through the summerwood of each layer. The colored zones were less than 1 mm. in thickness, while the unstained zones were from 2.5 to 4.5 mm. wide. In the alder (*Alnus oregona*) the same concentric zonation of water conduction is shown, but the movement of water in this stem does not take place through the late summerwood but is confined exclusively to the early formed springwood. In the walnut where heart- and sapwood are present all the conduction is in the sapwood through the late summer- and early springwood. In the pine (*Pinus radiata*) there is no localization of conduction in the xylem and the water moves through practically all parts of each annual ring.

Bowen (1931, 1933) found that the amount of water conducted over the external surface of moss plants, growing in either damp or dry areas, exceeded that conducted internally. The water that was conducted externally ascended in the form of capillary films between the leaves and the stem.

2. Water-conducting Cells.—The absorbing power of the epidermal cells of the root for water is always greater than that of the soil solution under the general conditions of plant growth. There is an increasing gradient in the absorbing power of the cortical cells for water from the epidermal cells of the root toward the endodermis. The absorbing power of the endodermis, however, in all the cases that have been investigated, drops considerably below that of the cortical cells bordering on it.

The cells concerned in the absorption and translocation of water across the root to the xylem elements of the central cylinder are all living cells. The tracheids and vessels of the root and stem with the exception of those which have been recently formed are, however, devoid of protoplasmic contents.

a. Gases in Conducting Cells.—The amount of water contained in these nonliving water-conducting elements has long been a matter of dispute. The observation of the contents of the tracheids and vessels under natural conditions is a difficult task, and in many of the observations the tissues examined were placed under abnormal conditions. It is thus impossible to determine whether the conditions of the cell contents observed were those normally prevailing or the result of the conditions imposed by the experiment. It has long been known that certain of the elements of the xylem contain gas bubbles with the liquid therein contained. At one time it was supposed that the presence of these bubbles lessened the total weight of a theoretical column of water and thus facilitated the ascent of sap in stems containing these so-called "Jaminian chains." It is now, however, almost universally recognized that the presence of a gas bubble in a tracheid or vessel at once eliminates that element from the conducting system. Those xylem elements which no longer function in the translocation of water are filled with gases, or the gases are intermingled as bubbles with the liquids yet remaining in them. The condition and composition of the gaseous mixture which fills these woody elements have been studied in detail by MacDougal (1925, 1926, 1927). He found that the gases of the central cylinder in the walnut may include carbon dioxide at a partial pressure 600 times as great as in the atmosphere, but with an oxygen pressure of only one-half that of the atmosphere. The gases in the central cylinder of the Monterey pine included carbon dioxide at a pressure 200 times as great as the atmosphere with the oxygen pressure reduced to one-fourth of that normally present in the air. The amounts of these gases vary during the season of year. Thus the carbon dioxide content of the oak was 3 per cent in October and 9 per cent in June. In the case of the willow the carbon dioxide content was 13.2 per cent in June and 8.4 per cent in August. The oxygen content of the walnut stem varied from 8.8 per cent in June to 12 to 13 per cent in July.

Gibbs (1930) found in jack pine that the heartwood contained approximately three times the amount of gas found in the sapwood. The heartwood, however, contained only 12 per cent of water as compared to 52 per cent in the sapwood. Chase (1934) in a study of the composition of gas drawn from the trunks of elm, oak, cottonwood, and white pine observed the following facts: (1) The percentage of carbon dioxide is highest during the growing season and is lowest in winter; (2) the percentage of oxygen varies inversely with the percentage of carbon dioxide at all times; (3) the volume or percentage of these two gases is less than their total in the atmosphere; and (4) the gas from the outer rings of sapwood is lower in carbon dioxide and higher in oxygen than that from the heartwood.

According to MacDougal, Overton, and Smith (1929), pressures and suctions on the gaseous system within the trunk of a tree are readily transmitted vertically for distances many times the length of the vessels. Tangential transmission, however, of suctions and pressures is at a very slow rate and even slower in a radial direction. The internal gaseous system of the tree is, however, in communication with the atmosphere radially through the cambium and vertically through the ends of the branches. The tensions in the air system have been shown to vary from less than ½ atmosphere to not to exceed 2 atmospheres. The relative volumes of the water and air systems within a tree are subject to variations during the season, so that specific conducting elements may at one time be partially or wholly filled with gas and at another time entirely filled with water.

Gibbs (1935) considered that the actual amount of gas in trees examined by him varied but little during diurnal changes but that it did vary with the seasonal fluctuation in water content. Priestley (1935) believed that under a deficit of water during the summer a large proportion of the water-conducting elements of trees become filled with gases. He did not propose, however, the means by which these vessels might be again refilled with water.

Using a specially devised method, Haines (1935) found that gases were sometimes present and sometimes absent in the conducting xylem of the twigs of elder, lilac, hawthorn, and ash. The presence or absence of gases depended largely on conditions external to the plant. He believed that a mechanism must exist in these plants for causing gas in bubble form to go again into solution.

b. Water in Conducting Cells.—According to MacDougal, Overton, and Smith (1929), it has been observed by Holle (1915) that when leaves were completely wilted the water columns remained intact in the conducting elements. These authors also stated that Bode (1923) in observations of the wilted plants of *Tradescantia, Cucurbita, Impatiens, Helianthus, Sambucus*, and *Phaseolus* noted that gas bubbles entered the conducting elements of these parts only when living cells in contact with them were touched or wounded. Bode thus concluded that bubbles may appear in conducting elements containing water only by being drawn in or by wounding. MacDougal, Overton, and Smith (1929) stated that in the stems examined by them the elements that are operative in conduction contain exclusively water, while those that are inoperative in conduction may contain gas exclusively or water and gas bubbles in the form of chains.

Gibbs (1935) considered his data indicated that there is a continuous water column in the conducting elements of the birch, larch, balsam, and spruce. The hardwoods examined by him showed a maximum water content in spring and a sharp drop in the summer. The degree of this variation differed from year to year. In some cases this loss of water from the woods during the summer was not replaced until the following spring. The softwoods showed no marked seasonal changes in their content of water.

It is thus apparently well established that the elements of the xylem concerned in conduction are completely filled with water and that on that account there is a continuous column of water through them from the endodermis to the chlorenchyma cells of the leaf. As mentioned in Chap. II, it has been shown that there is an increasing gradient of absorbing power for water in the chlorenchyma cells of the leaf from the fibrovascular bundles of the leaf toward its margins.

The evidence thus indicates that water columns are continuous in the plant from the soil particles through the epidermal cells, cortex, and endodermis of the root;

through the conducting xylem of the root and stem to the veins of the leaves; and thence through the cell walls and vacuoles of the chlorenchyma cells to the surfaces of the menisci formed in the submicroscopical cavities of the cell walls bordering on the intercellular spaces in the leaf. These water connections extending throughout the plant have been termed by MacDougal (1925 to 1926) the "hydrostatic system" of the plant. Woodhouse (1933) proposed the collective term "sap hydraulics" to designate the entrance of water into the plant, its passage through the stem, and its eventual loss from the stem and leaves.

3. Ascent of Water.—In attempting to determine the forces that are concerned in the ascent of water in plants, the investigator is handicapped because of the relatively small number of the external manifestations of the action of these forces as well as the difficulty encountered in measuring and correctly interpreting these manifestations. The external manifestations by which the action of the hydrostatic system of plants in elevating water may be estimated are, according to MacDougal (1925), (*a*) the amount of the intake of water, (*b*) the amount of transpiration, (*c*) the amount of water forced from the plant in the form of liquid, and (*d*) the registration of the daily reversible variations in stems as a result of the altered cohesive tensions in the ascending column of water. Some of these manifestations are more evident under conditions of low evaporation, while others are more evident under conditions of high evaporation. The forces that are concerned in the ascent of water in plants are apparently active all the time but are more manifest under one of these conditions than under the other. On that account, it is convenient in discussing the ascent of water to consider (*a*) the ascent of water in plants under conditions of low evaporation and (*b*) the ascent of water in plants under the conditions of high evaporation.

a. Low Evaporation.—Under conditions that are favorable for both low transpiration and the absorption of water from the soil, water is forced into and up through the xylem elements under considerable pressure. This fact is made manifest by the exudation of water from the stem and leaves of intact plants, the water so exuded being designated as "water of guttation." This action is also shown when bore holes are made into the stems and branches or when they are severed, since under these conditions water is forced out or exuded from the hole or stump apparently by some pressure within the stem or root. This exudation of water may also occur from perforations in the stem or from the stump of stems or branches under conditions of high transpiration, although a considerable period of time must elapse before the exudation develops. The force or pressure with which water is exuded from holes made in the stem or from the fresh stumps of stems or branches has been termed "root pressure," "sap pressure," "endodermal pressure," "bleeding pressure," or "exudation pressure." These terms have also been applied to the forces causing the exudation of water of guttation. The term "root pressure" has been the one most commonly used to designate not only the force with which water is being pushed up through the stem of the intact plant but also the force with which water is being exuded from some injured part. The term has been used because it has generally been considered that the force under consideration is generated in certain of the root cells. In some cases, however, as will be mentioned later, the exudation of water from cut or injured stems is brought about by forces generated in the living cells contiguous to the exuding region. In such cases "root pressure" is not the appropriate term to use. When water is exuded in considerable quantities from the stumps of cut stems or branches or from bore holes in the stem, as in the case of the grape vine, sugar maple, *Begonia*, or *Fuchsia*, the process has been termed "bleeding."

Lowry and Tabor (1931) reported that the stub of a single corn plant at the silking stage will exude in 3 days, under favorable conditions, more than 500 cc. of sap. By renewing the cut surface at frequent intervals, as much as 1,700 cc. of sap may be collected during a 15-day period. Some cut stems exude only a relatively small quantity of water, but there is no correlation between the amount of water exuded and the pressure with which it is being forced out of the stem.

The exudation pressures have been measured by manometers fitted to a bore hole in the side of large stems or to the stump, as in the case of the smaller stems or branches. Although many studies have been made of exudation pressures, the data concerning their value are very meager. Boehm (1892) found exudation pressures as high as 9 atmospheres in *Aesculus hippocastanum*, and Molisch (1902) reported pressures of over 6 atmospheres in the native trees of Austria. Clark (1874) reported root pressures of 2 to 2.5 atmospheres for *Betula lenta*, about 1 atmosphere for the sugar maple and 1.25 atmospheres for the grape vine. Eckerson (1908) studied the exudation pressures of the stumps of over 20 herbaceous plants growing under greenhouse conditions and found that they ranged from 0.43 to 1.4 atmospheres. MacDougal (1925) found that pressures of from 2 to 4 atmospheres developed in bore holes in the side of the trunk of Monterey pine. The pressures observed will depend to a considerable degree, however, upon the general conditions prevailing at the time the observations are made. White (1938) devised methods for measuring the root pressure of the excised roots of the tomato plant and found that pressures of 90 lb. per sq. in. were developed. This pressure is sufficient to raise water to a height of 200 ft. Root pressure thus may be a more important factor in sap movement than has generally been assumed.

It has generally been assumed that the exudation of water from cut stems and branches is due to the action of the cells of the root. There is considerable evidence, however, that exudation pressures are due, in many cases at least, to the action of cell masses contiguous to the cut surfaces (Molisch, 1902). It was considered by MacDougal (1925, 1926) that his work on the exudation pressures of the Monterey pine and the cactus *Carnegiea gigantea* indicated that the pressures developed were due to the local action of the living cells. Thus in the pine no root pressure could be observed, but exudation pressures as high as 4 atmospheres were observed in the stem. In the case of the cactus, exudation pressures as high as 2 atmospheres could be developed when no roots were present, and root pressure was thereby entirely absent. In the cases of guttation from epithem hydathodes, or in the case of continuous loss of water from a cut stem for a considerable period, as in the grape vine, the exudation is apparently caused by the action of certain cells of the root or stem other than those contiguous to the regions of water loss.

Heyl (1933) decided from experiments with the decapitated stems of *Ricinus* and *Brassica* that the intensity of bleeding closely paralleled soil temperatures. Experiments with narcotics showed definitely that the mechanism of bleeding is located in the roots and that it involves active secretion of water by living cells. The manner of exudation may not be the same for all plants, but this mechanism evidently depends upon complicated processes that are not purely osmotic in nature. Thus, James and Baker (1933) stated that there is no adequate evidence that the liquid escaping from the cut surfaces of plants issues from the vessels. They believed that any pressure on the sap that exudes from cut surfaces must be transmitted through the living cells and not through the vessels. On the other hand Ingold (1935) found indications that bleeding from birch trees occurs only when the vessels are injured.

In this connection the sap flow of the sugar maple should be mentioned. As is well known, large quantities of sap will escape from this tree during late winter and early spring through tap holes made in the lower part of the trunk. The causes for this escape of sap have long been a subject of discussion and are as yet little understood. Jones, Edson, and Morse (1903) found that the root pressure was very infrequent and that when it was observable it did not exceed 1.5 pounds. On the other hand, negative pressures were frequent during the time of flow and reached values as high as 6 pounds. They thus considered that the flow of sap cannot be attributed to root pressure, since the small positive pressure thus developed is too infrequent and slight to be the main causative factor. Gas expansion in the conducting elements occurs during the time of flow, but the force thus generated is considered to be inadequate to account for the results obtained. Jones, Edson, and Morse considered that the activities of the protoplasm of the living cells of the xylem are the dominant factors in causing this sap flow. They considered that the changes in temperature and the changes in gas pressure within the trunk are the stimuli that cause the protoplasm to become active.

Jones and Bradlee (1933) held that the periodic bleeding occurring at the close of winter, when the enclosing bark and the wood layer are tapped, represents an outpouring at the point of least resistance. In an untapped tree, the internal pressures aid in forcing the water to the extreme upper sections of the tree before transpiration begins.

Since the gradient of absorbing power for water increases from the epidermal cells of the root through the cortex to the endodermis, the migration of water from the epidermis to the endodermis can be attributed to osmotic and imbibitional forces, provided that a saturation deficit is maintained at the endodermis. In the leaves in the cases that have been examined, there is an increasing gradient of the absorbing power of the cells for water from the xylem elements of the veins across the chlorenchyma cells to the intercellular spaces. Here the movement of water from the veins toward the intercellular spaces may be explained on the basis of osmotic and imbibitional forces, provided a saturation deficit is maintained at the cell surfaces bordering upon the intercellular spaces.

The manner in which the water is transferred from the cortical cells of the root to the xylem elements and then forced upward is not known. The theories that have been proposed to account for root pressure are purely speculative or at the best are based upon very meager evidence. Little or nothing is known concerning the osmotic conditions of the parenchyma cells and xylem elements within the central cylinder of the root and in the conducting tissues of the stem. In the few studies that have been made, the suction force of the parenchyma cells of the central cylinder as well as the surrounding endodermal cells was found to be lower than that of the cortical cells. Whether this osmotic relationship is of common occurrence in plants, however, is not known. If it were, the entrance of water into the central cylinder by the ordinary methods of diffusion would not occur.

The structure of the endodermis suggests that it plays an important part in the entrance of water. The term "endodermal pressure," which is sometimes applied to the force that causes the exudation of water from cut stems, was so named because it was assumed that the endodermal cells force the water into the central cylinder of the root. If the endodermis is designated as the organ that forces water into the central cylinder and thence up through the stem, the nature of its action, according to our present knowledge of osmotic relationships, can be expressed only by saying that it is a process of "secretion" or "excretion." The use of these terms, however, is only an indirect way of stating that we know nothing concerning the manner in which the endodermis functions in the entrance of water into the central cylinder of the root.

Priestley (1920) and Priestley and Armstead (1922) formulated a theory of the mechanism by which exudation pressures might be developed. They considered that the solute concentration of the xylem, in certain regions at least, is maintained at a higher level than that of the adjoining cells, and that this supply of solutes is maintained by the endodermis, which prevents their outward leakage. They obtained evidence which indicated that the parenchyma cells around the xylem but within the central cylinder have specially permeable protoplasmic membranes that permit the exchange of organic substances between their cell sap and the solution in the xylem tubes. This exchange of solutes appears to involve both physical adsorption and physiological absorption. If the concentration of solutes in the xylem elements is greater than that in the cells adjoining them, it is easy to explain the entrance of water into them and the development of an exudation pressure.

Priestley and Armstead were able under suitable conditions to produce exudation pressures in stem and leaves as well as in the roots. In some instances, however, these pressures in the stem and leaves were only temporary, since the absence of a functional endodermis permits too rapid a leakage of the necessary solutes from the xylem for these pressures to be long maintained. In order to account for the fact that the water of guttation contains only a relatively small amount of solutes, these authors assumed that the solutes are absorbed from the xylem by the bordering cells during the upward passage of the sap. They thus assumed that in one region the sap in the xylem vessels may be receiving solutes from the surrounding cells, while in another region the same solute may be diffusing away from the xylem into the neighboring tissue.

b. High Evaporation.—As is well known, the greater portion of the water lost from plants is lost under conditions of relatively high evaporation. Under such conditions the water in the conducting elements of the xylem is under tension due to negative pressure, so that if a branch or stem is severed, water will be drawn into the cut ends instead of being forced outward. On that account, it has long been recognized that under conditions of high evaporation, root pressure or endodermal pressure is not the main force in causing the movement of water through the stem to the leaves. It has been almost universally considered that the forces which are operative in causing the upward movement of water reside in the stem and leaves, but the nature of these forces and even the cells concerned in the process have not, as a rule, been established with any degree of certainty.

The ascent of water in trees has especially attracted the attention of investigators, since the large quantity of water that is lost from them, the great distances that it has to be moved, and the great amount of energy that must be expended to bring this about are striking. On that account, most of the studies on the ascent of water have been confined to woody plants, and most of the theories proposed to account for this phenomenon have been based upon the conditions that are known to prevail or considered to prevail in woody stems. It is beyond the scope of this work to discuss in detail all the theories that have been proposed to account for the ascent of water in plants. Many of these theories are at present only of passing interest, since they were based upon conditions that are now known not to exist in the plants. A thorough and critical review of the literature on this subject may be obtained by consulting Copeland (1902) and Dixon (1910, 1914), and the student who is interested in this regard is referred to these sources for detailed information. It is the intention here to mention only briefly the earlier theories that have been proposed to explain the ascent of water in stems and to confine the discussion principally to the ideas that dominate the field in this regard at the present time.

The theories that have been proposed to explain the ascent of water in plants under conditions of high evaporation may be divided into two main classes: (1) vital and (2) physical. The authors of the vital theories assume that the forces that are active in raising water in plants are manifested, for the most part, by the living cells of the root, stem, and leaves, although most of them consider that purely physical factors may assume a definite, although a minor, part. Those who propose the physical theories contend that the major forces concerned in the elevation of water are purely physical in their nature and consider that their manifestation by the plant is identical with or similar to that which would occur in a purely physical apparatus under like conditions. Most of those who favor the physical explanation of the ascent of sap agree that the living cells of the plant play their part in the process, although their action is classed as of only minor importance. The matter of contention between the advocates of these two theories is thus apparently a question of whether vital or purely physical forces are the dominant ones in the process. It would seem that the forces which are dominant might depend upon the conditions surrounding the plant. Thus under some conditions, *e.g.*, low evaporation, vital forces might be the dominant ones, and physical forces might be exerting only a slight influence. Under another set of conditions, *e.g.*, under conditions of high evaporation, physical factors might dominate the process and thus overshadow the activity of the living cells.

1. *Vital Theories.*—Westermaier (1883 to 1884) formulated one of the first vital theories in regard to the ascent of water. He considered that

the upward passage of water in the stem was effected in the wood parenchyma, and that the tracheids and vessels acted as water reservoirs rather than conducting elements. Godlewski (1884) assumed a periodic change in the permeability of the protoplasmic membranes of the cells of the medullary rays of the xylem, on which account water was taken up on one side from the bordering xylem elements and then released into the xylem elements bordering upon the upper part of the opposite side. This was considered possible since it was assumed that the air pressure in the conducting xylem at the upper portion of the medullary ray was lower than the pressure inside those xylem elements in contact with the lower part of this ray. Janse (1887) supported the idea of Godlewski and was one of the first to note that if the lower portion of a branch is killed, the leaves above are affected within a few days. These theories, however, were proposed upon little or no experimental evidence and upon the assumption of certain conditions in the stem that are now known not to exist.

The later advocates of the vital theory of water ascent have been Ursprung (1905 to 1907), Ewart (1905, 1906, 1908), Schwendener (1892, 1909), Leclerc du Sablon (1910), Bose (1923) and Molisch (1928 to 1929). These advocates, however, base their assumptions upon a considerable amount of experimental evidence which, although convincing them of the nature of the forces involved in the ascent of water in plants, has not so appealed to many other workers in this field. Ursprung and Ewart, although agreeing that certain physical forces play a prominent part in the ascent of water, considered that the process is dependent to a certain extent upon the living cells of the stem. They considered that their experimental evidence indicates that the continuous ascent of water in the xylem is possible only in the presence of its living cells. The exact functions, however, that these living cells perform are not clearly set forth. Bose and Molisch, however, attribute the ascent of water practically entirely to the physiological activity of the living cells of the plant. The work of Bose (1923) has attracted much attention and on that account will be mentioned here in considerable detail, together with some of the adverse criticisms of the methods used and of the conclusions that he has deduced from his experimental evidence.

Bose observed by means of special apparatus that the cells of the lower and upper halves of the pulvinules of the lateral leaflets of the telegraph plant (*Desmodium gyrans*) execute alternate contractions, the period of a complete pulsation varying according to the circumstances from 1 to 4 min. Bose also claimed that by means of proper apparatus pulsating activities are noticeable in growing regions of the plant. These growth pulsations consist of a series of alternate expansions and contractions. It was ascertained that these pulsating activities can be modified in very definite directions by changing the physiological conditions. Thus pulsatory activity is depressed or arrested under diminished internal pressure. A rise of temperature stimulates it while a fall in temperature depresses it. A small amount of an anesthetic increases this rhythmic activity, while it is completely stopped by poisons.

Bose observed that the ascent of sap is modified by the same conditions that so strikingly affect the pulsatory activity. Thus, diminished internal hydrostatic pressure causes a depression in the ascent of sap, a rise of temperature increases the rate, and a fall in temperature depresses it. Small doses of an anesthetic increase the ascent of sap, while larger doses arrest it. Poisons also arrest the ascent of sap. From these observations Bose concluded that he had "proved conclusively that it is the pulsatory activity of living cells which maintains the ascent of sap in plants." He considered also that the response of transpiration to various factors indicates that it is "not a physical process but a physiological one carried on by a rhythmic tissue forming part of a rhythmic system continuous throughout the plant for the absorption and distribution of water." Thus he considered that transpiration is a physiological excretion of water by the leaf cells, the excreted water being removed by the physical process of evaporation. He also considered that the excretion of water can be shown when the leaves are covered with vaseline and considered that this is a further proof that transpiration is an active physiological process.

By electrical methods Bose considered that he demonstrated that a definite layer of tissue, the innermost layer of the cortex in dicotyledonous plants, is in a state of active pulsation consisting of alternate contraction and expansion. He thus concluded that these are the cells that carry the water upward. Since there is no such pulsatory activity to be observed in the wood, Bose considered that the xylem is by no means essential for the conduction of water and that it acts only as a reservoir, the water being pumped into or withdrawn from it according to different circumstances.

Molisch (1928, 1929) reported that he successfully repeated some of the experiments of Bose. In one experiment he employed a cut piece of stem of *Antirrhinum* which was held at the cut end by a piece of sponge. When the water supply was low the leaves on this stem became badly wilted. A few drops of a dilute solution of camphor applied on the sponge, however, caused the leaves to revive quickly and become fully erect in the course of 2 or 3 min. The results of this experiment are thought to be indicative of the fact that the movement of water in a plant is a physiological rather than a physical process. By means of one of the instruments of Bose, Molisch reported that the flow of sap along the stem was observed to consist of a series of pulsations. These pulsatory activities were greatly increased by drugs that enhance cardiac activity in the animal. Molisch stated: "There is an inherent activity in the stem itself, independent of those in the leaves and roots which cause the movement of sap. The sap can be made to flow either upward or downward according to different stimulation. The law of directing movement of sap is that it moves from the stimulated to unstimulated or depressed regions. A local depression or stimulation starts the machinery into action and makes the sap proceed toward the depressed or away from the overstimulated region."

Dixon and his workers (1924) repeated certain phases of Bose's work but were unable to verify his results. The work and conclusions of Bose have been adversely criticized by Shull (1923), MacDougal and Overton (1927, 1929), and Benedict (1927). One of the chief criticisms is that the relationship of the pulsations or rhythmic activity reported by Bose and Molisch to the ascent of sap has not been definitely established. It does not follow that because pulsation activities and the ascent of sap respond in a like manner to the same factors, one is the cause of the other. The method of determining the rate of water ascent by noting the time it took a wilted leaf to show signs of recovery when the stem is supplied with water is considered to be untenable. The fact that the leaf shows signs of recovery does not indicate that the water entering the stem has ascended to the leaf by that time any more than that water coming from the nozzle of a full garden hose is the identical water which at that instant is entering the distal end of the hose (MacDougal and Overton, 1927). It has been estimated that the pulsatory movement of sap claimed by Bose (1923) and Molisch (1928 to 1929) would imply a transfer through 200 to 400 living cells per second. Benedict (1927) measured the cross-sectional area of the living cells in the tree trunks of 10 different species and computed the maximum pumping capacity of the most rapid pulsating rates found by Bose. The actual rate of sap flow under maximal conditions of transpiration proved to be from 8,000 to 30,000 times as rapid as would be possible under the theory proposed by Bose.

Peirce (1934) believed that the movement of water from the roots through the stem and branches to the leaves is dependent upon the living cells between the root and the leaves. He stated that where living cells are present water ascends in stems, but where only dead cells exist there is no ascent of water to replace that lost by transpiration. He held, however, that the physical conditions of the soil and atmosphere influence to a considerable degree the amount of water translocated. Since each vessel, tracheid, and living cell is a constituent unit of the whole system, and since the proportion of these various parts may vary, the relative amount and speed of movement of water in one plant part may be different from those in another. Peirce (1935) showed that the rate of water movement in stems varied with temperature, thus indicating further that living cells are concerned in the process. Priestley (1935) noted that when the leaves are removed from woody stems, water continues to enter with as much force as if they were present. He thus concluded that the imbibitional forces in the leaves are not necessary for the ascent of water, but that living cells in the stem are the effective agents. Peirce (1936) from later experiments involving the relation of temperature to the ascent of sap concluded that living cells condition or affect but do not accomplish or effect the movement of water through vascular plants.

It appears evident under the conditions of high evaporation that all the forces which function in the ascent of water in plants cannot be attributed to the direct action of living cells. The living cells, however, evidently play an important part, especially those of the roots and leaves. Whether those in the conducting tissue play any part is evidently not definitely known. Dixon (1924) stated that the fact that water rises in stems of plants even when surrounded by a saturated atmosphere and the fact that dead leaves inadequately draw up water indicate that the living cells play some part in the ascent of sap. Smith, Dustman, and Shull (1931) believed that the rise of water under the conditions stated by Dixon can be explained solely on a basis of the saturation deficit that exists in leaf and stem tissues when the branches are cut. Although submerged in water, the amount of internal water continues to determine the intake of water until the saturation deficit is satisfied. They showed by experiments involving a number of branches of various plants that if the saturation deficit was satisfied there was no ascent of water.

It would be safe to state, however, that the living cells of the roots and leaves at least maintain good contact and thus keep intact the passageways across them for the movement of water even if they do not in themselves exert any vital force in the ascent of water in the stem.

2. *Physical Theories.*—Under this heading are grouped those theories which assume that the ascent of water is for the most part due to purely physical forces and that the living cells of the stem play only a minor part

in the process. Before discussing the physical theories of the ascent of sap, however, some of the experimental evidence which has been presented to show that the living cells are unessential in the ascent of water should be mentioned.

(a) *Absence of Living Cells.*—The general method of procedure that has been followed to demonstrate that the living cells of the stem are not concerned in the ascent of sap has been to kill a portion of the stem by exposing it to hot water, steam, or chemicals or to place the cut end of a branch into a solution of a chemical that is known to kill protoplasm when in contact with it. These methods, however, are subject to criticism, as will later be mentioned.

Boucherie (1840, 1841), while injecting timber with various compounds in order to test their effects upon its preservation, found that if a tree were cut across at its base and supplied with a poisonous solution, it drew this liquid up to its highest leaves. It would also draw up a second solution if supplied. The significance of this experiment in the ascent of sap, however, was apparently not recognized by this investigator or by contemporary botanists. Boehm (1889) noted that if a portion of a stem were killed, water would still ascend to the leaves, but the classical work of Strasburger (1891 to 1893) gave the first conclusive evidence that the living cells of the stem are at most of only minor importance in the ascent of water. He placed the base of cut stems in poisonous solutions, among which were copper sulphate and picric acid. In the case of branches set in a 5 to 15 per cent solution of copper sulphate, the liquid rose to a height of from 6 to 20 m. in from 16 to 20 days. In another experiment he killed the branches for a distance of 10 to 12 m. at the basal end with boiling water and observed that eosin rose in them from the cut end to a height of almost 17 m. The leaves at the uninjured end of the stem, however, remained alive for only a few days when they dried up and fell off. In another experiment Strasburger used a young oak tree 22 m. in height. This was sawed off near the ground and kept upright with the lower end in a solution of picric acid. After 8.5 days it was found that the highest branch was completely impregnated with picric acid, while fuchsin which had been added 3 days later than the picric acid was also found in the same regions. Roshardt (1910) performed experiments on 131 species of plants, including shrubs and herbaceous plants of 59 families, by killing portions of the stem, branches, or petioles by steam, ether, low temperatures, or other means and then determining the effect of this treatment upon the sap flow. A considerable number of his experiments were made upon grasses, among which were *Poa pratensis*, *Bromus sterilis*, *Bromus hordeaceus*, *Agropyrum repens*, and *Secale cereale*. He noted that water ascended the stem for a considerable time after certain portions of the lower part had been killed but that the leaves dried up after from 8 to 17 days. There was no distinct wilting, but the leaves began to dry at the tips.

Overton (1911) in experiments with the umbrella plant (*Cyperus alternifolius*) found that when 5 to 10 cm. of the stem was killed by treatment with picric acid, 95 per cent alcohol, or copper sulphate for 36 to 48 hr., sufficient water ascended through the poisoned portions to supply the transpiration needs for 90 days and to allow the development of new branches.

MacDougal, Overton, and Smith (1929) sawed off a willow tree 11 cm. in diameter and about 8 m. high and placed it upright with the basal end in a concentrated solution of picric acid. It was allowed to transpire for 4 days, after which it was transferred to acid fuchsin. In 4 days the tree absorbed 10 l. of the picric acid solution and about the same amount of acid fuchsin in the succeeding 4 days, both solutions being drawn up the tree for a distance of 6 m. This experiment was repeated with a walnut tree 10 cm. in diameter with results very similar to those observed for the willow.

The results obtained from the experiments that have just been mentioned indicate that the living cells of the stem play no important role in the ascent of sap. There has, however, been some question of the conclusions drawn from these experiments. It is pointed out that the killing of the cells in a given length of the stem does not show that those living cells above or below the injured portion are not concerned in the process. Some suggest that even though poisonous solutions are drawn up the stem, some of the living cells may not be injured. In the experiments in which the stems have been immersed in poisonous liquids, the leaves remain alive for only a relatively short period of time after the treatment. It has been contended that the leaves should remain alive for a long period of time after such treatments if the death of the living cells of the stem has no influence on the ascent of water. In the cases where a portion of the stem has been killed by the various means, Dixon (1905, 1909, 1915) and Overton (1911) considered that the leaves are not injured because of a lack of water being moved to them through the dead stem but by the transportation to them in the water stream of poisonous or plasmolytic substances that have been formed by the decomposition of the cells in the treated region. Both these investigators noted that the leaves showed injury in proportion to the length of the stem that was killed. This opinion was further substantiated by Dixon, who forced water through a steamed portion of the stem and found that it was tinged with a brown substance, while the water that was forced through an untreated stem emerged a clear liquid. Some of the investigators are of the opinion that the treatments of the stem cause the formation of substances which clog the lumina of the xylem elements and thus prevent the normal ascent of water. It has also been suggested that these treatments of stems might rupture the water columns and thus bring about a decrease in water movement. Roshardt (1910) considered that the decrease in water content under such treatment is due to the lack of energy which is normally furnished by the living cells and which the dead cells are unable to supply. It appears evident that the methods that have been used to render the living cells of the stems inactive in the study of the ascent of water bring about other changes in the stem which cannot be ignored.

(b) *Imbibition and Capillary Theories.*—The imbibition theory was proposed by Unger (1868) and championed by Sachs (1878, 1879), who became its chief exponent. This theory assumed that the water moved upward in the stem entirely in the walls of the xylem elements by the process of imbibition. It was considered that the creation of a saturation deficit in the cell walls at the top of a plant would cause a flow of water toward the point of loss from the adjoining portions of the cell wall and that this demand for water would thus be transmitted downward through the walls to the roots where it would be satisfied. The forces of imbibition are known to be sufficient to elevate the water in this manner much beyond the height of the tallest trees. The imbibition theory of the ascent of sap after the manner above mentioned, however, became obsolete when it was demonstrated beyond doubt that practically all the water moves upward in the lumina of the xylem elements. This fact, proved by numerous investigators among whom are Elfving (1882), Vesque (1884), Darwin and Phillips (1885), Errer (1886), Strasburger (1891), Dixon and Joly (1895), and Smith (1896), shows that if the lumina are blocked by gelatin, paraffin, ice, water vapor, or by compressing the stems until the lumina are closed, the supply of water to the upper portion of the plant is checked to the extent that severe wilting results. The ascent of water in the lumina is also demonstrated by means of dyes. There is no doubt that some water ascends in the walls of the xylem elements, but the amount is insignificant in comparison with that which moves upward through the lumina of the tracheids and vessels (Vines, 1896).

The elevation of water has been ascribed by some writers to the force of capillarity in the lumina of the xylem elements. The lifting power of a capillary is determined by the form and size of the meniscus of the liquid therein contained. The form of the meniscus is, in turn, influenced by the nature of the walls of the capillary. It was early recognized that the lumina of the xylem elements are of such a size that their lifting power by capillarity is not very large and cannot account for a rise of water, according to Strasburger (1891), to distances exceeding 3 m.

Stamm (1929) studied the capillary structure of the softwoods by determining the rate of electroendosmotic flow, the hydrostatic flow, and the force of surface tension in these woody tissues. The values he obtained in this manner agreed with the values obtained by microscopical measurements. Reed and Bartholomew (1927), and Reed (1930) believed that the transport of water in the fruits of oranges and lemons occurs through the hydrophylic layers of the cell walls rather than across the protoplasm and cell sap. Peirce (1934) believed that water moves along the inner surfaces of tracheids and vessels, and that the maintenance of a movable film of water on the inner surface of a vessel or tracheid depends upon the quantity of water in reserve in the cell wall. The size and relative proportion of vessels and tracheids, among other things, determine the heights that plants attain because these two factors determine the volume and speed of water movement. The terms "imbibition" and "capillarity" have been used more or less interchangeably by some writers. As here interpreted, however, capillarity is confined to small but well-defined lumina or tubes, while imbibition is confined to the submicroscopical pores or interstices of the cell wall.

(c) *Cohesion Theory.*—In 1894 Dixon and Joly published their cohesion theory of the ascent of water in plants. Askenasy (1895, 1896) independently developed a similar theory which was published shortly after the abstract of Dixon and Joly had appeared. This theory has been further elaborated by Dixon, and much work has been undertaken to substantiate it. It is so much in accord with the general conditions known to prevail in the plant and has been so well substantiated by experimental evidence that it has been almost universally accepted as the best explanation of the ascent of water in stems under conditions of high evaporation.

The cohesion theory of the ascent of water is based essentially upon the two following facts: (1) the cohesive force of water or the quality that water possesses of resisting tensile stress and (2) the imbibitional force of the evaporating cell walls of the leaf or the tractional force exerted by the water menisci of the submicroscopical cavities of these cell walls.

(1) *Cohesive Force of Water.*—The cohesive force of water has been known since about 1850, and, according to Copeland (1902), Nägeli (1866) first suspected its importance in the ascent of water. The publication of the cohesion theory of the ascent of water greatly stimulated interest in the cohesive properties of water, and studies in that regard have been reported by Dixon (1910, 1915), Ursprung (1913, 1915, 1916), and Renner (1915). These investigators all found that water has great cohesive power and demonstrated that the tensile strength of water containing dissolved air is at least 300 atmospheres. It has been shown that the xylem sap has a greater tensile strength than pure water and that when it is in position in the conducting xylem it may withstand tensions greater than 300 atmospheres without separating from the walls of these conducting elements.

(2) *Imbibitional and Osmotic Forces.*—The imbibitional forces developed by drying organic matter range from zero to 1,000 atmospheres, depending upon the degree of drying. The force of imbibition in the cell walls of the leaves is thus enormous and is much more than ample to lift a column of water to the height of the tallest tree. It has been estimated that the total force necessary to lift water to a height of 350 ft. through the xylem elements of a tree would not exceed 20 atmospheres. Only a small fraction of the possible imbibitional power of the cell walls of the leaves would be necessary to overcome this force. It is considered

that a saturation deficit of 2 per cent would be more than ample to lift water to the top of the tallest living tree, while a deficit of 0.1 per cent would exert a pull sufficient to raise water to the height of the ordinary annual plants. Ursprung (1925) proved that the suction power of the vacuoles of the leaf cells would furnish ample force to move the column of water in stems at an adequate rate to replace the water evaporated from the leaves, provided osmotic forces play a determining part in the lifting of the water in the stem.

It was calculated by Dixon (1897) that it would require a pressure of 100 atmospheres to tax the cell walls of the mesophyll cells of the leaf to their breaking point. Thus it is apparent that the osmotic pressures of the leaves never seriously tax the tensile strength of these cell walls.

There is some dispute as to what part the osmotic forces of the cells of the mesophyll of the leaves and the cortex of the roots play in the ascent of water. It is considered that these forces at least keep these cells turgid and thus prevent them from collapsing under the tension that pulls the water through them. The osmotic pressure of these cells, however, must equal or exceed the tension in the water that passes through them (Dixon, 1896).

(3) *Description of Theory.*—The cohesion theory of the ascent of water assumes that the water in the conducting tracts of the xylem exists in the form of unbroken columns and that these columns are continuous with each other both vertically and laterally through the cell walls. A continuous meshwork of water thus exists throughout the whole length of the stem, roots, and leaves. The terminals of this water meshwork are the menisci of the water in the submicroscopical cavities of the cell walls of the epidermal cells of the root and of the cells bordering upon the intercellular spaces of the leaf or the epidermal cells. This theory also assumes that the water in the conducting tracts is brought into a state of tension by the imbibitional forces in the evaporating cell walls of the leaf and that these water columns are able to withstand the tensile pull that is thus exerted upon them.

Thus in the process of transpiration, water particles are set free from the menisci of the submicroscopical cavities of the cell walls of the evaporating cells of the leaf. This loss of water particles produces incipient drying or a saturation deficit, the menisci become more concave, and their surface tension is increased. As a result of this, water is abstracted from the protoplasm of the cells which the evaporating walls enclose. The protoplasm in turn abstracts water from the vacuole, so, on account of this disturbance in water equilibrium, osmotic and imbibitional forces that abstract water from adjoining cells or cell parts are set in motion. This disturbance in equilibrium is transmitted eventually to the water of the xylem elements of the leaf. A pull is thus set up on the water in this conducting element, which, in turn, is carried down the petiole to the conducting xylem elements of the stem and root and then to the soil. Owing to the weight of the water columns and to their adhesion to the walls of the conducting tracts, the water is, under the conditions above mentioned, thrown into a state of tension which may in some cases be as great as 200 atmospheres (MacDougal, Overton, and Smith, 1929). As a result of this pull exerted by the water menisci of the evaporating cell walls of the leaf, water is moved upward through the root and stem to the leaves.

Under conditions of high transpiration the tensile stress of the water in the stem and roots must be transmitted across the cortex of the root to the submicroscopical cavities of the outer cell walls of the epidermal cells of the root. On this account, as the menisci of the evaporating cell walls of the leaf become more concave, the menisci of the exterior cell walls of the epidermal cells behave in a similar manner. The rendering of the menisci of the evaporating cell walls of the leaf more concave tends to bring them in equilibrium with a lower vapor pressure and consequently reduces evaporation, while the increased concavity of the menisci in the epidermal cells of the root renders them more ready to condense water vapor. The menisci of the cell walls of the epidermal cells of the root entrap the more active molecules of water, while the menisci of the evaporating cell walls of the leaf retain the less active ones. Dixon (1910) stated:

> Thus we may regard the flow of water up the highest tree as due to the evaporation and condensation produced by the difference between the vapor pressure at the surface of the epidermal cells of the root and that which obtains in the cell walls of the evaporating cells of the leaf.

Dixon (1914) considered that the loss of water by a plant under conditions of very low evaporation might be a process of secretion by the mesophyll cells of the leaves. If this is true, the energy that is necessary for the lifting of the water is considered to be supplied by respiration. Dixon estimated that the energy necessary to raise 1 cc. of water 100 m. in a tree and to overcome the resistance would amount to only 0.5 cal. To evaporate 1 cc. of water at 20°C. requires 592.5 cal.; hence, in order to obtain the energy needed to raise the water in a 100-m. tree the amount evaporated would be diminished by only 0.001 part. Even taking the highest and most excessive estimate of the resistance in the highest trees, the amount of water evaporated would be reduced by only one-hundredth to one-thirtieth if, in addition to evaporation, the energy absorbed by the leaf had to do the work of transporting the water from the roots. If under conditions of extremely low evaporation water is excreted from the leaf cells, ample energy can be derived from respiration to lift the water to the tops of the tallest trees.

Smith, Dustman, and Shull (1931) considered that respiration is not necessary for the ascent of water. They believed that the ascent of water is due entirely to physical and chemical forces, if osmotic action can be considered as purely physical and chemical in nature. Whether evaporation is exceedingly rapid or extremely slow, the cause of sap movement is considered to be due to surface tension and imbibitional and osmotic forces that are in no manner related to excretory and secretory activity of the living cells. Homan, Young, and Shull (1934) stated that the maximal tension or negative pressure in the tracheae depends on the relative humidity of the atmosphere. Since the relative humidities of natural atmospheres are frequently as low as 20 to 30 per cent, it is evident that the structure of plants enables them to prevent the evaporation of water to a marked degree. The epidermal cells with their cutinized surfaces and regulated stomata operate efficiently in maintaining the humidity of the atmosphere surrounding the mesophyll cells in an approximate state of saturation. Hungate (1934) believed that the cohesion theory offers the best explanation to date of the ascent of water in plants.

(4) *Evidence for Theory.*—Two general methods have been used to demonstrate the plausibility of the cohesion theory of the rise of water in plants. These methods have to do both with artificial systems and with living plant materials.

Renner (1911 to 1912) measured the suction force of transpiring twigs and found under the conditions of his experiment that it amounted to 10 to 20 atmospheres. He considered that the saturation deficit in the leaves was the cause of the water movement in the stem. By applying artificial suction to the leaves at a time when no saturation deficit existed, he determined the rate of flow produced by a given suction force. The rate of flow through the stem was then determined when rapid transpiration was proceeding, and the suction that was being exerted by the leaves was calculated from these data, since it was considered that the rate of flow through the stem was proportional to the suction forces exerted by the leaves. Nordhausen (1916, 1919, 1921) showed that suction forces exist in the leaves, although by his methods he did not obtain so high values as did Renner. In this connection it should be mentioned that Thut (1928) devised an apparatus that shows definitely the lifting power of evaporation under ordinary laboratory conditions. In 1932 he reported a special method for demonstrating the suction force of the twigs of *Acer negundo*, *Thuja occidentalis*, and *Platanus occidentalis*, among others.

If the water in the xylem elements is thrown into a state of tension by the suction forces of the leaves, these elements should show a contraction when transpiration is proceeding and should again expand when transpiration decreases or ceases entirely. MacDougal (1921, 1925 to 1926) and MacDougal and Shreve (1924) have observed that the trunks and branches of trees show daily alterations in diameter. This swelling and shrinking were attributed to changes in the wood due to changes in the transpiration rate. It was found that variations in diameter depended directly upon the depletion and repletion of the water supply of the wood, and the effect of increasing tension in the water upon the diameter of the conducting elements was noted. Bode (1923) made a direct microscopical examination of the vessels of wilted plants and found that they would shrink with increased transpirational pull and that they would expand again when the transpiring surfaces were removed from the plant. Haasis (1932) found that the fall and winter swelling of the trunks of Monterey pine and redwood was preceded by a progressive contraction during the dry season of the spring and summer. The relative humidity and the amount of available moisture in the soil most appreciably affected this shrinkage. He stated that a tree might be shrinking in one part and swelling in another. Gibbs (1935) found indications that water is continuous in the xylem of the birch, larch, balsam, and spruce. It has been shown by Dixon (1914) that the resistance of a current of water moving through wood at the velocity of the transpiration stream is equivalent to a head of water equal in length to the wood traversed. In a tree 100 m. in height, the tension of the ascending water would be approximately 20 atmospheres, a force that would in no way tax its cohesive powers.

Huber (1928) found that trees and nonsucculent xerophytes have the least resistance to the conduction of water per unit length of their stem. Their total resistance, however, may be high owing to the long stems of trees or to the difficulty of obtaining water under arid conditions. Groom (1910), Farmer (1918), Holmes (1918), Rivett and Rivett (1920), Coster (1931), Malhotra (1931, 1932), and Huber (1932) investigated rates of water flow in the conductive tissues of different plants. Coster (1931) studied the speed of the transpiration stream in herbs, shrubs, lianas, and trees in Germany and in the rain forests of West Java. Under optimum transpiration conditions the maximal speeds were 50 to 80 cm. per minute for herbs and shrubs, 100 to 200 cm. per minute for shrubs and trees, and 100 to 250 cm. per minute for lianas. The experimental values are lower than those which actually occur in nature since the severing of the plant slows the transpiration stream. In general, the speed is greater the larger the cross-sectional area of the elements and the taller the plants. Haines (1935) found that the greater the water deficit, the greater the resistance to flow by the walls and the protoplasm of the leaf cells. Pond (1903) concluded that the transpiration stream of the aquatic plants studied by him appeared to be caused primarily by forces in the basal part of the plant. This force seemed to be root pressure.

(5) *Objections to the Theory.*—The development of high tensions in the water columns of the plant introduces the greatest difficulty associated with the cohesion theory of the ascent of water, *viz.*, the permeability of the walls of the wood cells to gases. Why do gases not pass into the conducting elements when tensions are developed? Bailey (1916) found that the surface tension of the sap in the pit membranes of different conifers could be overcome by pressures that would be inadequate for the cohesion theory. Wright (1928) believed that the torus of these bordered pits could not exert a valve action and that pressures of 3 atmospheres would cause the passage of gases through them. Apparently many of the water-conducting elements fill with gas during the summer or during conditions of water deficiency and then are again filled with water during the winter or during conditions of an available water supply. Woodhouse (1933) believed that air enters the xylem cells through their walls before a sufficient tension is created to raise the water to the height of trees. He considered that the cohesion theory of sap ascent in itself is inadequate, but that combined with the imbibition theory a satisfactory hypothesis for the ascent of water is obtained. By this hypothesis the walls of the xylem imbibe and support the weight of water moving through it by tensile strength. He considered that the flow of water occurs through the largest vessels in which an imbibed tensile fluid can be maintained under the conditions of the moment. He believed that this explains the zonation of con-

ducting areas in the xylem, since it has vessels of varying diameters in definite seasonal zones.

Münch (1926, 1927, 1930) claimed that there is a downward stream of water containing osmotically active materials in the bark. When these dissolved materials are removed at the base of the tree, the water is forced into the xylem and refills the vessels emptied during the summer. He stressed the rapid streaming in the sieve tubes as significant.

(*d*) *Other Theories.*—A theory suggesting the manner of translocation of water in plants was advanced by Münch (1930). The basis of this theory lies in the fact that when two osmotic pressures operate against one another through a liquid connection, a movement of solution occurs from the direction of the greater pressure toward that of the lesser pressure until this movement has equalized the osmotic activities at the two ends of the path. Simultaneously, water will be taken in under the influence of the greater pressure and thrust out against that of the lesser.

In applying this principle, each living cell is not regarded as an osmotic unit, but the whole aggregation of living cells, the "symplast," is considered as being linked together by plasmatic fibrils into a single osmotic system. When a point in the system develops relatively high osmotic and hydrostatic pressures and external water is available, liquid movements will be set up from this position toward any other where these pressures are lower. If the root is intact there is no such easy escape; consequently the back pressure prevents such extensive movements, and the only movement that takes place is associated with the filling of the higher regions. This filling seems to consist mainly of passing fluid into the partially empty elements, distention of the living cells, and reduction to a limited extent of the tension in the vessels. When high hydrostatic tensions exist in the tracheae, positive hydrostatic pressures exist in the living cells surrounding them. The osmotic value of the cell sap of living cells is always higher than that of the tracheal sap, hence there will be a tendency for water to be withdrawn from the vessels across the semipermeable membranes of the surrounding cells. An easier line of escape for the moving fluid will be through the sieve tubes, and perhaps through the protoplasmic connections of living cells, since such movements do not involve the penetration of surface membranes.

Lund (1931, 1932) stated that an electric current flows downward in the outer cortex and upward in the wood axis. There is a possibility that one of the functions of this continuously upward electric current in the wood is to supply electrical energy for the electroendosmotic flow of sap in an upward direction in the conductive elements of the wood. This suggestion may apply to a downward flow in the cortex as well as transport across the stem.

4. Downward Movement of Water.—It has been known since the time of Hales (1727) that under certain conditions there may be a downward as well as an upward movement of water through the xylem. A few examples of this downward movement selected from the literature are cited herewith. Hales (1727) among other observations noted that a forked branch cut from a tree will remain fresh and continue to transpire for many days provided that one limb of the fork with its leaves is immersed in water. Dixon (1924) stated that if the tip of an upper leaf of a potato plant is cut under an eosin solution, the colored liquid is quickly drawn back into the conducting elements of the xylem of the leaf and thence into the same tissue in the petiole. From there it not only makes its way into the upper branches and leaves but also passes down the stem supporting these parts and into the xylem tissue of the tuber.

Mason and Maskell (1928) found in the case of the cotton plant that an eosin solution taken in through cuts in the leaf blades was in 6 hr. found to have penetrated both upward and downward in the xylem. The stain passed in with the foliar trace and occupied a median position in the wood, the peripheral region of the wood being unstained.

Arndt (1929) studied the movement of solutions through the xylem of the coffee tree. In one case the terminal internode was cut and an eosin solution applied through a funnel and rubber tube. In 15 hr., 50 cc. of the solution was absorbed and the corollas of flowers in lateral branches 80 cm. from the cut internode were stained. In 23 hr. after the beginning of the experiment all the leaves of the trees were stained and the total distance of eosin movement exceeded 200 cm. In one the stain descended 125 cm., passed horizontally 60 cm., and then ascended 100 cm. In none of the experiments did the eosin solution in moving either up or down the stem pass through the cambium into the phloem, and the cut surface of the phloem in contact with the dye did not conduct it. MacDougal (1925) noted that the outermost wood of the Monterey pine is not the path of an upward-moving current in the normal tree, but he could not, however, demonstrate a downward moving current in that part of the xylem. The path of reverse currents has been traced by Birch-Hirschfeld (1920) by means of lithium nitrate and eosin.

Harvey (1931) found that when suction was applied to a cut apple branch and a dye solution was applied simultaneously to a similarly decapitated twig located at a considerable distance from the former, the dye entered the main stem and streamed not only toward the source of suction but also rapidly away from it. Baker and James (1933) injected dyes into the wood of *Acer pseudoplatanus* and noted that they invariably moved both upward and downward. This movement never occurred in the bark or pith. The rate of upward movement exceeded the rate of downward movement only when transpiration was high.

The examples that have been cited above leave no doubt in regard to a downward current of water in the xylem. The cause of this movement, however, is not definitely known. Dixon (1924) considered that these reversals in the current of water are readily explained by the cohesion theory of the ascent of water. He argued that since transpiration causes the water in the conducting xylem to pass into a state of tension, this tension determines a flow from any source wherever situated and the continuous transpiration from the leaves draws the supply through the plant along the channels of least resistance. This path of least resistance, however, is not always the shortest path in the wood. Transverse motion across several xylem-conducting elements seldom occurs and the separate linear series of conducting elements are practically isolated from each other laterally. On that account, Dixon considered that this enables the tension developed by the transpiring cells of the leaves, although it raises a water column in one series of the conducting xylem, to draw a solution in a neighboring filament of xylem elements terminating above in some local supply. Smith, Dustman, and Shull (1931) believed that the downward translocation in the xylem, which occurs when leaves on wilted plants are truncated and supplied at the cut end with eosin, is a consequence of a root-saturation deficit that will be relieved from any direction by a free water supply. They claimed that this behavior is not connected with the ordinary transpiration or translocation currents in the plant.

Arndt (1929) noted that during the daytime when the leaves are actively transpiring, the eosin solution moved upward more rapidly than downward but that at late evening, at night, and in early morning the downward movement was not greatly reduced, while the upward movement was so greatly reduced that the downward movement exceeded it. Arndt considered, since only slightly reduced downward movement occurs at night and since topping a plant affects this movement but little, that it is independent of transpiration. If the postulation of Dixon (1924) that there are separate conduits for transportation in each direction is correct, Arndt considered that in the case of the coffee tree these conduits are so close together that a microscopical examination after short-interval experiments does not show stained and unstained areas as would be expected. He considered that the observations indicate that living cells other than those of the leaf must play an important role in sap conduction, since it does not seem probable that there could be simultaneously ascending and descending streams unless there are conduits that are hydrostatically isolated from each other by an impermeable partition of living cells.

5. Lateral Movement of Water.—The data concerning the lateral transfer of water in stems are very limited, and the observations in that regard have been confined almost entirely to woody stems. It is known that the lateral transfer is very slow and that translocation in a radial direction usually takes place more slowly than it does in a tangential direction. In many cases, the location of pits in the conducting xylem elements is such as to facilitate translocation in a tangential direction, while the medullary rays offer passageways for radial transfer. MacDougal, Overton, and Smith (1929) considered, however, that the movements which take place through the medullary rays are largely or wholly osmotic and that mass movements in a radial direction do not occur except possibly in some cases in the pines where the medullary-ray cells are of a tracheid nature.

Experimenting with a 13-year-old willow, they found that there was no radial transfer from one annual layer to another. Thus when dye was injected into the oldest 10 annual layers of this tree in such a manner as to exclude the dye from the outermost three layers, they found that to the height of the 10-year level there was no dye in the outer three annual layers. There was, however, a tangential movement of the dye in those annual layers to which the dye had access, and at a considerable distance above the bore hole the tangential movement in the summerwood was completely around the annual layer. The number of annual layers in which there was no dye remained constant from the bore hole to the 10-year level. The number of annual layers, toward the center of the tree, which contained the dye became progressively smaller until at the 10-year-old level only the innermost of the terminal portions of the wood of the tenth year showed the dye. These investigators considered that this experiment demonstrates that there is not a radial mass movement nor a radial transfer of the forces concerned in sap flow from one annual layer to another through the living cells of the medullary rays or through abutting xylem elements. They considered, however, that the possibility exists for an osmotic transfer of water through these structures.

It was noted by MacDougal, Overton, and Smith (1929) that, although the injection experiments showed no mass movement in a radial direction from the inner to the outer annual layers, the leaves at the tips of all branches above the bore hole showed the presence of the dye. Since these leaves are connected only with the wood of the current year and since there was no dye in this layer at the level of the bore hole, the question arises as to what path is followed by the transpiration stream in moving from the innermost layer of the xylem to the outmost one if there is no radial mass movement.

These authors explain this point by using as an example a 2-year-old unbranched stem. A stem of this type consists of a central column of xylem of the first year's growth whose terminal end is bluntly conical. This central column is surrounded by a thimble-shaped mass of wood whose upper half, which is the longitudinal growth of the present season, is solid. The lower portion, which is the lateral formation to the previous year's wood, is a hollow cylinder enclosing the last year's xylem. The leaves attached to the upper portion of the current year's growth exert a pull on the water in the xylem elements, and this is transmitted through the dome-shaped portion of the preceding year's growth to its conducting xylem. The transpiration stream thus moves from the xylem of the first year to that of the second year in a more or less vertical direction. The transpiration pull on the water columns of the first year may thus be transmitted to the columns in the xylem of the second year and thence to the xylem of the third year and so on to all the columns present in any annual layer.

It was noted by Arndt (1929), using a coffee tree, that, when a single lateral root was cut under an eosin solution, a one-sided staining was evident to the top of the tree after a period of 20 hr. In another case, this one-sided staining was still evident after 72 hr., the staining of the leaves of the laterals corresponding to the stained sector of the xylem. As the top of the tree was approached, the dye gradually diffused tangentially so that all the branches tended to become stained. In some experiments it was noted that as the dye passed upward it tended to diffuse laterally, but in all cases the radial and tangential diffusion was slow. Baker and James (1933) observed in *Acer pseudoplatanus* that radial transfer of water occurred from early to late wood and from one annual ring to another. They claimed to have demonstrated the radial and tangential anastomosing of vessels.

Auchter (1923) obtained evidence which suggested that water may move laterally in the peach tree. When the roots were removed from one side of the tree, the total water content of the whole plant was reduced and the growth checked to some extent. The moisture contents on the two sides of the tree, however, were identical and the leaves on the cut side did not wilt on bright, clear days. Apparently in these trees water moved through and around the stem without much difficulty. Furr and Taylor (1933) observed that the cross transfer of water was accomplished very readily in both old and young lemon trees, when irrigation water was applied on alternate sides of the tree on successive irrigation dates.

B. Translocation of Inorganic Compounds

1. Upward Transfer.—It has been commonly stated in the literature since the time of Sachs that the water and inorganic materials are carried upward for the most part through the xylem. Although it is apparently established beyond a doubt that the xylem is the pathway of the upward movement of water, the statements concerning the movement of solutes are apparently based upon inconclusive evidence.

The xylem has been considered as the pathway of the movement of the inorganic solutes on account of the following evidence: (1) It has been assumed that since the inorganic salts are dissolved in the water of the soil and since the pathway of the water is in the xylem, these inorganic materials would be absorbed and carried with it. (2) The inorganic salts have actually been found in the water-conducting elements and in the sap that has been obtained therefrom. (3) When cut stems are placed in solutions of dyes or salts, the solutes move for considerable distances in the xylem tissues. (4) Experiments in which the movement of dyes and salts has been studied in rooted and uninjured plants give indications that translocation of the solutes occurs in the xylem.

This evidence has been critically considered by Curtis (1923). He pointed out, as has been discussed in Chap. V, that it has apparently been definitely established that there is no direct relation between the absorption of water and the absorption of inorganic salts. He also considered that the fact that inorganic salts have been found in the sap of the xylem is not conclusive evidence that they are translocated in this tissue. He stated that unless it is known that the water moves by mass flow through open tubes and not by diffusion or through occasional membranes, or unless it is known that neighboring cells do not remove solutes from a passing stream perhaps transferring them and reintroducing them at a lower level, the mere presence of these solutes in the xylem sap is no proof that they are being moved therein.

The methods by which the samples of xylem sap have been obtained may be open to question. Curtis considered that sap obtained from a cut or bleeding stem may be different from that in a normal uninjured stem, since the movement of water through a series of cell membranes may be different from the flow that would occur if the membranes were cut open. Some of the methods for obtaining sap may, however, be free from this objection (Bennett, Anderssen, and Milad, 1927). Objections may also be made to the method of placing cut stems in dyes and salt solutions, since normally the water-conducting system may be considered as a closed system with no actual openings. Curtis considered that numerous sources of error have been overlooked in the experiments in which the translocation of salts and dyes has been studied in intact plants so that the results obtained have given no clear indications of the tissues concerned in the movement of the inorganic salts.

a. Through Xylem.—The observations of Anderssen (1929) in regard to the tracheal sap of the pear tree, however, indicate that the xylem functions in part in the translocation of the inorganic solutes. Thus he found that the individual inorganic constituents of the tracheal sap showed a large increase in concentration from early winter until early spring. From Nov. 10 to May 10 the calcium increased from 16.6 to 84.7 p.p.m., magnesium from 0.8 to 23.5 p.p.m., potassium from 23.6 to 59.6 p.p.m., and PO_4 from 10.6 to 25.2 p.p.m. On May 10 the calcium, magnesium, and PO_4 ions far exceeded their concentration in the soil extract. The total electrolytes were almost twice as concentrated in the outer annual ring as in the inner rings.

Bodenberg (1927) studied the tissues concerned in the translocation of inorganic salts in willow and *Acer negundo* by the use of lithium and cesium nitrates. These salts are not normally present in these plants and their presence can be detected spectroscopically. The removal of a section of xylem completely obstructed the upward movement of both salts, while the removal of a section of the phloem retarded only the movement into the parts above the ring. He believed that the upward and downward movement of mineral salts occurs principally through the xylem in the transpiration stream. He considered, however, that this movement may be aided by diffusion, by protoplasmic streaming, and by transfer along the threads of protoplasm.

Clements (1930) removed a ring of phloem from the stems of red and black raspberries, grape, apple, peach, plum, and cherry, and noted that the portions of the plant above the girdle continued to obtain mineral materials and nitrates from below. Since a ring of phloem was removed, it seemed apparent that the xylem served as a channel of transport for these materials. Clements considered that the distribution of the tracheae in the seasonal development of xylem is to a large extent the factor that determines the ease with which plants can tolerate girdling.

Crafts and Kennedy (1930), Morgan (1931), and Crafts (1933) studied the movement of toxic salts to the roots when the salts were applied as sprays to the leaves. After these salts entered the leaves they eventually entered the xylem and were transported downward to the roots where they slowly diffused from the xylem into the living cells, which were later killed. The rate of penetration of a toxic substance into plants and its translocation are influenced by temperature, incident radiation, humidity, and the rate of evaporation. Loomis, Smith, Bissey, and Arnold (1931, 1933) found that the movement of sodium chlorate within the plant is principally in the xylem and is most rapid in the direction of the transpiration stream. The removal of the phloem did not appreciably hinder the translocation of the sodium chlorate, but the checking of transpiration by vaselining the leaves did reduce this movement. Maskell and Mason (1930) and Mason and Maskell (1931) believed that in the cotton plant the evidence indicates that the bulk of the phosphorus, potassium, calcium, and inorganic nitrogen absorbed by the roots is carried in the transpiration stream through the xylem to the leaves. When a ring of bark, including the phloem, is removed between the foliage region and the roots, phosphorus, potassium, and inorganic nitrogen accumulated in the tissues of the stem and leaves above the ring and diminished in these parts below. They found that nitrogen, phosphorus, potassium, and other constituents of the ash ascend the stem mainly in the wood, are reexported from the foliage, and then move downward through the phloem toward the root. The calcium also ascends in the xylem, but there is no evidence that it moves downward from the leaves through the phloem. The ratios of nitrogen, phosphorus, and potassium in downward movement to the carbohydrate moving in the same direction appear to be in excess of the ratios required for the growth of the lower part of the plant. It is suggested that the excess of these elements may be liberated into the tracheal sap and thus again ascend the plant. They stated that this return toward the roots of many of the mineral nutrients that have been moved upward with the transpiration stream must influence the uptake of more nutrients by the roots. This may explain in part at least the reason why an increase in the transpiration rate does not affect the rate of the intake of salts.

The most serious objections that have been offered to ringing experiments is that the xylem may be altered by the ringing. Numerous authors have reported that tyloses or gums may plug the xylem elements in the region of a ring or other wound. However, in many of the cases that have thus been reported, no precautions were taken to protect the exposed xylem. Even when the ringed portion is protected by wax, vaseline, or paraffin, the xylem is in some way affected by the ringing, since it has been noted that under these conditions there is an increased resistance to the flow of water through it. The xylem, however, at the most, is only partially plugged in well-managed experiments, and this partial plugging cannot fully account for the hindrance in solute movement when the phloem is severed.

b. Through Phloem.—Curtis (1923 to 1925) conducted experiments to determine if the phloem is concerned in the translocation of the inorganic salts. The general method used by him was to ring twigs or branches by a complete or partial removal of the bark or to remove a portion of the xylem and to study the effect of this ringing upon the upward movement of nitrogen, ash, and carbohydrates. In these experiments all the ring wounds were covered with warm paraffin to prevent the entrance of bacteria and to prevent the tissue from drying out.

In the experiments with privet, peach, and lilac, the branches were selected in pairs, so the ringed and control branches were similar in size and position. Immediately after ringing, nitrogen as sodium nitrate was added to the soil. In some cases, calcium chloride and sodium chloride were also added. Observations were made upon the leaves above the ringed portions for area, dry weight, ash, and nitrogen content, and these quantities compared with those of the leaves from similar regions of the control branches. In the case of the peach, the total amount of nitrogen moving up the unringed stem was evidently much greater than that moving up the ringed stem, since there were from ten to twenty times as many leaves on the control stem and they were from 30 to 70 per cent larger than those upon the ringed twig. With the lilac the control leaves contained 1.8 to 2.2 times as much nitrogen per unit of dry weight as those from the ringed stem. The ash analysis also indicated that ringing hinders the movement of inorganic salts, since the check leaves contained from 1.5 to 2.0 times more ash than did those from the ringed stems.

In one series of experiments, Curtis (1925), by a technique that need not be described here, completely severed the xylem in one portion of a split stem, leaving the phloem intact. In another series, the xylem on one portion of the stem was left intact and the phloem severed by ringing. By the use of such divided stems where water was supplied by one set of roots and nitrogen by others, data were obtained which indicated that if the roots supplied with nitrogen were connected with the top by the xylem only, there was little transfer of nitrogen to the tops, while if they were connected by a strip of phloem only, considerable transfer of that element occurred.

Curtis also found that if a narrow band or strip of phloem containing from one-tenth to one-fourth of the normal amount was left intact in ringing, the transfer of solutes was normal. It has been shown many times that a small portion of the xylem cylinder will carry sufficient water to keep the leaves apparently normal. Curtis (1925) considered that, since a small amount of the phloem is sufficient to allow for an approximately normal solute transfer, while its complete severance causes a marked reduction, and since a small part of the xylem is sufficient to allow for approximately normal water conduction, while its complete severance practically stops water movement, it would appear that water moves chiefly in the xylem and solute movement occurs chiefly in the phloem. In later work MacDaniels and Curtis (1930) found that the phloem of the apple tree is the more important tissue concerned in the translocation of solutes since the severing of the phloem had the same effect as the severance of both the xylem and phloem.

There seems to be strong evidence which indicates that the inorganic solutes move upward in considerable amounts through the phloem. The cases of translocation, however, that have been studied are relatively few and have been confined to only a few species of plants. On that account, a much greater number and variety of plants must be studied before any general conclusion can be drawn.

2. Lateral Transfer.—The lateral transfer of inorganic solutes has received some attention. The most thorough work in this regard was done by Auchter (1923), who studied the lateral transfer of nitrates in 10-year-old peach trees, 5-year-old apple trees, 3-year-old privet bushes, and 5-year-old oak trees. In these experiments some plants received nitrates about all their roots, some had nitrates applied to only one-half of their roots, while others had all their roots removed from one side and nitrates applied evenly under both sides. In practically all cases the leaves increased in nitrogen content per unit of dry weight, per unit of green weight, and per unit of area on the nitrated sides. The leaves on those halves which had not received nitrate actually lost nitrogen in the cases where analyses were made before and after treatments. The check halves of the trees acted much the same as full-check trees, while the nitrated halves responded approximately the same as fully nitrated trees. In one experiment with Grimes Golden apple trees, the soil was cultivated on one side for 2 years, while the other side was not. The average terminal growth of the branches on the side that had been cultivated was 12 to 14 in., while that on the uncultivated side was only 3 to 4 in. It was considered that the additional nitrates in the cultivated soil were responsible for this difference in growth and that it indicated that they were used by the halves of the trees which were directly above the cultivated areas. Similar results in regard to fertilization and cultivation have been obtained by Blake (1921) for peach trees and by Knowlton (1921) for Rome Beauty apple trees.

In the case of dormant privet cuttings transplanted so that one-half of their roots were in quartz sand and the other half in fertile soil plus nitrate, Auchter (1923) found that the two halves of these stems broke their rest period unevenly. The sides above the fertile soil started to grow 3 weeks sooner and had made twice the amount of growth after a period of 5 weeks. Since nitrates are known to break the rest period of plants, it was considered that they had moved upward in the stem only on the side to which they were presented to the roots and that there was little or no lateral transfer.

Bodenberg (1929) believed that inorganic solutes cannot pass laterally through the stems of woody plants. He was unable to detect lithium nitrate spectroscopically in regions remote from the vertical line of flow in a willow stem after a period of 295 hr. MacDaniels and Curtis (1930) from their observations on the effect of the spiral ringing of apple trees reported that the lateral transfer of solutes in the stem of this tree is relatively slow, although it does occur when forced by spiral ringing. They stated that nutrient salts from below tend to move in straight lines parallel to the long axis of the conducting elements. Zimmerman, Hitchcock, and Crocker (1931) noted that ethylene gas enters a plant and moves rapidly in all directions from its point of entrance. It is not known, however, whether this gas diffuses through the intercellular spaces or whether it moves in the transpiration stream.

From these studies of woody plants it appears that the mineral nutrients absorbed by the roots on one side of a plant are, in a large measure, translocated to and used by the trunk, limbs, and leaves directly above them and that there is very little cross transfer of such nutrients in the plant. Since there are some indications of the cross transfer of water in stems, the mineral nutrients evidently travel independently of the water movement.

In the case of the corn plant, Gile and Carrero (1921) distributed the roots in three different flasks containing different nutrients. In one flask was a solution of potassium salts, in another the phosphorus salts, and in a third the nitrogen salts. There was a diminished assimilation of these nutrients under these conditions as compared with the plants growing in a solution that contained all these salts. It was considered that this diminished assimilation was not due to the inability of the roots to absorb the ions with sufficient rapidity but to the slowness with which the three ions were translocated to the cells where they were needed. These three ions thus each accumulated in different parts of the plant as a result of having been absorbed by different roots.

The same general results were obtained in a similar experiment performed by Carpenter (1936) with the roots of sugar cane. In this case all the roots of the plant except ten were removed. One of the intact roots was placed in distilled water and each of the other nine in individual solutions containing one of the elements necessary for the optimum growth of the plant. Henricksen (1933) found that the ash constituents in citrus trees move only through those conducting tissues which are connected to a definite portion of the root system.

White (1927) found that the entire vascular cylinder in the crown of the strawberry is composed of a network of short anastomosing bundles. These provide an efficient means of rapid transfer of water and solutes across the stem. Thus, in striking contrast to most plants, water and nutrients supplied to one side of a row of plants can apparently supply uniformly both sides of these plants.

3. Downward Transfer.—The downward movement of inorganic salts has received special study in leaves prior to their abscission. Before considering the migration of materials from the leaf in the autumn, a brief summary of the nature of leaf abscission will be given. According to Sampson (1918), Von Mohl (1860) was the first to state that previous to the fall of the leaf there is formed near the base of the petiole, a definite separation layer in which abscission always occurs by the separation of the cells from each other. The xylem elements are not included in this layer for they are eventually ruptured mechanically, following which the leaf falls. Von Mohl also stated that abscission and the formation of a protective tissue are two very distinct processes, and that the latter might either precede or follow the former.

The nature of the processes involved in the formation of the abscission layer and the fall of plant parts has been studied by Wiesner (1871, 1904, 1905), Molisch (1886), Tison (1900), Lee (1911), Fitting (1911), Hannig (1913), Lloyd (1914, 1916), Sampson (1918), and Heinicke (1919), and their observations may be summarized as follows: The formation of the abscission layer begins a short time before the maturity of the leaf and continues for a considerable period. Cell division begins to occur in two to four layers of cells across the base of the petiole. This begins in the epidermal and cortical regions and gradually extends inwardly to the phloem. Generally there results an abscission layer eight to twelve cells in thickness. These cells remain smaller and they have thinner walls than the adjacent cells of the petiole. After the formation of this abscission layer there is a disintegration of the middle lamella and of the cellulose of the secondary wall of the cells that compose this layer. The cells are thus no longer held together, and leaf fall results.

The external factors that accelerate leaf fall are extremely diversified. According to Lloyd (1914), they include high and low light intensities, high and low water supplies, high and low temperatures, low concentration of anesthetics, toxic condition of acids and salts, and wounding of the blade. Low concentrations of oxygen and high concentrations of anesthetics retard leaf fall.

The migration of the inorganic compounds from the leaves of deciduous plants previous to their fall has been studied by Tucker and Tollens (1900), Otto and Kooper 1910), Michel-Durand (1913, 1932), Swart (1914), Serex (1917), Rippel (1921), Combes and Kohler (1922), Seiden (1926), Deleano (1932), Deleano and Bordeianu (1932, 1934), Echevin (1932), Komatsu and Ozawa (1932), McHargue and Roy (1932), Deleano and Andreesco (1932), Murneek and Logan (1932), Thomas (1933), Denny

(1933), Polovrăgeanu (1933), Trandafirescu (1934), Sampson and Samisch (1935), and others. The leaves that have been studied included those from forest trees, fruit trees, and shrubs. With the exception of the findings of McHargue and Roy (1932) for certain forest trees, all the investigators give data that indicate the migration of nitrogen, phosphorus, potassium, magnesium, and iron from the leaves previous to their fall in autumn. The percentage of migration of a given element based on the maximal amount present during the life of the leaf has been termed the "reabsorption coefficient." The value of the reabsorption coefficient reported by various investigators has ranged from 1 to 90 per cent for the different leaves and elements considered.

It has been observed by Hornberger (1882) for corn, Snyder (1893) for wheat, Wilfarth, Römer, and Wimmer (1905) for a number of crop plants, Jones and Huston (1914) for corn, Burd (1919) for barley, and Miller (1937) for wheat that there is a migration of certain elements, especially potassium, from the stem and leaves to the roots and then to the soil. Penston (1935) has reviewed the work of several workers in regard to the loss of minerals from the plant.

4. Upward Migration.—There is evidence that during the development of new growth or of seeds that the vegetative parts are depleted of inorganic compounds to supply the raw materials for the synthesis of organic compounds in these developing parts. Thus Jones and Huston (1914) observed in the corn plant that the nitrogen in the stalks after the ears began to form decreased from 53.5 lb. per acre in August to 31.8 lb. in October, although the nitrogen content of the whole plant was increasing. The phosphorus decreased in the stalk from 15 to 6.3 lb. per acre during this same period, while it increased in the ear from 6.0 to 35.5 lb. per acre. The migration of potassium apparently was not so marked, since in the stalk it decreased only from 73.5 to 70 lb. per acre, while it increased in the ear from 10 to 25 lb. per acre.

Burke and Morris (1933) reported that the leaves and developing parts of an apple tree receive a considerable portion of their nutrients from the materials stored in the various regions of the plant, while the remainder is absorbed from the soil. Fagan and Watkin (1931) noted in the developing oat plant that there was a continuous migration of ash, phosphorus, potassium, iron, and nitrogen out of the straw into the spikelet from the time of its formation to the ripening of the grain. Miller (1937) noted that a portion of the minerals and nitrogen in the heads of wheat was obtained from the soil, while the remainder was drawn from storage forms in the stem and leaves of the plant.

C. Injection Experiments

A considerable number of injection experiments have been undertaken in an attempt to supplement the supply of water and nutrients to plants and to control fungous diseases and insect pests. These experiments should be mentioned here, as the factors involved are primarily those of translocation.

The injection of iron salts into the trunks of trees in order to overcome chlorosis has recently received considerable attention. Lipman and Gordon (1925) cured chlorosis, in lemon trees that had been chlorotic for several years, by the injection of a solution of ferrous sulphate. Bennett (1927) introduced iron salts directly into the trunk of pear trees that were affected with chlorosis. Over 6,000 trees were treated and a cure was effected in 90 per cent of the cases at a cost of 1.5 to 15 cents per tree, depending upon the size of the trees. Beneficial results were observed on those trees which did not fully recover from the chlorotic condition. Wann (1929) found that dry iron salts placed in an ordinary uncapped medicine capsule and inserted in holes bored in the stem of the plant proved effective in overcoming chlorosis of grapevines and of peach and apple trees. The iron salts used were ferric citrate, ferric phosphate, and ferric ammonium citrate, but the best response was obtained with the ferric phosphate. The leaves above the point of injection were the first to recover, after which the recovery spread gradually over the whole plant. Thomas and Haas (1928) were able to improve the chlorotic orange trees temporarily by the injection of iron salts. Subsequent injections, however, were necessary, and in a new place, since the wood was injured for some distance around the point of injection. Special methods for the injection of nutrient solutions into woody plants have been reported by Collison, Harlan, and Sweeney (1932); Roach (1934); and Thomas and Roach (1934) for apple trees. Collison and his workers believed that the injection method is a poor substitute for the natural intake of nutrients from the soil, but Thomas and Roach obtained increases in growth that were comparable to those obtained when the nutrients were applied to the soil.

Goff (1897) undertook to supply water in an artificial way to the roots of young transplanted trees. He placed a supply of water at a height equal to or greater than that of the tree and connected this with the cut end of a root. He reported that the buds on the treated plants developed more quickly than on the control plants and that the effects of this treatment were noticeable within 48 hr.

The injection of chemicals into plants for the control of fungous diseases has been undertaken by a considerable number of investigators. Bolley (1903 to 1907) found that trees would take up water solutions in large quantities when applied through a bore hole with an attached feeder. Chemical substances were carried to the most distal parts of the trees in a few hours. Thus in 10 hr. substances could be detected microscopically in the tops of cottonwood trees 30 to 45 ft. high. He was unable, however, to prevent fungous diseases by any injection experiments, but he considered that in some cases he held them in check. Mangin (1898), Morkrjetsky (1903), and Simon (1906) also considered that they obtained some control over fungous diseases by the injection of trees, potatoes, cauliflower, cabbage, and other plants. Injection experiments by Rankin 1917 and Rumbold (1915, 1920) using 80 or more chemicals with chestnut trees all gave negative results in the control of the chestnut-blight fungus. Scherer (1927) in injection experiments with white birch, gray birch, American elm, American chestnut, and apple trees obtained negative results in the control of fungous diseases and insect pests with the exception of apple trees that were injected with thymol. These seemed to show definite resistance to the progress of *Bacillus amylovorus*.

Injection experiments with potassium ferrocyanide have been tried for the control of insect pests, especially those in woody plants. The results that have been reported are extremely conflicting. Sanford (1914, 1915) and Shattuck (1915) have reported favorable results, while Surface (1914) and Flint (1915) could obtain no control whatsoever.

The conduction and behavior of the potassium cyanide when introduced into plants have been studied by Moore and Ruggles (1915) and Elliot (1917). Elliott found in smaller plants that the injury due to this chemical was very marked. On a hot, dry day the wilting of the leaves of herbaceous plants and small woody plants was noticeable under conditions of high evaporation within 10 min., and the fumes due to the potassium cyanide were readily detected coming from the leaves. Plants treated on cool, damp days showed little injury within 24 hr. The injury to these plants was also less extensive than to those treated on hot, dry days. The diffusion of potassium cyanide laterally is greatest when conduction is the slowest, and in woody stems Moore and Ruggles (1915) found that hydrocyanic acid passed through a particular area which had its point of departure on the upper side of the injection hole and that the chemical diffused but little in a lateral direction. Craighead and

St. George (1930) by an appropriate method allowed wood alcohol, copper sulphate, carbon disulphide, formaldehyde, mercuric chloride, potassium cyanide, and other compounds to penetrate the wood of pine trees for the control of the bark beetle. In a few hours, from 2 qt. to several gallons of liquid could be injected and distributed throughout all the outer annual rings to the uppermost branches and leaves. With some of the chemicals, complete brood mortality was obtained in some cases, provided that the application was made before the sapwood became stained blue by fungi and the ascending stream thus disturbed. The efficiency of the control of fungous diseases and injurious insects is limited apparently by the fact that a solution of any chemical that will injure the protoplasm of the cells of the fungus or insect will also injure the protoplasm of the host cells. The control of insect pests in woody stems by injections is further handicapped because the lateral conduction of solutions is very limited and the injected materials will come in contact for the most part only with those insects which are directly above the point of injection.

D. Translocation of Organic Compounds

1. Storage of Reserves.—Before discussing the translocation of organic compounds, a few facts should be stated in regard to their occurrence, storage, and depletion in the plant. The principal organic materials stored in plants are the proteins, carbohydrates, fats, and oils. The cells that are concerned in their storage are the medullary-ray cells, the parenchyma cells of the xylem and phloem, the parenchyma cells of the cortex, and the cells of seeds, bulbs, tubers, and rhizomes.

a. Formation of Reserves.—The place of the origin of these organic materials is an important point to consider in a discussion of their translocation. Are they manufactured at places other than that of their occurrence and then translocated to where they are found or are they manufactured in the cells in which they are stored?

Sugar is apparently the only organic compound for whose formation light is directly essential, although, as mentioned in Chap. VIII, the kind of sugar that is first formed is not known. From this primary sugar, all the other carbohydrates may be formed in any living cells regardless of light. The indications are that the proteins, fats, oils, and other organic compounds may also be formed in any living cell independently of light, provided that sugar, of one type or another, and the necessary inorganic materials are present. If such is the case, it is conceivable in the formation and storage of organic compounds that sugar of one kind or another is the only organic compound that needs to be translocated for long distances in the plant. Since it is formed, for the most part, only in the leaves and since it is found in either the free or combined form in every living cell, there is no doubt that it is moved farther and in greater abundance in the plant than any other organic compound.

There is evidence, however, in numerous cases that there is a movement of organic materials, other than the freshly manufactured sugars, from the vegetative part of the plant to the seeds and fruits in their formative stage. Thus Trowbridge, Haigh, and Moulton (1915) noted that the dry matter of the roots and stubble of wheat increased up to the milk stage, after which it decreased in amount, being moved to the parts of the plant above ground. The stalks and leaves contained their maximal amount of dry matter at blooming time, after which it decreased until the ripening of the grain. The protein in the stems and leaves decreased from 240 lb. per acre at blooming to 120 lb. at the maturity of the grain. The most rapid depletion during that period was from the milk stage to ripening. During that time the total carbohydrates decreased from 1,700 to 1,400 lb. per acre in the leaves and stems. Data quoted by Jones and Huston (1914) from German experiments showed during a 22-day period of grain formation that the nitrogen in the straw decreased from 83.5 to 25.0 lb. per acre, while the nitrogen content of the head increased from 23.4 to 60.8 lb. per acre. The dry matter in the straw decreased from 77.9 to 60.1 lb. per acre, while it increased in the head from 11.9 to 32.6 lb. per acre. The work of Jones and Huston (1914) with maize indicates that there is little depletion of the organic materials in the vegetative parts of this plant during the formation of the grain. Thus they found that the dry matter, fats, and carbohydrates in the stalk and blades remained practically constant, although they rapidly increased in the ear. Although the carbohydrates remained constant in amount in the vegetative parts, they increased from 4 to 40 lb. per acre during a 40-day period of grain formation.

Murneek (1928, 1932) found in the apple after flowering that the direction of movement of most organic compounds seems to be primarily if not exclusively toward the fruit that had set. The relative amount of foliage on a branch determines partially at least the size and quality of the apple fruit, but it has little or no effect on the weight and composition of the seed. The apple fruit draws heavily upon the products of photosynthesis during the latter part of its development. It was observed by Weinberger and Cullinan (1932) in peach trees that the average size of fruit on branches that had been ringed was greater by 20 per cent than those grown on branches not ringed. The organic materials that normally would have been translocated down the stream were prevented from moving and were thus available to the developing fruit.

Jones and Bisson (1932), and Bisson and Jones (1932) observed in the development of the garden pea that in the pod there was an early maximum of the absolute weight of nitrogen, total sugars, starch, and ash, after which there was a decline. In the peas there was an increase in the absolute weight of nitrogen, starch, crude fiber, and ash throughout the entire growth of 48 days. The absolute weight of sucrose reached its maximum in 32 days, after which it declined. Rewald and Riede (1933) noted that the seeds of the soybean during their development grew at the expense of the other parts of the plant. Fagan and Watkin (1931), and Miller (1937) found that a portion of the carbohydrates and nitrogenous materials in the seeds of wheat was obtained by depleting the stem and leaves of these substances.

The distribution of sugars and dry matter in the maturing corn stem from tasseling to maturity was studied by Welton, Morris, and Hartzler (1930). They found in the varieties Clarage and Burr Leaming that the percentages of dry matter and sucrose were larger in the upper than in the lower part of the stem. The percentage of dry matter increased and the percentage of reducing sugars decreased progressively from the tasseling stage to maturity in both varieties. Sayre, Morris, and Richey (1931), and Brunson and Latshaw (1934) found in corn that protein and carbohydrates tended to accumulate in other portions of the plant when the set of grain was prevented. The composition of the cobs was influenced more than any other part. They were higher in protein, fat, and ash, and lower in fiber in poorly filled plants than in those which had normal grain development. A sweet-corn plant usually does not produce so great a weight of grain as a dent-corn plant of equal size. By pollination studies Henson (1932) showed that this behavior results in part from the fact that the elaborated carbohydrates are more completely transferred to the grain from the vegetative parts of dent corn than from those of sweet corn.

In the development of the endosperm of corn, Lampe (1931) stated that the reducing sugars decreased as the cells approached maturity. The sucrose content of the endosperm in the several varieties rose gradually to its maximum at about the fifteenth day after pollination and then decreased but not so rapidly as the reducing sugars. In the nonsweet corns the disappearance of sugars from the endosperm was essentially complete at maturity, while in the sweet types a small amount of sugar, mostly sucrose, remained in the mature endosperm. This was mostly confined to the central region where the dextrin was located. Kretovitch (1933) stated that, in the wheat grain, sucrose, to the extent of 3 per cent, is present in the embryo. The embryo makes up 19 per cent of the weight of the grain. There is no sugar in the aleurone layer, and glutenin and albumin compose the major portion of the protein in this part of the seed.

b. Time of Storage.—The time of the storage of organic compounds as well as the tissues in which they are laid down should be noted, since considerable work has been done in order to determine the relation of the amount of these reserves to fruit formation, to the amount and quality of vegetative growth, to the conservation of plants, or to their eradication.

In timothy (Trowbridge, Haigh, and Moulton, 1915) the dry matter in the bulbs increased throughout the growing period, but the amount became constant before the ripening of the seed.

Working with alfalfa, Graber and his coworkers (1927) did not find any exact period of growth when the storage of organic food reserves began in the roots but noted that such storage was greatly accelerated during the periods of blooming and seed formation, as shown in the following table:

DISTRIBUTION OF ORGANIC MATERIALS IN THE TOPS AND ROOTS OF ALFALFA AT DIFFERENT PERIODS DURING THE GROWING SEASON

Adapted from Graber, Nelson, Leukel, and Albert, 1927

Stage of growth	Per cent of moist weight		Per cent of dry weight									
	Dry matter		Total sugars		Soluble starches, dextrins		True starch		Hemicellulose		Total nitrogen	
	Roots	Tops	Roots	Tops	Roots	Tops	Roots	Tops	Roots	Tops	Roots	Tops
Early-spring stage, 4/13	26.8	20.5	11.2	6.1	2.8	1.9	6.0	1.9	11.3	10.9	2.1	3.8
Flower-bud stage, 6/21	28.0	20.1	5.3	3.9	5.5	1.5	17.8	2.1	12.1	12.0	2.3	2.5
Full-bloom stage, 7/10	36.9	24.6	5.1	3.1	14.4	2.2	17.6	1.9	12.5	14.5	2.2	2.2
Seed-pod stage, 8/7	56.0	43.0	5.0	1.6	6.7	1.4	20.0	2.0	9.6	12.6	2.7	2.6
Fall dormancy, 10/22	44.8	26.8	4.6	6.1	4.7	1.6	24.9	3.1	8.2	11.9	2.5	2.9

Aldous (1929) found that in the stems of the sumac (*Rhus glabra*) and buckbrush (*Symphoricarpos vulgaris*) the largest amount of reserve materials was reached about Aug. 15 in both cases. In the former the minimum amount of stored materials was present about June 10, while in the latter the minimum was reached as early as May 10. The location of the food reserves in the two plants is of interest. In the sumac the major portion is stored in the parenchyma cells of the bark, while in the buckbrush practically all of it is stored in the medullary rays and parenchyma cells of the xylem.

In the 2- to 3-year-old portions of apple twigs, Traub (1927) found that the total reserve carbohydrate of the xylem tissues was higher than that of the cortex phloem at all times of the year and that this reserve reached its maximum after leaf fall. The pentosan content formed a relatively large proportion of the total reserve carbohydrates. They made up from 45 to 68 per cent of the total carbohydrates of the cortex phloem, from 65 to 93 per cent of the carbohydrates of the outer xylem, and from 58 to 83 per cent of those of the inner xylem. The acid hydrolyzable hexosans made up from 2.3 to 26.5 per cent of the carbohydrates of the cortex phloem and from 10.1 to 31.8 per cent of those of the outer xylem.

During the summer this hexosan content increased, but with the approach of cold weather there was a change of hexosans into sugar beginning about October. The total sugars in these parts also increased at this time, owing to translocation from the leaves. During the winter the hexosans disappeared almost entirely from the cortex phloem and were reduced more than 50 per cent in the xylem tissues. There was a general depletion of hexosans in May in all the tissues, owing to the translocation of sugar used in growth. The total sugar content of the phloem cortex was always markedly higher than that of the xylem tissues.

The nitrogen maximum in all parts was reached in March and April preceding growth in length. After a decline during the growth in length, the nitrogen content was relatively constant during the summer. From the fall of leaves through the winter it remained practically constant. The amino nitrogen was highest during the growing season, decreased as the rest period approached, and reached a minimum during the dormant period. The protein nitrogen was highest during the dormant season, decreased as the active growing season approached, and reached a minimum in June. The protein nitrogen was consistently higher in the cortex phloem than in the xylem tissues. Gardner (1925) observed that starch deposition in the shoots of the Bartlett pear begins very shortly after the cessation of length growth. Starch is deposited first at the tip of the shoot and then progressively downward toward the base. In the various tissues of the current season's growth, starch appears in the following order: medullary rays, wood parenchyma, and bark. Microchemical tests for sugar showed that the bark is much higher in sugar content than is the wood. The cortical parenchyma contains the largest amount of sugar.

Gäumann (1927) noted that the carbohydrates in the sapwood of *Picea excelsa* were at the maximum in October and April, and at the minimum in February and July. In *Abies pectinata* they were at the maximum in November and April, and at the minimum in February and June.

Kraybill, Sullivan, and Miller (1930) noted that in the Stayman apple tree there was a maximal content of starch in all parts of the tree after leaf fall in the autumn. This was utilized in the new growth in the spring, and starch storage began anew after petal fall in May. Grandfield (1930) found that the organic food reserves in *Solidago rigida*, *Verbena stricta*, and *Veronia baldwinii* were at their low point just before these plants began to show buds. Aldous (1930) noted that the organic reserves in the roots of little bluestem, side oats, Indian grass, hairy grama, buckbrush, sumac, ironweed, and vervain decreased in organic reserves to about the time of flowering after which these plants began to restore these reserves. Arny (1932) reported that there was a decline in the reserves of carbohydrates and nitrogenous compounds in the underground storage organs of leafy spurge, Australian field cress, and sow and Canada thistle from spring until blooming. Storage of these compounds then began in these parts and continued at a moderate rate until the close of the season. It was observed by Martin (1934) that in *Melilotus alba* the growth of tops predominated during the first of the growing season, while root growth was more prominent during the latter part. Grandfield (1935) stated that the total nitrogen and carbohydrate contents reached their maximal accumulation in the roots and crowns of alfalfa at or near the time of blooming. Long (1935) believed that the strawberry plant behaves in general like a biennial plant in that it stores large quantities of carbohydrates and nitrogen during the summer and autumn, and uses them the following spring for new growth. Culpepper, Caldwell, and Moon (1935) found that as the strawberry ripened it showed a marked increase of sugar, which makes up about 70 to 80 per cent of the soluble solids and about 50 per cent of the total solids of the ripened fruit.

Werner (1935) found in the potato that whenever carbohydrates were manufactured in excess of the capacity of the plant to build new tissues they accumulated in the potato tuber after a transitory period of storage in the leaves, stems, and stolons. Such accumulations occurred when the day length was shortened, temperature lowered, and the nitrogen supply reduced.

c. Factors Influencing.—Studies have been made by numerous investigators concerning the frequency and time of cutting of the aerial parts of plants on the amount of stored reserves in the underground parts. In some cases, the purpose of these studies has been to determine the methods of agricultural practice that will be conducive to the storage of a reserve in these parts to produce an optimum growth and development of the plants the following season. In other cases the purpose has been to determine the best practice to follow in order to obtain a maximal yield of hay or forage.

It was noted by Winkler (1929) that the nonpruned vines of *Vitis vinifera* contained from three to four times as much total carbohydrate at the beginning of the growing season as the normally pruned vines. In the case of kudzu (Pierre and Bertram, 1929), the fewer the number of cuttings the greater is the production of food reserves. The roots of plants receiving six cuttings per season decreased in weight during a period of 2 years. Those receiving four cuttings increased in weight 150 per cent, those receiving two cuttings increased about 400 per cent, while those from plants receiving only one cutting increased in weight about 1,250 per cent. The percentage of reserve starch and nitrogen was found to be less than one-half as much in the roots from plants receiving six cuttings as in the roots of those receiving four or a fewer number of cuttings.

It has been found by Aldous (1930) for certain prairie grasses, Shutt, Hamilton and Selwyn (1930) for meadow foxtail, Graber (1931) for blue grass, Parker and Sampson (1931) for *Bromus hordeaceus* and *Stipa pulchra*, Robertson (1933), and Biswell and Weaver (1933) for various grasses that the yield of hay or forage from these plants was least from those which had been cut or clipped frequently during the growing season and highest from those which were cut only at their maturity. Janssen, McClelland, and Metzger (1930) reported that the amount of sugar in the stalks of sorghum was slightly depressed by stripping them of the leaves 3 or 4 days previous to cutting.

The effects of defoliation or injury to the leaves upon the production of grain has been extensively studied in corn. The dominant idea in such studies has been to approximate "hail injury" and to determine the loss in yield that such injury might cause. It was reported by Culpepper and Magoon (1930) that defoliation reduced the yield of grain to the greatest extent when it was performed 10 to 15 days after silking, and that beyond this time the undesirable consequences of defoliation were lessened progressively as the treatment was delayed. Hume and Franzke (1929) and Dungan (1930, 1934) found that the removal of blades or portions of blades most severely affected the yield of corn when the treatment was applied during the early silk stage. Leaf removal before or after this stage is progressively less harmful to yields. Tearing the sides of the blades from the midribs but leaving them intact at the base resulted in a yield that was 70 per cent higher than the one obtained after the complete removal of the blades. The reduction in yields corresponded roughly at all stages with the percentage of the leaf area removed.

It has been reported by Sturkie (1930) for *Sorghum halapense*, Graber and Ream (1931) for bluegrass, Robertson (1933), Biswell and Weaver (1933) and Sprague (1933) for various grasses, Thornton and Nicol (1934) and Grandfield (1935) for alfalfa, Leukel and Barnette (1935) for Bahia grass, and Virtanen and Nurmia (1936) for sweet clover that frequent cutting or clipping of the tops during the growing season resulted in a marked decrease in the storage reserves of the root. Thornton and Nicol (1934) stated that the clipping of alfalfa did not significantly alter the number of nodules on the roots.

According to Thomas (1932) apple trees supplied in the spring with different combinations of sodium nitrate, potassium sulphate, and monohydrogen calcium phosphate in low concentrations showed marked differences in their storage reserves. Thus the concentrations of simple sugar in the trees that did not receive any sodium nitrate was much lower than in nitrogen-treated trees.

d. General Facts.—It was observed by Jacobson (1934) that the stems of Leoti Red sorgo had 39.9, 40.1, 38.6, 45.3, and 44.3 per cent sugar on a dry basis at the milk, soft dough, early hard dough, late hard dough, and ripe stages, respectively, while for the same stages, Atlas sorgo had 30.2, 40.2, 38.6, 35.4, 38.6, and 36.5 per cent sugar.

Gerry and Hall (1935) stated that the gum or oleoresin exudes from cuts on the living pine tree from its earliest seedling stage. This exudate does not come from the ordinary sap-conducting tissues but from the areas of parenchyma cells in the outer sapwood. These cells contain actively functioning protoplasm for a longer period of time than the associated cells that conduct water. Resin ducts are spaces that arise in the midst of clusters of parenchyma cells by a separation or splitting apart of those cells. In the slash pine there are more than 450 horizontal resin passages per inch of tangential surface, while the vertical ones number about 200 per square inch of cross section. The oleoresins are products of the protoplasm, but as yet not one fact is known with certainty concerning the process by which they are elaborated. In the oleoresinous exudate there are present glucosidal complexes of resin acids, volatile oils, and reactive sugars.

2. Utilization of Reserves.—These reserve organic compounds are used by the plant in respiration and in the formation of new plant parts, after a rest period as in the case of renewed growth in the spring or for the production of new organs and parts when these have been removed. Thus it has been found by Salmon, Swanson, and McCampbell (1925), Nelson (1925), Graber (1927), Albert (1927), and Leukel (1927) that in alfalfa new top growth is initiated in a large measure at the expense of previously deposited root reserves and that unless such reserves are sufficiently replenished during the period of successive cuttings, a reduction in the reserve contents of the root occurs, which diminishes the amount of new top and root growth, following each cutting. The effects of too frequent cutting at immature stages are to prevent the plants from forming organic food reserves in sufficient quantities for storage in the roots for their future development, as well as that of the tops. When alfalfa is cut too frequently, the vigor of succeeding growths decreases rapidly, and many plants may die during the growing season. It has also been observed that the amount of winter injury to these plants may be correlated with the time and number of the cuttings during the summer. The vigor of bluegrass, redtop, and timothy is also influenced by the number of cuttings during their growing season.

Gerhardt (1929) noted that in the common milkweed the leaf, stem, and root yielded carbohydrates and nitrogenous compounds to the developing seed. The rubber in this plant amounts to as much as 3.5 per cent, and it is stored for the most part in the leaves. Pentosans and hemicelluloses are stored in the stem, while starch and nitrogen are the main reserves in the root. Crafts and Kennedy (1930) believed that the large quantities of starch stored in the roots of *Convolvulus arvensis* makes possible the persistent vegetative activity of this plant. The rapidity with which this reserve can be depleted or replenished indicates the presence of a very efficient conducting system. Davis (1931) found that in the nonbearing sugar prune trees, the amount of starch was consistently higher than in the bearing trees. The roots of bearing trees contained a relatively small amount of starch, whereas those of the nonbearing trees were high in this carbohydrate. During the greater part of the summer the amount of total nitrogen in the bearing trees was lower than that in the nonbearing ones. Stuart and Appleman (1935) reported that potato sprouts were richer in total nitrogen than the tubers from which they grew. The nonprotein nitrogen was much higher in the medulla of the potato than in the cortex. It was noted by Vickery, Pucher, Leavenworth, and Wakeman (1935) in the tobacco plant that during the development of the seeds there was evidence that organic and inorganic substances were being transferred from various parts of the plant to them.

Bartel, Martin, and Hawkins (1935) found that tillers increased the development of the main stalks of dwarf hegari when the leaves of the main stalk were removed previous to heading. The longer the tillers were left on the plant, the greater was the weight of the main stalk. Although each tiller develops a root system of its own, a vascular connection exists between the main stalk and the stalk of the tiller.

In practice it has long been contended that if obnoxious brush and briars are cut at certain seasons of the year, few if any sprouts will arise from the roots and stump and that thus the plants would be eradicated. If, however, these were cut at other seasons of the year, numerous and vigorous sprouts would be developed, so that within a year or two the growth would be as heavy as or heavier than before. It seems very probable that this growth of sprouts is correlated with the amount of organic reserves stored in the stump and roots. If the cutting is done when the reserves are almost depleted, few sprouts develop; while if an abundant supply is present, new growth is very vigorous. Brown (1930) found under the conditions of Connecticut that July was the best time to mow pasture land to kill the undesirable plants.

3. Distance Moved.—Although it is definitely known that the reserve organic materials are utilized by the plant, the question arises as to how far these materials are moved from their place of storage to their regions of utilization. This question is of special interest in the case of woody plants where the distances between the various plant organs are relatively great. Since foods are stored in large amounts in stems and roots in late summer and autumn and since these disappear when growth is renewed in the spring, the general conclusion was early formed that foods are translocated from the roots and lower parts of the stem to the developing shoots and leaves.

It has been stated by Hartig (1858), Leclerc du Sablon (1906), and Butler, Smith, and Carey (1917) that the percentage of carbohydrates in the roots of various woody plants may be higher than that of the stem. This has been considered to indicate that the roots might act as storage organs for foods to be supplied to the developing parts of the stem. Curtis (1920), however, pointed out that the actual amount of carbohydrate and other reserves present in the roots may be less than that in the stem, since the total mass of stem is much greater than that of the roots, as shown by Chandler (1919), who calculated from the percentage concentrations given by Butler the relative amounts of foods available in the roots and tops of the apple tree. Aldrich (1932) considered that the roots of the Imperial apple obtained nitrogen from the soil during the winter months. This was also the opinion of Sullivan and Kraybill (1930) for Stayman apple trees in Indiana, and of Weinberger and Cullinan (1934) who noted that nitrogen might be absorbed by the roots of peach trees in autumn and early winter. This nitrogen was held in the roots until growth started in the spring, when it was rapidly translocated to twigs and buds. Murneek (1933) found that although the roots of an 18-year-old apple tree weighed only a third as much as the tops, the total quantity of starch present was about equal to that in the aerial parts.

Curtis (1920) ringed stems of *Acer saccharum* in early spring at different distances from the tips. In one series the rings were in the first-year wood, in another in the second- and third-year wood, and in another in wood ranging from 5 to 15 years of age. Of the 15 twigs ringed in the 1-year-old wood, the average growth in length during the following month was 0.84 cm., while that of the check stems was 2.22 cm. The terminal growth of those stems which were ringed in the second and third year's growth was 2.03 cm. as compared with 2.25 cm. for the corresponding checks. Those twigs ringed in the 5- to 15-year-old wood showed no difference in growth from the controls. Similar results were also obtained for the pear tree.

In apple twigs it was found that shoot growth was fairly vigorous when no food farther back than that obtained from a branch about

1 cm. in diameter was available. When the ring was no farther back than the 5- to 10-year-old wood, the growth of the shoots above the ring approached that of the unringed twigs. Soon after the shoots have started, much or all the food necessary for continuing growth is apparently produced by the leaves of the same shoot, so that in some cases the growth of ringed twigs with leaves was fully as great as that of twigs not ringed.

Haller (1930) studied the relation of the distance and direction of the fruit from leaves to the size and composition of apples. Branches were ringed, leaving above each ring one fruit and 30 leaves. These were left intact at varying distances from the fruit. It was found that apples can draw upon elaborated foods that are synthesized at a considerable distance from them. There was no decrease in the size or sugar content of the fruit when no leaves were nearer than 4.5 ft. in Grimes Golden, 6 ft. in Ben Davis and York Imperial, and 10 ft. in Baldwin, compared with the fruit produced when the same number of leaves were immediately adjacent.

Winkler (1932) stated that the average cluster of grapes requires all the food that has been synthesized by at least 16 leaves. This indicates that the fruit must use the foods that have been manufactured at a distance considerably removed from the region of its utilization. He considered that food moved upward and downward, in this case, for a distance of at least 3 ft.

Zimmerman and Connard (1934) by the arch method grafted the branches of tomato, coleus, salvia, and tobacco, and showed that both mineral salts and elaborated foods moved upward and downward through the stem.

Using the Bartlett pear, Gardner (1925) found that when the growth of twigs began in the spring, the stored starch disappears first from the tip and then from the lower regions of the shoot successively. In the various tissues of the shoot, the stored starch disappears in the following order as growth begins: bark, medullary rays, wood parenchyma, and pith. In some cases, at least, root growth occurs before shoot growth begins in the spring, as shown by Goff (1898) for a large number of woody plants. The observations of Jones, Edson, and Morse (1903) on the amount of water in the trunk of the sugar maple would also indicate that root growth precedes stem growth. They found that the amount of water in the stem increased from 36.5 per cent on Mar. 1 to 47 per cent on Apr. 28, at which time the buds opened and the water content of the stems fell off. Crider (1928) and Rogers (1929) have noted that roots of weeds grow under conditions that are unfavorable for the development of the aerial portions. Since apparently the roots of woody plants begin to grow earlier in the spring and continue later in the autumn than the twigs, it would seem probable that they would utilize a considerable portion of their stored foods. Since a twig ringed in that part of a stem from 5 to 15 or more years old produces a growth above the ring which approximates that of a normal stem, the indications are that the upward movement of food from points below the ring is not essential for the development of the twig. This statement is made on the assumption that the major portion of the organic materials is translocated through the phloem. Curtis considered that the distance from which food is withdrawn back of the tip to be used in shoot growth may depend upon the different species, individual plants, and the conditions of the present and previous seasons. The indications are that there is normally no movement upward to the tips of the branches from the roots and little or none from the main trunk.

4. Tissues Concerned.—No uniformity of opinion regarding the tissues that are concerned in the translocation of organic materials in plants has been reached by investigators. Some claim that the major movement is through the xylem and consider that the phloem is totally inadequate for that purpose. Others consider that the phloem carries the major portion of these compounds. Both sides present considerable experimental evidence that is worthy of brief consideration here.

a. The Xylem.—Those who consider the xylem as the main pathway for the translocation of organic materials base their opinion upon two general observations: (1) the physical unsuitability of the phloem for their translocation; (2) the composition and amount of the organic substances in the tracheal sap.

1. *Physical Unsuitability of the Phloem.*—Dixon and Ball (1922) and Dixon (1922 to 1924) presented theoretical evidence to show that the phloem is unsuitable for the translocation of organic substances. They mentioned that the phloem tissue is always relatively small in cross sections; that it is formed of short cells and comparatively short, narrow tubes with many cross partitions that must be traversed by the stream carrying the organic substances if this tissue is used as a pathway for their translocation. They considered that the resistance to a translocation stream must be high owing to the fact that a large proportion of the cross section of the phloem is occupied by viscous contents of protoplasm and proteins. They stated that the velocities of transport comparable with diffusion alone could be maintained only under such conditions, and that such velocities could not account for the quantity of material that is moved through the stem and roots.

To illustrate these points they cited as an example the translocation of carbohydrates from the leaves of the potato to its tubers. They selected a branch connecting the stem with a tuber that contained 50 g. of carbohydrate that had been deposited during a period of 100 days and determined that the area of the cross section of the phloem tissue of this branch amounted to 0.422 sq. mm., not allowing for the cell walls and nonfunctional elements. Assuming that all the carbohydrates in the tuber passed through the phloem tissue and that they were carried in a solution whose concentration was 10 per cent, 500 cc. of this solution would have passed in 100 days. The velocity of the solution in its passage would thus have been $\frac{500}{0.0042 \times 100 \times 24}$ or approximately 50 cm. per hour. The concentration of the solution, however, probably never exceeds 5 per cent; hence the rate of translocation must have approximated 100 cm. per hour. In the yam, Mason and Lewin (1924) believed that higher values than those for the rate of diffusion would have to be assumed. Dixon and Ball assumed that if a 10 per cent sugar solution were supplied by the leaves to the phloem elements and this were converted into an insoluble carbohydrate in the tuber 50 cm. distant as soon as it reached this destination, then a rate of transport by diffusion alone of only 0.2 mm. per day would be expected. They therefore considered that the observed rate of translocation of carbohydrates is so high that it cannot be accounted for by diffusion alone through the phloem elements, even though protoplasmic streaming is taken into account. They considered that these facts force one to conclude that the phloem elements are not adequate to transmit the amounts of carbohydrates actually known to travel downward in the stems and petioles of plants. Dixon and his students maintained that the major portion of the organic compounds are moved upward and downward through the conducting elements of the xylem in the xylem sap. Dixon considered that his observations on the seedlings of *Vicia* strengthens this view. He noted (1923) that the tracheal elements in these seedlings are differentiated in the petioles of the cotyledons much earlier than are the sieve tubes, so that before the latter are formed, considerable transport of organic substances to the embryo takes place. He believed that this is evidence that the tracheae and not the sieve tubes carry the organic substances.

Dixon and his pupils believed that the phloem is a means for the transmission of organic substance in a radial direction and considered that the medullary rays receive their organic materials from it. The phloem elements with the medullary rays and wood parenchyma present a large surface to the conducting xylem, which thus facilitates the interchange of the organic substances between the living cells and these nonliving conducting elements.

Curtis (1925), however, calculated the cross-sectional areas of the conducting tissues of potato rhizomes and found that the xylem in this case has an even smaller cross-sectional area than the phloem. The small area of the cross section of the phloem cannot thus be considered as evidence that the sugar travels in the xylem. Furthermore, according to Curtis, sugars have never been demonstrated to be present in the conducting elements of the potato xylem. It should be noted in this regard that the relative areas of the conducting xylem elements and the living cells of the stem differ widely for different plants, as shown by Dixon and Marshall (1915).

2. *Composition and Amount of Organic Compounds in Tracheal Sap.*—Hartig (1858) ringed a number of trees and noted after a time that the starch below the ring had disappeared. He concluded from this that

the reserve food products must have moved up through the xylem. As pointed out by Curtis (1920), however, he overlooked the fact that the foods stored below the ring might have been used locally for growth or have been moved downward through the phloem and used in root growth. Hartig considered that the materials formed in the leaves are passed down in the bark and stored in the wood parenchyma and medullary rays and in the spring are brought into solution and passed upward through the conducting elements of the xylem.

Schroeder (1871) made some observations on the sap of *Acer platanoides* and *Betula.* He noted that the sugar content of the sap from the birch decreased from early to late spring. The sugar concentration of the maple sap was greater in the upper part of the stem than in the lower, while the opposite condition prevailed in the birch.

Fischer (1888, 1890) by microchemical means demonstrated the presence of reducing substances in the conducting xylem elements of numerous trees and found that they were less in amount in the summer and autumn than in the spring. Dixon and Atkins (1915, 1916) examined the tracheal sap of a number of deciduous woody plants and found that the concentration of the soluble carbohydrates in the transpiration stream was greater than that of the electrolytes. The sugars were both the reducing and nonreducing types, and, estimated as glucose, their concentration ranged from 0.3 to 2 per cent depending upon the date of examination. The greatest concentration of sugar occurred in the early spring. This was followed by a rapid dilution in spring and early summer with a minimum occurring in the summer or autumn. A rise in concentration took place during the winter, culminating in a spring maximum just prior to the opening of the buds. These authors concluded that the conveyance upward of sugars is a continual and primary function of the xylem. The sheath of wood parenchyma around the vessels functions as a storage place from which sugars pass into the rising transpiration stream. They believed that the medullary rays convey carbohydrates from the bark to these sheaths.

The sap of the sugar maple has a sugar concentration of between 2 and 3 per cent. This flow of sap is considered good evidence that the conducting elements of the wood are able to carry this sugar both upward and downward at a comparatively rapid rate. No sap in this case comes from the bark (Adams, 1923). MacDougal (1926) found considerable amounts of reducing substances in the sap of the xylem of Monterey pine, the more concentrated solutions being in the outer rings.

Anderssen (1929), in the case of the tracheal sap of the pear, found that the free reducing substances and sucrose showed a high concentration in late winter and early spring, a rapid decrease in late spring, and a very low amount during the summer. These substances were found to be limited to the outer annual ring in 3-year-old pear branches during the late winter and spring. He thought this showed that the starch in the inner rings when hydrolyzed does not pass into the adjacent tracheae but is transferred along the medullary rays toward the outer tracheae into which it passes.

The nitrogen in the tracheal sap of the pear was found to consist almost entirely of organic nitrogen, and no nitrates could be detected until after the sap had been concentrated. On Nov. 10 the sap contained 60 p.p.m. of total nitrogen and on May 10, 163 p.p.m., of which 110 p.p.m. were present as amino nitrogen and 49 p.p.m. as amide nitrogen.

Some evidence has been presented that indicates that the living cells of stems and petioles are not necessary for the translocation of organic materials. This has been taken as proof that the organic matter moves through the xylem rather than through the phloem. The work of Deleano (1911) has been the most frequently quoted in proof of this. He chloroformed the petioles of grape leaves and found that the translocation of organic materials from the leaves continued, although at a reduced rate. Some movement of organic materials from the leaves also took place after the petioles had been killed by heat.

b. The Phloem.—It is the intention to state here in a general way the type of experiments performed to determine the relation of the phloem to the translocation of organic materials and the theories that have been proposed to account for their movement.

1. *The Kind of Experiments.*—Most of the experiments may be classified under the following headings:

(a) Ringing.—The view that the phloem elements are the channels for the movement upward and downward of organic compounds in the plant rests mainly upon the results obtained from ringing experiments in which the phloem portion of the stems has been severed by removing a ring of bark extending to the cambium. Experiments of this type were some of the earliest performed in plant physiology, and in general they have indicated that the downward passage of organic material is prevented or greatly hindered by the ringing. The death of a tree due to girdling is attributed to the fact that the downward movement of organic materials manufactured in the leaves is prevented to the extent that the roots die from starvation.

It has generally been assumed that the removal of the phloem in no way injured or affected the functioning of the xylem, but recent observations indicate that it does so unless special precautions are taken. On that account, the evidence obtained in the earlier ringing experiments cannot be taken as conclusive proof that the results obtained were due alone to the severance of the phloem tissue. It would be well now to review some of the more recent ringing experiments and consider the evidence thus obtained in regard to the part played by the phloem in the movement of organic material.

Curtis (1920) by ringing the twigs of *Philadelphus pubescens*, *Acer saccharum*, *Crataegus sp.*, *Pyrus malus*, and others found that the defoliated stems did not continue to grow after the starch supply above a ring was depleted. If the stem above a ring, however, was not defoliated, the leaves were able to furnish a sufficient supply of material to allow for considerable growth. The carbohydrates stored in the xylem below the ring can apparently not be removed through the xylem but may be transferred radially to the phloem, where they may be carried downward if there is no second ring below.

Mason (1922) in the case of the bitter cassava (*Manihot utilissima*) found that the final weight of the tuberous roots of the ringed plants was only about one-fourth that of the unringed stems. Gardner (1925) in a study of the conductive tissues in the shoots of Bartlett pear concluded that the phloem is the tissue largely concerned in the longitudinal movement of foods. By using potassium hydroxide a ring was made which did not injure the xylem, yet it did not conduct the food stored between two rings past them, while the starch above and below the rings was removed. Both old and new tracheal tubes of excised pear branches allowed water and solutions of glucose and asparagine to pass through under pressure. In his opinion there is little reason to believe that the outer and inner xylem act differentially in the conduction of foods and water, because only the outer xylem conducts foods, while the whole cross section is used for water transport. He found that ringing did not induce the formation of tyloses within the tracheae of the pear shoots but that it did affect the xylem by actual mechanical injury by clogging and by drying the outer vessels. The inner xylem, comprising 75 to 80 per cent of the ringed shoots, was uninjured and was available for carbohydrate conduction but was evidently not used for that function. In *Robinia pseudoacacia*, tyloses effectively blocked the passage of water and solutes of glucose and asparagine through the innermost xylem. The conduction of foods in this species is thus apparently limited to the phloem or to the outermost xylem.

Loomis (1934, 1935) reported that in woody plants ringing stopped the upward movement of the organic nitrogenous compounds formed in the roots and the downward movement of the carbohydrates formed in the leaf. Under ordinary conditions, the protoplasm of the phloem cells is impermeable to these organic compounds so that they do not pass into the xylem. Their movement is thus confined to the plasmodesmal connections between the living cells of the phloem. Loomis noted that, after 4 to 5 weeks, ringing lost its effectiveness in checking the upward and downward movement of organic compounds. He suggested that the cells of the ringed segments became increasingly permeable with time so that these compounds diffused into the xylem, and were thus moved past the interruption in the phloem. Phillis and Mason (1936) in ringing experiments upon the cotton plant observed that in intact stems, nitrogen traveled upward in considerable amounts from the basal to the apical region, while little or none entered the apical region of the ringed group. They inferred that some of the nitrogen at least which entered the apical region of normal plants traveled in the phloem, and they concluded that nitrogen and carbohydrate could travel simultaneously in opposite directions in this tissue.

(*b*) *Insects*.—The position taken by certain aphids on the vascular bundles of leaves has been considered an indicator of the tissue in which the organic materials travel. Thus Davidson (1921) by fixing the aphids *in situ* found that they seek the phloem part of the vascular bundle when tapping the plant for nutrients. The observation of Schneider-Orelli (1909) on the influence of a leaf miner upon translocation is worthy of note. He found that when it had bored through veins the storage of starch in the tissues was affected. It was observed that the destruction of the xylem and the greater portion of the bundle sheath could be brought about without causing an accumulation of starch in the distal part of the leaf, but that an injury of phloem resulted in an accumulation of starch proportional to the extent of the injury.

(*c*) *Protoplasm*.—It has previously been mentioned in this chapter that the sieve tubes, companion cells, and phloem parenchyma are living cells and have a high colloidal content compared with the vessels and tracheids of the xylem. It has also been mentioned that the presence of the contents of these cells has been considered by some to prevent the translocation of organic matter in sufficient amount, by simple diffusion, to account for the total quantity that is actually moved. Some, on the other hand, consider that the protoplasm plays some role in this translocation, and considerable experimental work has been done to determine if it is actually concerned in this process.

DeVries (1885) pointed out that diffusion alone is apparently too slow to account for the movement of organic compounds that actually occurs. He suggested that the rotation and circulation of the protoplasm might aid in the translocation process, since he and others had noted protoplasmic streaming in the phloem tissue. Curtis (1929) observed streaming in the companion cells and phloem parenchyma in numerous plants but in no sieve tubes except those of *Elodea*. He considered that injury in preparing sections might account for the failure to observe streaming in the mature sieve tubes and believed that it is very likely that rotation and streaming of some sort do actually occur in them under normal conditions.

Czapek (1897) and Deleano (1911) found that chloroforming the petioles without killing the cells delayed starch removal from the leaves. Czapek did not consider the streaming of the protoplasm as essential but seemed to favor some secretive action by the protoplasm. Steward (1928), however, believed that streaming of the protoplasm might be the cause that facilitates the diffusion of the organic compounds through protoplasmic membranes.

Curtis (1928 to 1929) studied the effects of chilling upon the translocation of carbohydrates. In the case of leaves, he encased the petioles with rubber tubing through which water of a desired temperature could be run. He found that the total carbohydrate content in the leaves with the chilled petioles was always higher than that in the controls, the ratio of the total carbohydrate of the chilled leaves to the checks being 1.37. The ratio of the sugar content of the chilled leaves to the checks was 1.49. The chilling of the petioles to 2 to 3°C. clearly interfered with the removal of sugar from the leaf, but chilling to 7 to 9° did not cause a significant retardation. Using plants grown in sand and watered with known chemicals and treating the stems as mentioned above, Curtis found that chilling interfered with the upward movement of inorganic nutrients. Thus the gain in total ash of the unchilled plants was 58.3 per cent of that at the beginning, while that of the chilled stems was only 35 per cent.

The temperatures at which chilling interferes with the translocation of materials is approximately that at which streaming in protoplasm ceases. The evidence indicates that living cells take an active part in both the upward and downward movement of materials and that this movement is probably facilitated by the streaming of the protoplasm. It should be noted that chilling apparently has little or no effect upon water movement, since plants bled as freely from the cut ends of chilled stems as they did from the unchilled.

Schumacher (1930, 1932) concluded that the sieve tubes are the channels of transport for organic materials. He observed the location of the potassium salt of fluorescein by means of ultraviolet light. He found that this compound is taken up by leaves and is transported by the protoplasm into the sieve tubes and then downward.

2. *Theories of Movement.*—It is apparently well established that the translocation of the major portion of the organic compounds in plants is through the phloem. The means, however, by which this movement of these compounds is accomplished are as yet not known (Kruseman, 1931).

A rather elaborate theory of transport in the phloem was proposed by Münch (1926, 1927, 1930). According to Curtis (1935) this theory suggests that there is a unidirectional flow of phloem contents that is caused by a pressure gradient. This pressure gradient is established and maintained by differences in osmotic concentrations of supplying and receiving tissues. In applying this theory it is assumed that a semipermeable membrane separates the living cells of the phloem from the xylem tissues. The plasmodesms are supposed to allow for the flow of solutions from one living cell to another, while the sieve pores allow for similar flow through the specialized sieve tubes. Thus, according to this theory, if the supplying cells or tissues have a high osmotic concentration, the receiving cells a low concentration, and the plasmodesms and sieve pores allow for mass flow, there should be a transport of solution from the cells with high concentration and high turgor to those with low.

The theory of Münch has been adversely criticized by Weevers and Westenberg (1931), Curtis (1935), and Mason, Maskell, and Phillis (1936). Three of the numerous objections as stated by Curtis are as follows: (*a*) The theory does not account for the observed simultaneous movement in both directions through the phloem, (*b*) the pressure gradients available are insufficient to cause a flow of solution through the conducting tissues into receiving cells and exudation of water from the receiving cells, and (*c*) existing osmotic and turgor gradients have been observed in many cases to lead in a reverse direction from that required by the proposed hypothesis. Curtis (1920), Fernald (1925), and Curtis and Scofield (1933) found in certain cases that the osmotic relationships of plant tissues were directly opposite to those demanded by the hypothesis of Münch.

Crafts (1931, 1932, 1933, 1936) considered that the movement of organic materials in plants occurs through the entire cross section of the phloem including the cell walls. The phloem exudation in cucurbits indicates that the movement is not of local origin but is a direct manifestation of a process occurring throughout the plant. As a result of this process, organic foods are moved over long distances in a relatively short time. It appears that this conduction is of the nature of an elongated, somewhat elastic osmometer working under a positive pressure and activated by a concentration gradient of its osmotically active contents. Crafts stated that the mechanism of xylem exudation appears to be of a different nature for the ordinary osmotic action cannot account for the rapid flow obtained from plants. When root pressure is manifested, the xylem vessels have their total contents replaced many times during the day. Any theory of root pressure that fails to explain the simultaneous absorption and secretion of water and inorganic ions into nonliving vessels is inadequate. Although the movements of organic and inorganic materials may not be directly correlated, they are both in progress at the same time and involve the same general tissues.

Phloem sap may be coagulated in the stem by heat, and if a heated portion of a stem is cut no phloem exudation occurs, while heating has no visible effects upon xylem exudation. Exudation from the cut xylem depends upon conditions affecting water loss, since the water deficit must be satisfied before an actual exudation occurs. Phloem exudation will occur from severely wilted plants (Mothes, 1931), and, although the volume of the exudate is reduced in proportion to the water deficit, the dry weight composition tends to increase.

Steward and Priestley (1932) adversely criticized the theory propounded by Crafts that there is a movement of organic materials through the cell walls. Mason and Maskell (1934) reported that their experimentation strengthened the hypothesis that movement of materials along the phloem is determined independently for each material by the concentration gradient of its mobile form in the channel of transport. Mason, Maskell, and Phillis (1936) considered that neither the Münch theory of mass flow, nor the theory of protoplasmic streaming gives a satisfactory explanation of the movement of organic materials in the cotton plant. The student is referred to the monograph by Curtis (1935) for a detailed discussion of the movement of solutes in the plant.

5. Lateral Translocation.—The lateral transfer of organic materials in plants is for the most part in a radial direction. This movement takes place principally through the medullary rays but is never so marked as the longitudinal movement. Translocation around the stem is apparently almost negligible. It has long been observed that the annual rings of trees on the side directly under the larger limbs or on the sunny side of trees bordering on a forest are much larger than those on the opposite side. Czapek (1897) in ringing experiments found that ringing could be nullified only by vertical bridges connecting the upper and lower portions of the bark. Oblique and zigzag bridges were ineffective in the transference of organic materials across the ring. This indicates that there is little lateral translocation. Auchter (1923) obtained evidence in the case of fruit trees which indicates that the foods manufactured on one side of these are used and stored mainly in that side or are translocated for the most part, to the roots directly beneath. Caldwell (1930) noted that the removal of the leaves from one side of the crown of the Swede turnip during the early stages of growth caused the development of a root that was flattened on the defoliated side. The removal of the leaves from one side of a sunflower plant was followed by an asymmetrical development of it. When differentiated treatment of the roots of 2-year-old privet cuttings was performed, there resulted a considerable difference in the foliage on each side.

6. Translocation of Carbohydrates.—It has previously been mentioned that sugars are translocated farther and in greater quantities than any other organic compound and are almost without exception the form in which carbohydrates are moved. Mason and Maskell (1928) in their thorough and detailed study of the transport of carbohydrates in the cotton plant have given us the most extensive information concerning carbohydrate translocation that is at present available. On that account their results will be reviewed here in considerable detail.

They found that the concentration of the total sugars in the cell sap of the bark was greatly in excess of that of the wood or leaf. Thus in one 24-hr. period the sugar content of the xylem varied from 0.05 to 0.07 per cent, the sugar content of the bark from 3.0 to 3.5 per cent, and the sugar content of the leaf from 1.5 to 1.8 per cent. In another observation it was found that the total sugar concentration in the boll was approximately 0.5 per cent higher than that of the bark, the concentration ranging from 5 to 6 per cent in both. Under the same conditions the sugar concentration of the leaves varied from 1.5 to 2.5 per cent. The concentration curves of these three plant parts, however, ran parallel, for the most part, but those for the bark and boll lagged behind that of the leaf. The sugar concentration rose in all three of these parts during the day and fell during the night, but the concentration in the wood was not appreciably affected by the change from day to night. The diurnal variations in the concentration of total sugars in the leaf were more highly correlated with variations of the concentrations in the bark than in the wood. In two experiments in which samples of the bark were taken from different levels on the stem it was shown that the concentration of the total sugars in the upper part of the stem was higher than that in the lower portion.

Reducing sugars were present in the leaves greatly in excess of sucrose. In the boll, the excess of reducing sugars over sucrose was especially marked. In the stem, however, sucrose was in excess of the reducing sugar, the concentration of sucrose in the bark increasing during the day and decreasing during the night, while the reducing sugars remained relatively constant. Phillis and Mason (1933) noted that sucrose is the only sugar in the cotton plant that shows well-marked and consistent diurnal changes in the lamina and petiole of the leaf accompanied by corresponding changes in the rate of transport.

Mason and Maskell (1928) believed that their chemical analyses suggest that in the cotton plant the translocation of sugar occurs in the bark. As the major part of the fluctuations in the total sugars of the bark was found to be due to changes in the sucrose, they believed that the bulk of the carbohydrate translocated through this stem travels in that form. They considered that sugar moves from the leaf to the bark by a process similar to physical diffusion in that concentration changes in the leaf are followed by similar changes in the bark. They assumed that even though the exchange of the solute may take place against a concentration gradient, the possibility of a definite causal relationship between concentration changes in the leaf and bark is not excluded. Since a concentration gradient exists in the bark down the stem, the movement of sugars down the stem toward the roots resembles movement by simple diffusion, in that the direction of movement is from a region of high to one of low concentration.

Since the sucrose concentration of the bark was found to be on an average 2 per cent higher than that of the boll and since the rate of transport of sugar into the boll tended to increase as the sucrose gradient from bark to boll increased, it was concluded that the sucrose gradient between bark and boll is probably an important factor in determining variations in the rate of carbohydrate transport through the pedicel to the boll. It was also shown that the rate of transport of sugar into the boll was approximately four times greater by day than by night, a fact that is in harmony with the preceding conclusions, since the variation in the sucrose concentration in the bark follows closely in the changes in sugar concentration in the assimilating organs due to alternation of day and night.

By well-controlled ringing experiments, Mason and Maskell obtained further evidence that the transport of carbohydrates in the cotton plant occurs in the bark. (1) Ringing caused an accumulation of carbohydrates, not only in the bark and wood just above the ring but also in the leaves 2 ft. distant from it. (2) Ringing interrupted the downward transport of carbohydrate in the stem but did not interrupt the downward movement of a dye in the wood. It was thus concluded that the movements in these two parts are fundamentally different. (3) When the wood and bark were completely separated from each other, the movement of sugar took place through the bark at nearly a normal rate, but no movement of sugar through the wood could be detected. Baldwin (1934) found that from 3 to 4 years were required for girdled hardwoods to die. From 20 to 300 times as much sugar accumulated above the cut as below, and from two to three times as much as in normal trees. Two years after girdling, the sugar reserves above the cut were lower in amount than in the trunk of normal trees. In order to determine what portion of the bark is concerned in the translocation of sugar, Mason and Maskell subdivided the bark into three zones—outer, middle, and inner of approximately equal cross-sectional area—and analyzed each of these portions. They found that sucrose increased in concentration from outside to inside, while reducing sugars increased in the reverse direction. The radial gradient of reducing sugars was relatively stable, but the gradient in sucrose fluctuated markedly, increasing or decreasing as the sugar supply in the bark increased or decreased. There was a high positive correlation between the percentage of sieve tubes in any zone of bark and the sucrose concentration in that zone, and a high negative correlation between the percentage of sieve tubes and the concentration of reducing sugars.

The sugars, in particular the sucrose, in the sieve tubes fluctuated markedly with changes in the sugar supply to the bark. The changes in the sugars of the parenchyma were, however, very small. The gradient in sugar concentration from sieve tubes to bark parenchyma was found in some cases to be 500 times as steep as the vertical gradient in the sieve tubes. There was a high correlation between the concentration of total sugars in the whole bark and the concentration in the sieve tubes, so that the gradient in the whole bark could be used as an estimate of the gradient in the sieve tubes. These observations appear to offer conclusive proof that the longitudinal translocation of sugars for the most part takes place through the sieve tubes.

It should be kept in mind, however, that all the individual carbohydrates in plant material are capable of conversion into each other, and that in being transported from the leaves to the roots, a given carbohydrate may have been transformed numerous times (Virtanen and Nordlund, 1934).

It has been noted by Jost (1907), Mangham (1910 to 1922), Dixon and Ball (1922), and others that physical diffusion alone seems much too slow to account for the large quantities of carbohydrates that are moved longitudinally in plants. Mangham (1917) proposed a theory to account for the translocation of carbohydrates through the phloem elements that does not involve the physical diffusion of sugars through the protoplasmic membranes. He assumed that in the plant protoplasm there are present constituents capable of adsorbing sugars from solution; that for any given concentration of sugar present in the liquid phase of the protoplasm and the cell sap continuous with it there would be a definite concentration of sugar present at the adsorbing surface; and that any alteration of concentration in any region would lead to a readjustment of concentration equilibrium which would be transmitted or propagated as a wave through the system of these adsorbing particles and the solution immediately in contact with them. The connecting threads of the sieve tubes are assumed to provide a continuous protoplasmic pathway so the sieve plates would cause little if any obstruction to the progress of the wave of readjustment of concentration equilibrium. He considered that the sieve tubes would thus permit the passage of the waves with more effect than would be possible in any other type of cell except laticiferous cells.

The work of Mason and Maskell (1928), however, indicates that sugars move through the sieve tubes by a process similar to diffusion. Thus if sugars move in this tissue by diffusion, it would seem possible that the rate of movement of sugar at any point would be proportional to the concentration gradient in the channels of transport and to the cross-sectional area of the channels. Mason and Maskell investigated this point by examining the changes in rate produced by reducing the area of the channels of transport. It was found that a reduction by ringing in a cross-sectional area of the bark of approximately 45 per cent caused a reduction in the rate of sugar translocation of only 22.4 per cent. Thus the rate of transport across unit cross-sectional area of the bark in the constricted area was increased by reducing the area available for transport. This increase in rate was accompanied by and could be explained in terms of an increase in the gradient of sugar concentration across the constricted region.

The rate of sugar movement through the sieve tubes is, however, much higher than that which ordinarily occurs in purely physical diffusion. Mason and Maskell (1928) stated in the case of the cotton plant that the observed diffusion constant of sugar in the sieve tubes is about 40,000 times as great as the diffusion constant for sugar in a 2 per cent solution of sucrose in water and that it is almost identical with the diffusion constant for molecules of the size of the sucrose molecule diffusing in air. Although there seems to be good evidence that the carbohydrates travel as sugars in the sieve tubes, no satisfactory explanation has been given of the manner in which these products are moved with adequate speed in the required amount (MacDougal, 1926). The rate of movement of sugar from bark to wood appears, on the other hand, not very much greater than would be accounted for by purely physical diffusion.

The translocation of carbohydrates in germinating seeds and in seedling plants has been extensively studied by Sachs (1862), Detmer (1880), LeClerc and Breazeale (1911), Boysen-Jensen (1915), Choate (1921), Toole (1924), Yocum (1925), and others. Only a few of these observations will be discussed here, as detailed information may be obtained by consulting the references mentioned above. It should be kept in mind that the transformation of carbohydrates from one form to another is very frequent and rapid in plants, especially in seedlings, so that any given molecule of sugar passes no considerable distance without changing into other sugars or into insoluble carbohydrates for temporary storage.

In the wheat seedling, Choate (1921) detected reducing sugars in the coleorhiza, root, endosperm, coleoptile, plumule, and scutellum after 12, 18, 24, 36, 48 and 96 hr., respectively, while starch appeared in the root cap in 12 hr. after the beginning of germination. In the endosperm, reducing sugar is first found near the basal end of the embryo but is eventually found throughout the whole tissue. At the end of 7 days, starch was still present in the endosperm, although all the grains were corroded. Yocum (1925) noted that the greatest amount of reducing sugar in the plumules and roots was present at the ninth day. There was no evidence of sucrose formation taking place in the seed. When the starch showed a rapid loss the dextrins increased and then dropped off at about the same rate as the total sugars, but at no time were there sufficient dextrins in the roots and plumules to indicate that translocation of dextrins occurs. The starch content of the seed and seedling dropped off rapidly the first 7 days, the decline being extreme from the third to the sixth day. The amount of sugar was highest at 6 days. The proportional increase of plumule and roots in dry weight was greatest at this period, so the greatest starch hydrolysis and the greatest sugar content accompanied the greatest growth. Starch was found in the plumules and roots at all stages. It appears that if the seedling has good growing conditions, it will store temporary starch in all its tissues. The seedlings of wheat depend entirely upon the endosperm for their carbohydrate supply for 6 days. From the sixth to the twelfth day they depend only partially on this source, after which they depend entirely upon photosynthesis for this supply. Toole (1924) in the case of germinating corn could find no reducing sugars in the cells of the scutellum, although sucrose was present in about the same amount as in the dry embryo. He thought that sucrose might represent the translocation form of carbohydrates in the corn seedling.

7. Translocation of Fats and Oils.—Our knowledge concerning the translocation of fats and oils is very meager and is confined almost entirely to observations of the germination of fatty seeds. The known facts may perhaps be best presented by reviewing the results that have been reported.

Sachs, as early as 1859, observed considerable quantities of oil in the hypocotyls of seeds and concluded that it had moved out from the seed in that form. He believed, however, that the greater portion of the reserve oil in seeds is transformed to carbohydrates for transport and use.

Schmidt (1891) reported extensive experiments on the translocation of oil in seedlings. He believed that in many cases oils are transported as such to the different regions of the seedling, where it is broken up into the necessary products for growth. His conclusions were based on the following experiments: Pea seedlings wer grown in the dark until the stored foods of the cotyledons were exhausted. A longitudinal incision was made just above the ground and a strip of filter paper saturated with oil was inserted in the incision. By using colored oils, their presence in the intercellular spaces and in the cytoplasm could be detected. When almond oil mixed with oleic acid was used, he observed that the rise of the oil was increased as the amount of fatty acid was increased. He reported that the neutral oil with which he worked did not enter the cell unless there was present at least 10 per cent free acid. When pure fatty acid was used, its presence could be detected in the cytoplasm within a few hours. Schmidt explained these results by assuming that soaplike compounds were formed with the free acid and that the emulsification of the neutral fat by these soapy substances occurred. He thus assumed that the neutral oil passed through the cell walls from cell to cell as a fine emulsion.

Rhine (1926) repeated the work of Schmidt. He found that the movement of oil in plants treated after the manner of Schmidt was through the intercellular spaces of the outer region of the cortex. Contrary to the observations of Schmidt, however, he found that a fatty acid in the pure state did not rise so rapidly as did the neutral fats of these acids. Thus linseed oil rose at the rate of 2 cm. per hour, while the fatty acids of this oil rose at the rate of only 1 cm. per hour. The fatty acids entered the cytoplasm of the cells from the intercellular spaces and appeared in the cytoplasm in the form of small droplets. Linseed oil, however, with a free acid content of less than 2 per cent penetrated the cell walls as well as the free acid.

Rhine believed that since the oil in the cells during translocation is in the emulsified state, while the oil on the outside of the cells in these experiments was pure and homogeneous, the cases could not be considered comparable. He believed that the movement of oils through the cell walls must be demonstrated from the emulsion state with water if the continuous phase of the experimental results is to be taken as an indication of what actually occurs in the plant. When pea seedlings were placed in an emulsion of linseed fatty acids, no intake whatsoever of the oils took place.

Since the seedlings that had been used in the foregoing experiments were starved and etiolated, it was considered that an abnormal condition might prevail in the walls bordering on the intercellular spaces. In order to investigate this point, two sets of seedlings were treated. One set was well supplied with water and the other partially desiccated by drying with a fan. In the seedlings that had an abundance of water, the rise of the oil in most cases was zero and no penetration of the cell walls occurred. The increased drying effects of air currents increased the rate of the ascent of the oil and facilitated its penetration. It was thus definitely shown that the observed intake of oil by plants was due to an unnatural drying out of the cell walls—a condition that does not prevail in the cells of germinating fatty seed through which oil has been supposed to pass.

It was suggested by Schmidt (1891) that soaps formed from the fatty acids might facilitate the movement of oils through the cells. If soaps are found to function in that manner, the tissues must be alkaline in order that they may not decompose. Rhine (1926) determined the pH value of the hypocotyls of a number of seedlings of fatty seeds with the following results:

Hypocotyls of	Range of pH
Peanut	5.4 to 5.8
Sunflower	5.2 to 5.6
Castor bean	5.2 to 5.6
Hemp	6.2 to 6.6

The phloem of these tissues was examined with special care, since it has been reported to be alkaline, but its acidity was found to be the same as that of the adjoining tissues. Under the conditions of acidity that were found, it would be difficult to conceive of the existence and movement of soaps through the tissues.

The respiratory quotient of the hypocotyls or other growing parts of the seedling should serve as an index to the nature of the food that is being translocated through them. If oil is being translocated as such and moved into the hypocotyls, it would be expected that oil would be used in respiration, and in that case the respiratory quotient would be relatively low, approximately 0.6. If the hypocotyl is receiving sugar, it would be used in respiration, and the respiratory quotient of the hypocotyl would be approximately 1.00. Rhine found that the respiratory quotient of the hypocotyls of the seedlings of sunflower and cotton averaged 0.770, while that of the pea, wheat, barley, and buckwheat averaged 0.775. The respiratory quotient of the hypocotyl of these germinating oily seeds was thus almost the same as that of starchy seeds. This indicates that the food supplied to the hypocotyls by the storage region of fatty seeds reaches them in a form that yields the same value for the respiratory quotient as the food supplied by the typical carbohydrate storage seeds. This food is commonly supposed to be sugar.

Rhine studied the oil gradient from the source of supply to the growing tip in the case of the seedlings of cotton, sunflower, squash, cocklebur, castor bean, peanut, and flax. In no case did he find a gradient from the storage region to the growing tip, but in every case there was a distinct gradient in the other direction; *i.e.*, the amount of fat increased with the approach toward the tip of the hypocotyl, as shown in the following table:

Seedling	Percentage of ether extract on wet basis		
	Tips	Middles	Bases
Cotton	5.75	2.24	1.33
Sunflower	5.31	1.34	0.66
Squash	2.59	0.60	0.22

If oils are moved as such from cell to cell, some would be expected to be present in the cell walls. Rhine, however, was unable to observe in either cross or longitudinal sections any evidence of its passing through the walls.

Rhine concluded that from a physical standpoint the movement of oils as such through the cells is improbable. In moving downward into the hypocotyls, the oil must pass through the cytoplasm whose specific gravity is 1.00 or greater. The various extracts of the hypocotyls made by Rhine showed in all cases a specific gravity lower than that of water. Oil movement downward if it occurs would thus have to take place against the pressure of a denser medium, a procedure that would seem rather improbable.

The presence of oil in the hypocotyls of seedlings is no proof that it has been moved there as such. Starch is present in most cases in the growing regions of plants, yet no one considers that it has been moved from its place of storage in that form.

The evidence thus indicates that oils, especially in seedlings, are converted into sugars at their place of storage and translocated to the different parts of the plant in that form. After the arrival of sugar at these regions, a portion of it at least may be converted into oil as a temporary storage product. It seems to be proved that oil as such is rarely if ever moved from cell to cell or from one region of the plant to another.

Knight, Chamberlin, and Samuels (1929), Ginsburg (1931), Rohrabaugh (1934), and Young (1934, 1935) have studied the penetration of petroleum oils applied as sprays.

Ginsburg (1931) noted that emulsified oils penetrated the leaves of apple, peach, and tomato much more slowly than did pure oils. Only those oils of very low viscosity penetrated the upper surface of these leaves. Young (1934, 1935) found that petroleum oil sprayed on apple leaves entered the parenchyma cells and tracheae of the veins and petioles. The oil in the twigs, however, was outside the parenchyma cells of the cortex, rays, pith, and tracheae. The petroleum oil placed upon or injected into the leaves of rutabaga, cucumber, potato, turnip, and onion passed downward into the stems, roots, bulbs, and other plant parts.

Some of the oil was in the tracheae but most of it was between the parenchyma cells of the cortex, where apparently under the conditions of the observations most of it was conducted.

Knight, Chamberlin, and Samuels (1929) noted in *Citrus* that petroleum oils applied as sprays began to appear in the palisade layer of cells in about 4 hr. These oils eventually accumulated in the cells along the larger veins and passed into the vascular system. They next appeared in the phloem, rays, pith, and in the old wood fibers of the xylem, but none was observed in the wood of the current season's growth. They considered that most of these oils were deposited in the smaller twigs and limbs. Rohrabaugh (1934), however, found in his experiments with *Citrus* that oil sprays entered by capillarity, and only penetrated under the most favorable circumstances to the depth of five or six cells beneath the epidermis. The less volatile oils remained in the citrus leaves throughout their life, which was more than 2 years.

Van den Honert (1932) studied the rate of transport of potassium oleate along the interface between water and ether by measuring the resulting pH change at the distal end of the interface. The precipitation of oleic acid on the interface sets free potassium hydroxide, which is allowed to neutralize hydrochloric acid in a bulb into which the interface extends. The potassium hydroxide under these conditions is transported at a rate 68,000 times that at which it could have diffused through the water layer. He suggested that in the sieve tubes the protoplasm-vacuole interface is the one in which transport occurs, and that this fact is probably the explanation for the very rapid transport of organic materials in plants.

8. Translocation of Proteins.—Maskell and Mason (1930), and Mason and Phillis (1934) stated that in the cotton plant the movement of nitrogen through the bark from the leaves to the roots occurs against a negative gradient of organic, crystalloidal nitrogen. The indications are that there is no well-defined positive gradient in either amino acids or residual nitrogen in the sieve tubes. It should be noted, however, that a positive gradient in the sieve tubes may be masked by a negative one in the companion cells. The gradient in the sieve tubes may consist of two components, a dynamic gradient in translocatory nitrogen, superimposed upon a steeper reverse gradient of relatively immobile, storage nitrogen. There is apparently little or no movement of proteins as such through cell membranes or through the special conducting elements of the plant. Only a few proteins are soluble in water, and, since they are slowly diffusing substances, their movement must necessarily be very slow. The contents of the sieve tubes give the protein reactions and have been considered on that account to be the elements in which proteins are moved longitudinally in the plant. These elements are rich in protoplasm and colloidal material, but how much of this if any is protein in transit cannot be determined.

The proteins can apparently be formed in any region of the plant, provided that a supply of sugar and the necessary inorganic materials are present. It seems probable, therefore, that proteins are not, as a rule, manufactured in superabundance in any special portion of the plant and then translocated for considerable distances for utilization or storage, but that they are formed in the immediate region in which they are found. The frequently observed fact that, when a nitrogen supply is abundant in the tops of plants the carbohydrates are used for top growth in larger proportions than when the nitrogen is more limited, is indicative that protein formation proceeds wherever the necessary materials are present and that the proteins are utilized more or less locally (Chandler, 1919). It was mentioned in Chap. IX that proteins are manufactured in the leaves in considerable amounts during the day and moved from them especially at night, but to what distances they are moved is not known. Amino acids and amides have been detected in both the xylem and phloem elements, but whether they are of local origin or are in transit from regions more remote has not been definitely determined.

The movement of protein from their place of storage to their point of utilization has been studied for the most part in seedlings. The richest stores of proteins occur in seeds and are rapidly broken down and moved to the growing portions of the seedling during germination. The general procedure has been to study the kind and quantity of the hydrolytic products in the original storage parts and in the newly formed organs of the seedlings. A few of the observations that have thus been made will now be reviewed.

Schulze (1900) and Schulze and Castoro (1903) found in the seedlings of *Lupinus albus* and other leguminous seedlings that amino acids appeared as leucine, tyrosine, and hexone bases. In the seedlings of the Alaska pea, Thompson (1915) noted that the amount of the total nitrogen in the amino form increased from 6.58 per cent of the total nitrogen in the 1-day-old seedlings to 28.27 per cent in those 7 days old. The distribution of the nitrogen in the 7-day seedlings was as follows:

Part of seedling	Percentage on a dry basis of		Percentage of total nitrogen in the amino form
	Total N	Amino N	
Plumule.............	7.02	0.254	36.18
Cotyledons...........	3.36	0.421	12.53
Root..............	5.33	1.705	32.00

Stark (1927) found in the seedlings of Manchu soybeans grown at 20°C. that the α-amino nitrogen increased from 1.86 per cent of the total nitrogen at the age of 1 day to 23.2 per cent at the end of 39 days. In the case of the Midwest variety, the increase was from 2.1 to 17.8 per cent.

These data indicate that the proteins in the seed are broken down into amino acids, which are translocated as rapidly as they are formed to the plumules and roots. In corn seedlings (Pettibone and Kennedy, 1916) the total nitrogen in the seed decreased from 75.0 mg. per 20 seedlings when the plumules were ½ in. in length to 23.1 mg. when they had reached a length of 10 in. The total nitrogen in the plumules during the period increased from 12.3 to 57.5 mg., while in the roots it had increased from 8.2 to 307 mg. Amino acids were present in considerable amounts in all three parts. An analysis of the sap exuded from the stumps of several seedlings showed that amino acids and other protein decomposition products resembling proteoses and peptones were present. Observations similar to these have been observed by Choate (1921) working with wheat seedlings.

Jodidi (1925 to 1927) observed also in etiolated corn seedlings that 48 per cent of the proteins present in the seed are converted into water-soluble diffusing nitrogen compounds within 8 days. During this period there is a rise in the amino nitrogen and a diminution in the peptide nitrogen. This is taken to mean that amino acids increase at the expense of polypeptides, which, along with proteoses, are among the first degradation products of proteins.

Since nitrogen increases in the plumules and roots as it disappears from the seed, and since amino acids are present in these parts and in the flowing sap, it seems evident that the reserve proteins of the seeds are transported to their points of utilization in the seedling in the form of amino acids.

Bibliography

Adams, J., 1923: The translocation of carbohydrates in the sugar maple, *Nature,* **112:** 207.

Albert, W. B., 1927: Studies on the growth of alfalfa and some perennial grasses, *J. Am. Soc. Agron.,* **19:** 624–655.

Aldous, A. E., 1929: The eradication of brush and weeds from pasture lands, *J. Am. Soc. Agron.,* **21:** 660–666.

———, 1930: Effect of different clipping treatments on the yield and vigor of prairie grass vegetation, *Ecology,* **11:** 752–759.

———, 1930: Relation of organic food reserves to the growth of some Kansas pasture plants, *J. Am. Soc. Agron.,* **22:** 385–393.

Aldrich, W. W., 1932: Nitrogen intake and translocation in apple trees following fall, winter, and spring sodium nitrate applications, *Proc. Am. Soc. Hort. Sci.,* **28:** 532–538.

ANDERSSEN, F. G., 1929: Some seasonal changes in the tracheal sap of pear and apricot trees, *Plant Physiol.*, **4**: 459–477.

ARNDT, C. H., 1929: The movement of sap in *Coffea arabica*, *Am. J. Bot.*, **16**: 179–190.

ARNY, A. C., 1932: Variations in the organic reserves in underground parts of five perennial weeds from late April to November, *Minn. Sta. Tech. Bull.* 84.

ASKENASY, E., 1895; Ueber das Saftsteigen, *Verh. naturh.-med. Ver. Heidelberg*, **5**: 325–345; *Bot. Cent.*, **62**: 237–238.

———, 1896: Beiträge zur Erklärung des Saftsteigens, *Verh. naturh.-med. Ver. Heidelberg.*, **5**: 429–448.

ATKINS, W. R. G., 1916: "Some Recent Researches in Plant Physiology," Chap. XI, The Function of the Wood, pp. 186–215.

AUCHTER, E. C., 1923: Is there normally a cross transfer of foods, water, and mineral nutrients in woody plants? *Md. Sta. Bull.* 257.

BAILEY, I. W., 1916: The structure of the bordered pits of conifers and its bearing upon the tension hypothesis of the ascent of sap in plants, *Bot. Gaz.*, **62**: 133–142.

BAKER, H., and W. O. JAMES, 1933: The behavior of dyes in the transpiration stream of sycamores (*Acer pseudoplatanus*), *New Phytol.*, **32**: 245–260.

BALDWIN, H. I., 1934: Some physiological effects of girdling northern hardwoods, *Bull. Torrey Bot. Club*, **61**: 249–257.

BARNES, C. R., 1930: "A Text Book of Botany. Vol. II. Physiology." Revised and enlarged by C. A. Shull. American Book Co., N. Y. and Cincinnati.

BARTEL, A. T., J. H. MARTIN, and R. S. HAWKINS, 1935: Effect of tillers on the development of grain sorghums, *J. Am. Soc. Agron.*, **27**: 707–714.

BARTHOLOMEW, E. T., 1926: Internal decline of lemons. III, Water deficit in lemon fruits caused by excessive leaf evaporation, *Am. J. Bot.*, **13**: 102–117.

BENEDICT, H. M., 1927: Application of Bose's theory of sap rise to ten species of trees, *Am. J. Bot.*, **14**: 623.

BENNETT, J. P., 1927: Chlorosis studies, *Calif. Sta. Rept.*, **1927**: 85–87.

———, F. G. ANDERSSEN, and Y. MILAD, 1927: Methods of obtaining tracheal sap from woody plants, *New Phytol.*, **26**: 316–323.

BIRCH-HIRSCHFELD, L., 1920: Untersuchungen über die Ausbreitungsgeschwindigkeit gelöster Stoffe in der Pflanze, *Jahrb. wiss. Bot.*, **59**: 171–262.

BISSON, C. S., and H. A. JONES, 1932: Changes accompanying fruit development in the garden pea, *Plant Physiol.*, **7**: 91–106.

BISWELL, H. H., and J. E. WEAVER, 1933: Effect of frequent clipping on the development of roots and tops of grasses in prairie sod, *Ecology*, **14**: 368–390.

BLAKE, M. A., 1921: Effect of an early application of nitrogen on peach trees deficient in vigor, *Proc. Amer. Soc. Hort. Sci.*, **18**: 139–145.

BODE, H. R., 1923: Beiträge zur Dynamik der Wasserbewegung in den Gefässpflanzen, *Jahrb. wiss. Bot.*, **62**: 92–127.

BODENBERG, E. T., 1927: Tissues involved in the transfer of mineral salts in plants, *Pub. Puget Sound Biol. Sta.*, **5**: 231–244; *Exp. Sta. Rec.*, **62**: 122.

———, 1929: Lateral transfer of lithium nitrate in *Salix*, *Am. J. Bot.*, **16**: 229–237.

BOEHM, J. A., 1878: Les causes de l'ascension de la sève, *Ann. Sci. Nat. Bot.*, VI, **6**: 223–236.

———, 1889: Ursache des Saftsteigens, *Ber. bot. Ges.*, II, **7**: 46–56.

———, 1890: Ueber Ursache der Wasserbewegung in transpirirenden Pflanzen, *Verh. Zool.-Bot. Ges. Wien*, **40**: 149–159.

———, 1892: Ueber einen eigenthümlichen Stammdruck, *Ber. bot. Ges.*, **10**: 539–544.

———, 1893: Capillarität und Saftsteigen, *Ber. bot. Ges.*, **11**: 203–212.

Bolley, H. L., 1904, 1905, 1907: Tree feeding and tree medication, *N. Dak. Sta. Ann. Rept.*, **14**: 55–58 (1904); **15**: 56 (1905); **17**: 35 (1907).

Bose, J. C., 1923: "Physiology of the Ascent of Sap," Longmans, Green and Company, New York and London, 277 pp.

Boucherie, A., 1840: Rapport sur un Mémoire relatif à la conservation des bois, *C. R. Acad. Sci. Paris*, **11**: 894–898.

———, 1841: Nouvelles recherches sur la conservation des bois, *C. R. Acad. Sci. Paris*, **12**: 337–339.

Bourdouil, C., 1931: The formation and transformation of the starch in bananas, *Rev. bot. app. Agr. trop.*, **11**: 656–660; *Chem. Abst.*, **26**: 753.

Bowen, E. J., 1931: Water conduction in *Polytrichum commune*, *Ann. Bot.*, **45**: 175–200.

———, 1933: The mechanism of water conduction in the *Musci* considered in relation to habitat. I, Mosses growing in wet environments, *Ann. Bot.*, **47**: 401–422.

———, 1933: The mechanism of water conduction in the *Musci* considered in relation to habitat. II, Mosses growing in damp situations, *Ann. Bot.*, **47**: 635–661.

———, 1933: The mechanism of water conduction in the *Musci* considered in relation to habitat. III, Mosses growing in dry environment, *Ann. Bot.*, **47**: 889–912.

Boysen-Jensen, P., 1915: Über synthetische Vorgänge im pflanzlichen Organismus. II, Vorkommen, Bedeutung und Bildung des Rohrzuckers bei der Keimung von *Pisum sativum*, *Jahrb. wiss. Bot.*, **56**: 431–446.

Brown, B. A., 1930: Effect of time of cutting on the elimination of bushes in pasture, *J. Am. Soc. Agron.*, **22**: 603–605.

Brunson, A. M., and W. L. Latshaw, 1934: Effect of failure of pollination on composition of corn plants, *J. Agr. Res.*, **49**: 45–54.

Burd, J. S., 1919: Rate of absorption of soil constituents at successive stages of plant growth, *J. Agr. Res.*, **18**: 51–72.

Burke, E., and H. E. Morris, 1933: Nutrient elements used by leaves and growth of apple trees, *Plant Physiol.*, **8**: 537–544.

Butler, O. R., T. O. Smith, and B. E. Carey, 1917: Physiology of the apple. Distribution of food materials in the tree at different periods of vegetation, *N. H. Sta. Tech. Bull.* 13.

Caldwell, J., 1930: Studies in translocation. II, The movement of food materials in plants, *New Phytol.*, **29**: 27–44.

———, 1930: Studies in translocation. I, Movement of food-materials in the Swedish turnip, *Proc. Roy. Soc. Edinburgh*, **50**: 130–141; *Biol. Abst.*, **5**: 2657.

Capen, R. G., and J. A. LeClerc, 1933: Chemical composition of native Alaskan hays harvested at different periods of growth, *J. Agr. Res.*, **46**: 665–668.

Carpenter, C. W., 1936: Notes on Pythium root rot. VIII, Absorption of essential elements by segregated roots of sugar cane, *Hawaiian Planters' Rec.*, **40**: 137–142.

Chandler, W. H., 1914: Sap studies with horticultural plants, *Mo. Sta. Res. Bull.* 14.

———, 1919: Some results as to the response of fruit trees to pruning, *Proc. Amer. Soc. Hort. Sci.*, **16**: 88–101.

Chase, W. W., 1934: The composition, quantity and physiological significance of gases in tree stems, *Minn. Sta. Tech. Bull.* 99.

Choate, H. A., 1921: Chemical changes in wheat during germination, *Bot. Gaz.*, **71**: 409–426.

Clark, W. S., 1874: The circulation of sap in plants, *Mass. Bd. Agr. Ann. Rept. Sec.*, **21**: 159–204.

Clements, H. F., 1930: The upward movement of inorganic solutes in plants, *Res. Studies State Col. Wash.*, **2**: 91–106.

COLLISON, R. C., J. D. HARLAN, and M. P. SWEENEY, 1932: Direct tree injection in the study of tree nutrition problems, *N. Y. Sta. (Geneva) Tech. Bull.* 192.

COMBES, R., and D. KOHLER, 1922: Ce que déviement les hydrates de carbone quand meurent les feuilles des arbres, *C. R. Acad. Sci. Paris,* **175**: 590–592.

———, 1926: Émigration des substances azotées des feuilles vers les tiges et les racines des arbres au cours du jaunissement automnal, *Rev. Gén. Bot.,* **38**: 430–448, 510–517, 565–579, 632–645, 673–686.

COPELAND, E. B., 1902: The rise of the transpiration stream: an historical and critical discussion, *Bot. Gaz.,* **34**: 161–193, 260–283.

COSTER, C., 1931: Die Geschwindigkeit des Transpirationsstromes, *Planta,* **15**: 540–566; *Chem. Abst.,* **8**: 1659.

CRAFTS, A. S., and P. B. KENNEDY, 1930: The physiology of *Convolvulus arvensis* (Morning glory or bindweed) in relation to its control by chemical sprays, *Plant Physiol.,* **5**: 329–344.

———, 1931: Movement of organic materials in plants, *Plant Physiol.,* **6**: 1–43.

———, 1931: The function of the sieve tube in conduction, abst., *Am. J. Bot.,* **18**: 893.

———, 1932: Phloem anatomy, exudation and transport of organic nutrients in cucurbits, *Plant Physiol.,* **7**: 183–226.

———, 1933: The use of arsenical compounds in the control of deep-rooted perennial weeds, *Hilgardia,* **7**: 361–372.

———, 1933: Sieve-tube structure and translocation in the potato, *Plant Physiol.,* **8**: 81–104.

———, 1936: Further studies on exudation in cucurbits, *Plant Physiol.,* **11**: 63–79.

CRAIGHEAD, F. C., and R. A. ST. GEORGE, 1930: A new technique in tree medication for the control of bark beetles, *Sci.,* **72**: 433–435.

CRIDER, F. J., 1928: Winter root growth of plants, *Sci.,* **68**: 403–404.

CULPEPPER, C. W., and C. A. MAGOON, 1930: Effect of defoliation and root pruning on the chemical composition of sweet corn kernels, *J. Agr. Res.,* **40**: 575–583.

——— and J. S. CALDWELL, 1932: Relation of age and of seasonal conditions to composition of root, petiole and leaf blade in rhubarb, *Plant Physiol.,* **7**: 447–479.

———, ——— and H. H. MOON, 1935: A physiological study of development and ripening in the strawberry, *J. Agr. Res.,* **50**: 645–696.

CURTIS, O. F., 1920: The upward translocation of foods in woody plants. I, Tissues concerned in translocation, *Am. J. Bot.,* **7**: 101–124.

———, 1920: The upward translocation of foods in woody plants. II, Is there normally an upward transfer of storage foods from the roots or trunk to the growing shoots? *Am. J. Bot.,* **7**: 286–296.

———, 1923: The effect of ringing a stem on the upward transfer of nitrogen and ash constituents, *Am. J. Bot.,* **10**: 361–383.

———, 1925: Studies on the tissues concerned in the transfer of solutes in plants. The effect on the upward transfer of solutes of cutting the xylem as compared with that of cutting the phloem, *Ann. Bot.,* **39**: 573–585.

———, 1928: Studies indicating that translocation is dependent on the activity of living cells and the possible bearing of this on solute distribution, *Am. J. Bot.,* **15**: 630–631.

———, 1929: Studies on solute translocation in plants. Experiments indicating that translocation is dependent on the activity of living cells, *Am. J. Bot.,* **16**: 154–168.

——— and H. T. SCOFIELD, 1933: A comparison of osmotic concentrations of supplying and receiving tissue and its bearing on the Münch hypothesis of the translocation mechanism, *Am. J. Bot.,* **20**: 502–512.

——— 1935: "The Translocation of Solutes in Plants," McGraw-Hill Book Company, Inc., New York and London, 273 pp.

Czapek, F., 1897: Über die Leitungswege der organischen Baustoffe im Pflanzenkörper, *Sitzungsber. Akad. Wiss. Wien*, **106**: 117–170.

Darwin, F., and R. W. Phillips, 1885: On the transpiration stream in cut branches, *Proc. Cambridge Phil. Soc.*, **5**: 330–367.

———, S. H., Vines, and J. Joly, 1896: A discussion on the ascent of water in trees, *Ann. Bot.*, **10**: 630–661.

Davidson, J., 1921: Biological studies of *Aphis rumicis*, *Ann. App. Biol.*, **8**: 51–65.

Davis, L. D., 1931: Some carbohydrate and nitrogen constituents of alternate-bearing sugar prunes associated with fruit bud formation, *Hilgardia*, **5**: 119–154.

Deleano, N. T., 1911: Über die Ableitung der Assimilate durch die intaken, die chloroformierten und die plasmolysierten Blattstiele der Laubblätter, *Jahrb. wiss. Bot.*, **49**: 129–186.

——— and M. Andreesco, 1931: L'absorption et la migration negative de l'azote dans les feuilles de *Salix fragilis*, *Bull. Soc. Sci. Cluj Roumanie*, **6**: 209–220.

———, 1932: Die Anwendung der biometrischen Methode zum Forschen der Rolle von Mineralstoffen im Leben der Pflanze, *Bull. Cult. si Fermentarii Tutunului*, **21**: no. 2.

———, 1932: Die Rückwanderung der Stickstoffsubstanz aus den Laubblättern und ihre Speicherung im Stamme und Wurzel, *Bull. Soc. Sci. Cluj Roumanie*, **7**: 45–49.

——— and M. I. Andreesco, 1932: Beiträge zum Studium der Rolle und Wirkungsweise der Mineral- und organischen Stoffe im Pflanzenleben. I, Der quantitative Stoffwechsel der Mineral -und organischen Substanzen in den *Salix fragilis*-Blättern während ihrer Entwicklung, *Beitr. Biol. Pflanzen*, **19**: 249–286; *Biol. Abst.*, **8**: 2170.

——— and C. Bordeianu, 1932: Beiträge zum Studium der Rolle und Wirkungsweise der Mineral- und organischen Stoffe im Pflanzenleben. II, Der quantitative Stoffwechsel der Mineral- und organischen Substanzen in den Blättern und geschälten Samen von *Aesculus hippocastanum* während Entwicklung, *Beitr. Biol. Pflanzen*, **20**: 179–197.

——— and ———, 1934: Beiträge zum Studium der Rolle und Wirkungsweise der Mineral- und organischen Stoffe im Pflanzenleben. II. *Acad. Romana Mem. Sec. Stiint Mem.* 10, Bucuresti.

Denny, F. E., 1933: Changes in leaves during the period preceding frost, *Contr. Boyce Thompson Inst.*, **5**: 297–312.

Detmer, W., 1880: "Vergleichende Physiologie des Keimungsprocesses der Samen," Fischer, Jena.

DeVries, H., 1885: Ueber die Bedeutung der Circulation und der Rotation des Protoplasma für den Stofftransport in der Pflanze, *Bot. Ztg.*, **43**: 2–6, 17–26.

Dixon, H. H., and J. Joly, 1894: On the ascent of sap, *Proc. Roy. Soc.*, **57**: 3–5.

——— and ———, 1894: On the ascent of sap, *Ann. Bot.*, **8**: 468–470.

——— and ———, 1895: On the ascent of sap, *Phil. Trans. Roy. Soc.*, 186 B, I, 563–576.

——— and ———, 1895: The path of the transpiration-current, *Ann. Bot.*, **9**: 403–420.

—, 1896: Note on the role of osmosis in transpiration, *Proc. Roy. Irish Acad.*, **3**: 767–775.

———, 1897: The tensile strength of cell walls, *Ann. Bot.*, **11**: 585–588.

———, 1903: The cohesion theory of the ascent of sap, *Sci. Proc. Roy. Dublin Soc.*, 10, No. 4.

———, 1905: Note on the supply of water to leaves on a dead branch, *Sci. Proc. Roy. Dublin Soc.*, **11**: 7–12.

———, 1906: On the transpiration current in plants, *Proc. Roy. Soc.*, B, **79**: 41–57.

———, 1909: Vitality and the transmission of water through the stems of plants, *Notes Bot. School Trinity Col. Dublin*, **2**: 5–18; *Sci. Proc. Roy. Dublin Soc.*, **12**: 21–34.

———, 1910: Transpiration and the ascent of sap, *Progressus rei botanicae*, **3**: 1–66.

———, 1910: Transpiration and the ascent of sap, *Ann. Rept. Smithsonian Insi.*, **1910**: 407–425.

———, 1914: "Transpiration and the Ascent of Sap in Plants," Macmillan & Company, Ltd., London, 216 pp.

——— and W. R. G. Atkins, 1915: On the constituents and concentration of the sap in the conducting tracts and on the circulation of carbohydrates in plants, *Sci. Proc. Roy. Dublin Soc.*, **14**: 374–392.

——— and E. S. Marshall, 1915: A quantitative examination of the elements of the wood of trees in relation to the supposed function of the cells in the ascent of sap, *Proc. Roy. Dub. Soc.*, **16**: 358; *Notes Bot. School Trinity Col. Dublin*, **2**: 264.

———, 1915: Notes on the spread of morbid changes through plants from branches killed by heat, *Notes Bot. School Trinity Col. Dublin*, **2**: 221–224.

———, 1915: Changes produced in the sap by the heating of branches, *Notes Bot. School Trinity Col. Dublin*, **2**: 225–229.

———, 1915: On the tensile strength of sap, *Notes Bot. School Trinity Col. Dublin*, **2**: 230–235.

——— and W. R. G. Atkins, 1916: Osmotic pressure in plants. IV, On the constituents and concentration of the sap in the conducting tracts, and on the circulation of carbohydrates in plants, *Notes Bot. School Trinity Col. Dublin*, **2**: 275–293.

——— and J. Joly, 1916: On the composition of the sap in the conducting tracts of trees at different levels and at different seasons of the year, *Sci. Proc. Roy. Dublin Soc.*, **15**: 51.

——— and N. G. Ball, 1922: Transport of organic substances in plants, *Nature*, **109**: 236–237.

———, 1922: Transport of organic substances in plants, *Nature*, **110**: 547–551.

——— and ———, 1923: On the channels of transport from the storage organs of the seedlings of *Lodoicea*, *Phoenix* and *Vicia*, *Sci. Proc. Roy. Dublin Soc.*, **17**: 185–196.

———, 1923: Transport of organic substances in plants, *Notes Bot. School Trinity Col. Dublin*, **3**: 207–215.

———, 1924: "The Transpiration Stream," University of London Press, 80 pp.

———, 1933: Bast sap, *Sci. Proc. Roy. Dub. Soc.*, **20**: 487–494.

Doyle, J., and P. Clinch, 1927: Seasonal changes in conifer leaves with special reference to enzymes and starch formation, *Proc. Roy. Irish Acad.*, B, **37**: 373–414.

Dungan, G. H., 1930: Relation of blade injury to the yielding ability of corn plants, *J. Am. Soc. Agron.*, **22**: 164–171.

———, 1934: Losses to the corn crop caused by leaf injury, *Plant Physiol.*, **9**: 749–766.

Dustman, R. B., and L. C. Shriver, 1931: The chemical composition of *Ambrosia trifida* at successive growth stages, *J. Am. Soc. Agron.*, **23**: 190–194.

Eames, A. J., and L. H. MacDaniels, 1925: "An Introduction to Plant Anatomy," McGraw-Hill Book Company, Inc., New York, 363 pp.

Echevin, R., 1932: Autumn migration of phosphorus from tree leaves, *C. R. Acad. Sci. Paris*, **194**: 2160–2162.

Elfving, F., 1882: Ueber die Wasserleitung im Holz, *Bot. Ztg.*, **40**: 707–723.

ELLIOT, J. A., 1917: The conduction of potassium cyanide in plants, *Phytopath.*, **7**: 443–448.

ERRERA, L., 1886: Ein Transpirationsversuch, *Ber. bot. Ges.*, **4**: 16–18.

EWART, A. J., 1904: The ascent of water in trees, *Proc. Roy. Soc.*, **74**: 554–556.

———, 1906, 1908: The ascent of water in trees, *Phil. Trans. Roy. Soc.*, B, **198**: 41–85; **199**: 341–392.

FAGAN, T. W., and J. E. WATKIN, 1931: The distribution of the nitrogenous and mineral constituents in the oat plant at different stages of growth, *Welsh J. Agr.*, **7**: 229–246.

FARMER, J. B., 1918: On the quantitative differences in the water conductivity of the wood in trees and shrubs. I, The evergreens. II, Deciduous plants, *Proc. Roy. Soc.*, B, **90**: 218–250.

FERNALD, E. I., 1925: The imbibition of bud-development as correlated with the osmotic concentration of cell sap, *Am. J. Bot.*, **12**: 287–305.

FISCHER, A., 1888: Glycose als Reservestoff der Laubhölzer, *Bot. Ztg.*, **46**: 405–417.

———, 1890: Beiträge zur Physiologie der Holzgewächse, *Jahrb. wiss. Bot.*, **22**: 73–160.

FITTING, H., 1911: Untersuchungen über die vorzeitge Entblätterung von Blüten, *Jahrb. wiss. Bot.*, **49**: 187–263.

FLINT, W. P., 1915: The effect of cyanide on the locust-borer and the locust tree, *Sci.*, **42**: 726–727.

FREAR, D. E., 1931: A negative correlation observed between the nitrate nitrogen in the juice of beet leaves and the weight of the leaves, *J. Agr. Res.*, **42**: 53–56.

FURR, J. R., and C. A. TAYLOR, 1933: The cross-transfer of water in mature lemon trees, *Proc. Am. Soc. Hort. Sci.*, **30**: 45–51.

GARDNER, F. E., 1925: A study of the conductive tissues in shoots of the Bartlett pear and the relationship of food movement to dominance of the apical buds, *Calif. Sta. Tech. Paper* 20.

GÄUMANN, E., 1927: Der jahreszeitliche Verlauf des Kohlehydratgehaltes in Tannen- und Fichtenstamm, *Ber. bot. Ges.*, **45**: 591–597.

GERHARDT, F., 1929: Propagation and food translocation in the common milkweed, *J. Agr. Res.*, **39**: 837–853.

GERRY, E., and J. A. HALL, 1935: Biochemical phases of oleoresin production, *Plant Physiol.*, **10**: 537–543.

GIBBS, R. D., 1930: Sinkage studies. II, The seasonal distribution of water and gas in trees, *Can. J. Res.*, **2**: 425–439; *Biol. Abst.*, **5**: 2658.

———, 1935: Studies of wood. II, On the water content of certain Canadian trees and on changes in the water-gas system during seasoning and flotation, *Can. J. Res.*, **12**: 727–760.

———, 1935: Studies of wood. III, On the physiology of the tree with special reference to the ascent of sap and the movement of water before and after death, *Can. J. Res.*, **12**: 761–787.

GILE, P. L., and J. O. CARRERO, 1921: Assimilation of nitrogen, phosphorus and potassium by corn when nutrient salts are confined to different roots, *J. Agr. Res.*, **21**: 545–573.

GINSBURG, J. M., 1931: Penetration of petroleum oils into plant tissue, *J. Agr. Res.*, **43**: 469–474.

GODLEWSKI, E., 1884: Zur Theorie der Wasserbewegung in den Pflanzen, *Jahrb. wiss. Bot.*, **15**: 569–630.

GOFF, E. S., 1897: The application of artificial root pressure to recently transplanted trees, *Wis. Sta. Ann. Rept.*, **14**: 272–282.

———, 1898: The resumption of root growth in the spring, *Wis. Sta. Ann. Rept.*, **15**: 220–228.

GRABER, L. F., N. T. NELSON, W. A. LEUKEL, and W. B. ALBERT, 1927: Organic food reserves in relation to the growth of alfalfa and other perennial herbaceous plants, *Wis. Sta. Res. Bull.* 80.

———, 1931: Food reserves in relation to other factors limiting the growth of grasses, *Plant Physiol.*, **6**: 43–72.

——— and H. W. REAM, 1931: Growth of blue grass with various defoliations and abundant nitrogen supply, *J. Am. Soc. Agron.*, **23**: 938–944.

GRANDFIELD, C. O., 1930: The relation of organic food reserves to the effect of cutting pasture weeds at different stages of growth, *J. Am. Soc. Agron.*, **22**: 709–713.

———, 1935: The trend of organic food reserves in alfalfa roots as affected by cutting practices, *J. Agr. Res.*, **50**: 697–710.

GROOM, P., 1910: Remarks on the ecology of *Coniferae*, *Ann. Bot.*, **24**: 241–269.

HAAS, A. R. C., and F. F. HALMA, 1929: Chemical relationship between scion and stock in *Citrus*, *Plant Physiol.*, **4**: 113–121.

——— and L. J. KLOTZ, 1935: Physiological gradients in citrus fruits, *Hilgardia*, **9**: 181–217.

HAASIS, F. W., 1932: Seasonal shrinkage of Monterey pine and redwood trees, *Plant Physiol.*, **7**: 285–295.

———, 1933: Shrinkage and expansion in woody cylinders of living trees, *Am. J. Bot.*, **20**: 85–91.

———, 1934: Diametral changes in tree trunks, *Carnegie Inst. Wash. Pub.* 450.

HABERLANDT, G., 1914: "Physiological Plant Anatomy." Translation by M. Drummond. Macmillan & Company, Ltd., London, XV + 777 pp.

HAINES, F. M., 1935: Observations on the occurrence of air in conducting tracts, *Ann. Bot.*, **49**: 367–379.

———, 1935: Transpiration and pressure deficit. II, *Ann. Bot.*, **49**: 521–565.

HALLER, M. H., 1930: The relation of the distance and direction of the fruit from the leaves to the size and composition of apples, *Proc. Am. Soc. Hort. Sci.*, **27**: 63–68.

HANNIG, E., 1913: Untersuchungen über das Abstossen von Blüten u. s. w., *Z. Bot.*, **5**: 417–469.

HARTIG, T., 1858: Über die Bewegung des Saftes in den Holzpflanzen, *Bot. Ztg.*, **16**: 329–335, 337–342.

HARVEY, E. M., 1931: Movement of water in plants as affected by a mutual relation between the hydrostatic and pneumatic systems, *Plant Physiol.*, **6**: 495–506.

HARVEY, T. B., 1930: Tracing the transpiration stream with dyes, *Am. J. Bot.*, **17**: 657–662.

HAWKINS, R. S., R. L. MATLOCK, and C. A. HOBART, 1933: Physiological factors affecting the fruiting of cotton with special reference to boll shedding, *Ariz. Sta. Tech. Bull.* 46.

———, S. P. CLARK, G. H. SERVISS, and C. A. HOBART, 1934: Varietal differences in cotton boll shedding as correlated with osmotic pressure of expressed tissue fluids, *J. Agr. Res.*, **48**: 149–157.

HEINICKE, A. J., 1919: Concerning the shedding of flowers and fruits and other abscission phenomena in apples and pears, *Proc. Am. Soc. Hort. Sci.*, **16**: 76–83.

HENRICKSEN, H. C., 1932: Absorption and movement of plant nutrients, *Puerto Rico Sta. Agr. Notes*, **60**: 1–3; *Exp. Sta. Rec.*, **70**: 167.

———, 1933: Absorption of ash constituents by citrus trees, *Ann. Rept. Puerto Rico Exp. Sta.*, **1932**: 18–19.

HENSON, P. R., 1932: The effect of the su factor on translocation and storage of carbohydrates in the corn plant, abst., *25th Ann. Meeting Am. Soc. Agron.*

HEYL, J. G., 1933: The influence of external factors on sap of plants, *Planta*, **20**: 294–353; *Exp. Sta. Rec.*, **71**: 23.

HILL, A. W., 1901: The histology of the sieve-tubes of *Pinus*, *Ann. Bot.*, **15**: 575–613.

———, 1908: The histology of the sieve-tubes of angiosperms, *Ann. Bot.*, **22**: 245–291.

HOLLE, H., 1915: Untersuchungen über Welken, Vertrocknen und Wiederstraffwerden, *Flora*, **108**: 73–126.

HOLMES, M. C., 1918: A study in the anatomy of Hazel-wood with reference to conductivity of water, *Ann. Bot.*, **32**: 553–567.

HOMAN, C., T. F. YOUNG, and C. A. SHULL, 1934: Evaporation in its thermodynamic relation to the cohesion theory and to imbibitional and osmotic pressures of cell constituents, *Plant Physiol.*, **9**: 653–661.

HORNBERGER, R., 1882: Chemische Untersuchungen über das Wachsthum der Maispflanze, *Landw. Jahrb.*, **11**: 359–523.

HUBER, B., 1923: Theoretische Betrachtungen zur Kohäsionstheorie der Wasserbewegung in der Pflanze, *Biol. Zent.*, **43**: 30–49.

———, 1924: Beiträge zur Kenntnis der Wasserbewegung in der Pflanze. II, Die Strömungsgeschwindigkeit und die Grösse der Widerstände in den Leitbahnen, *Ber. bot. Ges.*, **42**: 27–32.

———, 1928: Weitere quantitative Untersuchungen über das Wasserleitungsystem der Pflanzen, *Jahrb. wiss. Bot.*, **67**: 859–877.

———, 1932: Beobachtung und Messung pflanzlicher Saftströme, *Ber. bot. Ges.*, **50**: 89–109.

HUME, A. N., and C. FRANZKE, 1929: The effect of certain injuries to leaves of corn plants upon weights of grain produced, *J. Am. Soc. Agron.*, **21**: 1156–1164.

HUNGATE, R. E., 1934: The cohesion theory of transpiration, *Plant Physiol.*, **9**: 783–794.

INAMDAR, R. S., and K. V. VARADPANDE, 1930: The problem of carbohydrate transport in plants. A glucose effect on the permeability of cell membranes to sugar molecules as studied by the intensity of respiration when leaves of *Artocarpus integrifolia* were injected with varying concentrations of glucose solution, *Proc. Indian Sci. Cong.* (*Calcutta*), **17**: 300–301; *Exp. Sta. Rec.*, **66**: 621.

INGOLD, C. T., 1935: Note on exudation and exudation pressures in birch, *New Phytol.*, **34**: 437–441.

JACOBSON, L. A., 1934: Determination of total sugars in stalks of sorghum varieties, *unpub. data, Hays branch, Kans. Sta.*

JAMES, W. O., and H. BAKER, 1933: Sap pressure and the movements of sap, *New Phytol.*, **32**: 317–343.

JANSE, J. M., 1887: Die Mitwirkung der Markstrahlen bei der Wasserbewegung im Holz, *Jahrb. wiss. Bot.*, **18**: 1–69.

JANSSEN, G., 1929: The relationship of organic root reserves and other factors to the permanency of alfalfa stands, *J. Am. Soc. Agron.*, **21**: 895–911.

———, C. K. MCCLELLAND, and W. H. METZGER, 1930: Sap extraction of sorghum and the localization of juice and sugars in internodes of the plant, *J. Am. Soc. Agron.*, **22**: 627–639.

JODIDI, S. L., 1925: Nitrogen metabolism in etiolated corn seedlings, *J. Agr. Res.*, **31**: 1149–1164.

———, 1927: Isolation and identification of some organic nitrogenous compounds occurring in etiolated corn seedlings, *J. Agr. Res.*, **34**: 649–656.

JONES, C. H., A. W. EDSON, and W. J. MORSE, 1903: The maple sap flow, *Vt. Sta. Bull.* 103.

——— and J. L. BRADLEE, 1933: The carbohydrate contents of the maple tree, *Vt Sta. Bull.* 358.

JONES, H. A., and C. S. BISSON, 1932: Changes in the composition of the garden pea after harvest, *Plant Physiol.*, **7**: 273–282.

JONES, W. J., and H. A. HUSTON, 1914: Composition of maize at various stages of its growth, *Ind. Sta. Bull.* 175.
JOST, L., 1907: "Lectures on Plant Physiology," English translation, Clarendon Press, London, pp. 167–172.
KENNEDY, P. B., and A. S. CRAFTS, 1927: The application of physiological methods to weed control, *Plant Physiol.*, **2**: 503–506.
KNIGHT, H., J. C. CHAMBERLIN, and C. D. SAMUELS, 1929: On some limiting factors in the use of saturated petroleum oils as insecticides, *Plant Physiol.*, **4**: 299–321.
KNIGHT, R. C., 1920: Storage and translocation, *Sci. Prog.*, **15**: 200–206.
KNOWLTON, H. E., 1921: A preliminary experiment on half tree fertilization, *Proc. Am. Soc. Hort. Sci.*, **18**: 148–149.
KOMATSU, S., and S. OZAWA, 1932: Biochemical study of the autumn fall of the leaf, *Mem. Col. Sci. Kyoto Imp. Univ.*, **A15**: 57–65; *Chem. Abst.*, **26**: 4629.
KRAYBILL, H. R., J. T. SULLIVAN, and L. P. MILLER, 1930: Seasonal changes in the composition of Stayman apple trees. I, Carbohydrates, *Proc. Am. Soc. Hort. Sci.*, **27**: 206.
KRETOVITCH, W., 1933: Distribution of sugar and nitrogenous substances in wheat grain, *Biochem. J.*, **27**: 1687–1690.
KRUSEMAN, W. M., 1931: De Invloed van Temperatuur en Narcose op Het Transport der Assimilaten. Purmerend.
LAMPE, L., 1931: A microchemical and morphological study of the developing endosperm of maize, *Bot. Gaz.*, **91**: 337–377.
LECLERC DU SABLON, M., 1904, 1906: Recherches physiologiques sur les matières de réserves des arbres, *Rev. Gén. Bot.*, **16**: 341–368; **18**: 5–25, 82–96.
———, 1910: Sur le mécanisme de la circulation de l'eau dans les plantes, *Rev. Gén. Bot.*, **22**: 125–136; *Bot. Cent.*, **116**: 87.
LECLERC, J. A., and J. F. BREAZEALE, 1911: Translocation of plant food and elaboration of organic plant material in wheat seedlings, *U. S. Bur. Chem. Bull.* 138.
LEE, E., 1911: The morphology of leaf-fall, *Ann. Bot.*, **25**: 51–106.
LEONARD, O. A., 1936: Seasonal study of tissue function and organic solute movement in the sunflower, *Plant Physiol.*, **11**: 25–61.
LEUKEL, W. A., 1927: Deposition and utilization of reserve foods in alfalfa plants, *J. Am. Soc. Agron.*, **19**: 596–624.
——— and J. M. COLEMAN, 1930: Growth behavior and maintenance of organic foods in Bahia grass, *Fla. Sta. Bull.* 219.
———, J. P. CAMP, and J. M. COLEMAN, 1934: Effect of frequent cutting and nitrate fertilization on the growth behavior and relative composition of pasture grasses, *Fla. Sta. Bull.* 269.
——— and R. M. BARNETTE, 1935: Cutting experiments with Bahia grass grown in lysimeters, *Fla. Sta. Tech. Bull.* 286.
LI, H. W., and T. N. LIU, 1935: Defoliation experiments with kaoliang, *J. Am. Soc. Agron.*, **27**: 486–491.
LIPMAN, C. B., and A. GORDON, 1925: Further studies on new methods in the physiology and pathology of plants, *J. Gen. Physiol.*, **7**: 615–623.
LLOYD, F. E., 1914: Abscission, *The Ottawa Naturalist*, **28**: 41–75.
———, 1916: The abscission of flower buds and fruits in *Gossypium* and its relation to environmental changes, *Trans. Roy. Soc. Canada*, **10**: 55–61.
———, 1916: Abscission in *Mirabilis Jalapa*, *Bot. Gaz.*, **61**: 213–230.
LONG, J. H., 1935: Seasonal changes in nitrogen and carbohydrate content of the strawberry plant, *Proc. Am. Soc. Hort. Sci.*, **33**: 386–388.
LOOMIS, W. E., R. BISSEY, and E. V. SMITH, 1931: Chlorates as herbicides, *Sci.*, **74**: 485.

———, E. V. Smith, R. Bissey, and L. E. Arnold, 1933: The absorption and movement of sodium chlorate when used as an herbicide, *J. Am. Soc. Agron.*, **25**: 724–739.

———, 1934: The translocation of nitrogen in woody plants, *Proc. Am. Soc. Hort. Sci.*, **32**: 61–64.

———, 1935: Translocation and growth balance in woody plants, *Ann. Bot.*, **49**: 247–272.

Lowry, M. W., and P. Tabor, 1931: Sap for analysis by bleeding corn plants, *Sci.*, **73**: 453.

Lund, E. J., 1931: Electric correlation between living cells in cortex and wood in the Douglas fir, *Plant Physiol.*, **6**: 631–652.

———, 1932: Control of the flux equilibrium of electro-chemical processes and electric polarity in the Douglas fir by temperature, *Plant Physiol.*, **7**: 297–307.

———, 1932: Comparison of the effects of temperature on the radial and longitudinal electric polarities in wood and cortex of the Douglas fir, *Plant Physiol.*, **7**: 505–516.

MacDaniels, L. H., and O. F. Curtis, 1930: The effect of spiral ringing on solute translocation and the structure of the regenerated tissue of the apple, *N. Y. Sta. (Cornell) Mem.* 133.

MacDougal, D. T., 1921: Growth in trees, *Carnegie Inst. Wash. Pub.* 307.

——— and F. Shreve, 1924: Growth in trees and massive organs of plants, *Carnegie Inst. Wash. Pub.* 350.

———, 1925: Absorption and exudation pressures in plants, *Proc. Am. Phil. Soc.*, **64**: 102–130.

———, 1925: Reversible variations in volume, pressure and movements of sap in trees, *Carnegie Inst. Wash. Pub.* 365.

———, 1926: The hydrostatic system of trees, *Carnegie Inst. Wash. Pub.* 373.

———, 1925–1927: Composition of gases in tree trunks, *Carnegie Inst. Wash. Year Book* **25**: 160–162; **26**: 162–163.

——— and J. B. Overton, 1927: Sap flow and pressure in trees, *Sci.*, **65**: 189–190.

———, ———, and G. M. Smith, 1929: The hydrostatic-pneumatic system of certain trees: movements of liquids and gases, *Carnegie Inst. Wash. Pub.* 397.

Malhotra, R. C., 1931: A contribution to the physiology and anatomy of tracheae, with special reference to fruit trees. I, Influence of tracheae and leaves on the water conductivity, *Ann. Bot.*, **45**: 593–620.

———, 1932: A contribution to the physiology and anatomy of tracheae with special reference to fruit trees. II, Water conductivity in higher plants and its relation to tracheae, *Ann. Bot.*, **46**: 11–28.

———, 1932: The distribution of some reserve substances in hard wheat at successive growth stages and their possible utilization, *J. Agr. Sci.*, **22**: 485–496.

Mangham, S., 1910–1911: The translocation of carbohydrates in plants, *Sci. Prog.*, **5**: 256–285, 457–479.

———, 1915: Observations on the osazone method of locating sugars in plant tissues, *Ann. Bot.*, **29**: 369–391.

———, 1917: On the mechanism of translocation in plant tissues. An hypothesis, with special reference to sugar conduction in sieve-tubes, *Ann. Bot.*, **31**: 293–311.

———, 1922: Transport of organic substances in plants, *Nature*, **109**: 476–477.

Mangin, L., 1898: Sur la nutrition et la défense de la vigne par injection, *J. Prak. Agr.*, **1898**: 918–920.

Martin, J. N., 1934: The relative growth rates and interdependence of tops and roots of the biennial white sweet clover, *Melilotus alba*, *Am. J. Bot.*, **21**: 140–159.

Maskell, E. J., and T. G. Mason, 1930: Studies on the transport of nitrogenous substances in the cotton plant. I, II, III, *Mem. Cotton Res. Sta. Trinidad*,

2; *Ann. Bot.*, **43**: 205–231, 615–652 (1929); **44**: 1–29; 233–267, 657–688 (1930).

MASON, T. G., 1922: A note on the growth and the transport of organic substances in bitter cassava (*Manihot utilissima*), *Sci. Proc. Roy. Dublin Soc.*, **17**: 105–112; *Bot. Abst.*, **12**: 759.

——— and C. J. LEWIN, 1924: On the rate of carbohydrate transport in the greater yam, *Dioscorea alata*, *Sci. Proc. Roy. Dublin Soc.*, **18**: 203–205.

——— and E. J. MASKELL, 1928: Studies on the transport of carbohydrates in the cotton plant. I, A study of diurnal variation in the carbohydrates of leaf, bark and wood, and of the effects of ringing, *Ann. Bot.*, **42**: 189–253.

——— and ———, 1928: Studies on the transport of carbohydrates in the cotton plant. II, The factors determining the rate and the direction of movement of sugars, *Ann. Bot.*, **42**: 571–636.

——— and ———, 1928: Studies on the transport of carbohydrates in the cotton plant. I, II, *Mem. Cotton Res. Sta. Trinidad*, Ser. B, *Physiology*.

——— and ———, 1931: Further studies on transport in the cotton plant. I, Preliminary observations on the transport of phosphorus, potassium and calcium, *Ann. Bot.*, **45**: 125–173.

——— and ———, 1934: Further studies on transport in the cotton plant, *Ann. Bot.*, **48**: 119–141.

——— and E. PHILLIS, 1934: Studies on the transport of nitrogenous substances in the cotton plant. VI, Concerning storage in the bark, *Ann. Bot.*, **48**: 315–333.

———, E. J. MASKELL, and E. PHILLIS, 1936: Further studies on transport in the cotton plant. III, Concerning the independence of solute movement in the phloem, *Ann. Bot.*, **50**: 23–58.

MAXIMOW, N. A., and M. A. KROTKINA, 1930: Investigations on the after effect of low temperature on the length of vegetative period, *Trudy pro Prikladnoi Botanike Genetike i Selekcii*, **23**: 427–478.

MCCARTY, E. C., 1932: Some relations between the carbohydrates and the growth rate in the wild oat, *Avena fatua*, *Occasional Papers Riverside Jr. Col. Calif.*, **6**: 1–32; *Chem. Abst.*, **26**: 4081.

MCHARGUE, J. S., and W. R. ROY, 1932: Mineral and nitrogen content of the leaves of some forest trees at different times in the growing season, *Bot. Gaz.*, **94**: 381–393.

MCLEAN, D. M., 1933: The effect of harvesting at different stages of maturity upon the yield and chemical composition of barley, *Sci. Agr.*, **13**: 698–727.

MCLEAN, R. C., and L. R. HUTCHINGS, 1935: Streamline flow and the movement of solutes in the transpiration stream, *Plant Physiol.*, **10**: 773–780.

MICHEL-DURAND, E., 1913: Variations des substances hydrocarbonée des feuilles au cours du développement, *C. R. Acad. Sci. Paris*, **156**: 1926–1929.

———, 1932: Sur l'azote, le soufre, le phosphore, le potassium des feuilles de *Prunus laurocerasus* au moment de lur chute, *C. R. Acad. Sci. Paris*, **195**: 1419–1420.

MILLER, E. C., 1937: A physiological study of hard winter wheat at different stages of growth, Unpub. data, *Kansas Exp. Sta.*

MOLISCH, H., 1886: Untersuchungen über Laubfall, *Bot. Cent.*, **25**: 393–394.

———, 1902: Ueber localen Blütungsdruck und seine Ursachen, *Bot. Ztg.*, **60**: 45–63.

———, 1928, 1929: The movement of sap in plants, *Nature*, **122**: 168–169; *Sci.*, **69**: 217–218.

MOORE, W., and A. G. RUGGLES, 1915: The action of potassium cyanide when introduced into the tissues of a plant, *Sci.*, **42**: 33–36.

MORGAN, A., 1931: Experiments on the control of hoary cress, *Victoria J. Agr.*, **29**: 504–515.

MORKRJETSKY, A., 1903: Über die innere Therapie der Pflanzen, *Z. Pflanzenkr.*, **13**: 251–265. Quoted by Rumbold, 1920.

MOTHES, K., 1931: Zur Kenntnis des N-Stoffwechsels höher Pflanzen, *Planta*, **12**: 686–731.

MOYER, L. S., 1937: Recent advances on the physiology of latex, *Bot. Rev.*, **3**: 522–544.

MÜNCH, E., 1926: Dynamik der Saftströmungen, *Ber. bot. Ges.*, **44**: 68–71.

———, 1927: Versuche über den Saftkreislauf, *Ber. bot. Ges.*, **45**: 340–356.

———, 1930: "Die Stoffbewegungen in der Pflanze," Jena, 234 pp.

MURNEEK, A. E., 1928: Nitrogen and carbohydrate distribution in organs of bearing apple spurs, *Mo. Sta. Res. Bull.* 119.

———, 1932: Relation of leaf area to fruit size and food reserves in apple seeds and branches, *Proc. Am. Soc. Hort. Sci.*, **29**: 230–234.

——— and J. C. LOGAN, 1932: Autumnal migration of nitrogen and carbohydrates in the apple tree, *Mo. Sta. Res. Bull.* 171.

———, 1933: Carbohydrate storage in apple trees, *Proc. Am. Soc. Hort. Sci.*, **30**: 319–321.

NÄGELI, C., 1866: Die Theorie der Capillarität, *Bot. Mitth.*, **2**: 471–501. Quoted by Copeland, 1902.

NELSON, N. T., 1925: The effect of frequent cutting on the production, root reserves and behavior of alfalfa, *J. Am. Soc. Agron.*, **17**: 100–113.

NICHOLS, P. F., 1930: Variations in content of sugars and related substances in olives, *J. Agr. Res.*, **41**: 89–96.

NICOL, H., 1934: Yield, duration and drought-resistance of lucerne as influenced by frequency and time of cutting, *Empire J. Exp. Agr.*, **2**: 380–390.

NIGHTINGALE, G. T., 1933: Effect of temperature on metabolism in tomato, *Bot. Gaz.*, **95**: 35–58.

NORDHAUSEN, M., 1916: Ueber die Saugkraft transpirierender Sprosse, *Ber. bot. Ges.*, **34**: 619–639.

———, 1919: Zur Kenntniss der Saugkraft und der Wasserversorgung transpirierender Sprosse, *Jahrb. wiss. Bot.*, **58**: 295–335.

———, 1921: Weitere Beiträge zum Saftsteigeproblem, *Jahrb. wiss. Bot.*, **60**: 307–353.

NORMAN, A. G., 1933: A preliminary investigation of the development of structural constituents in the barley plant, *J. Agr. Sci.*, **23**: 216–227.

OTTO, R., and W. D. KOOPER, 1910: Beitrag zur Abnahme bezw. Rückwanderung der Stickstoffverbindungen aus den Blättern während der Nacht sowie zur herbstlichen Rückwanderung von Stickstoffverbindungen aus den Blättern, *Landw. Jahrb.*, **39**: 167–172.

——— and ———, 1910: Untersuchungen über Stickstoffassimilation in den Laubblättern, *Landw. Jahrb.*, **39**: 999–1004.

OVERTON, J. B., 1911: Studies on the relation of the living cells to transpiration and sap flow in Cyperus. I, II, *Bot. Gaz.*, **51**: 28–63, 102–120.

———, 1925–1926: Stem anatomy and sap conduction, *Carnegie Inst. Wash. Year Book*, **25**: 155–158.

——— and J. B. SMITH, 1926–1927: Additional observations on stem anatomy and sap conduction, *Carnegie Inst. Wash. Year Book*, **26**: 159–160.

PARKER, K. W., and A. W. SAMPSON, 1931: Growth and yield of certain *Gramineae* as influenced by reduction of photosynthetic tissue, *Hilgardia*, **5**: 361–381.

PEIRCE, G. J., 1929: To demonstrate the course of sap ascent in plants, *Sci.*, **69**: 186–187.

———, 1934: Observation on sap hydraulics, *Am. J. Bot.*, **21**: 211–227.

———, 1935: The ascent of sap, *Madrono*, **3**: 1–3.

———, 1936: Are living cells involved in the ascent of sap? *Am. J. Bot.*, **23**: 159–162.

PENSTON, N. L., 1935: Return of mineral elements to the soil by plants, *Nature*, **136**: 268–269.

PETTIBONE, C. J. V., and C. KENNEDY, 1916: Translocation of seed protein reserves in the growing corn seedling, *J. Biol. Chem.*, **26**: 519–525.

PHILLIS, E., and T. G. MASON, 1933: Studies on the transport of carbohydrates in the cotton plant. III, The polar distribution of sugar in the foliage leaf, *Ann. Bot.*, **47**: 585–634.

——— and ———, 1936: Further studies on transport in the cotton plant. IV, On the simultaneous movement of solutes in opposite directions through the phloem, *Ann. Bot.*, **50**: 161–174.

PIERRE, W. H., and F. E. BERTRAM, 1929: Kudzu production with special reference to influence of frequency of cutting on yields and formation of root reserves, *J. Am. Soc. Agron.*, **21**: 1079–1101.

PLATENIUS, H., 1934: Chemical changes in carrots during growth, *Plant Physiol.*, **9**: 671–680.

———, 1934: Physiological and chemical changes in carrots during growth and storage, *N. Y. Sta. (Cornell) Mem.* 161.

POLOVRĂGEANU, I., 1933: Sur la variation quantitative des substances minérales au cours du développement des feuilles de *Populus pyramidalis* et *Robinia pseudo-acacia, Inst. Art. Grafice E. Mavin, Bucuresti VI, Univ. Bucuresti.*

POND, R. H., 1903: The biological relation of aquatic plants to the substratum, *Contrib. Biol. Great Lakes, Rept. U. S. Fish Comm.*, 483–526.

PRIESTLEY, J. H., 1920: The mechanism of root pressure, *New Phytol.*, **19**: 189–200.

——— and D. ARMSTEAD, 1922: Physiological studies in plant anatomy. II, The physiological relation of the surrounding tissue to the xylem and its contents, *New Phytol.*, **21**: 62–80.

———, 1935: Sap ascent in the tree, *Sci. Prog.*, **30**: 42–56.

RANKIN, W. H., 1917: The penetration of foreign substances introduced into trees, *Phytopath.*, **7**: 5–14.

REED, H. S., and E. T. BARTHOLOMEW, 1927: Water translocation in young fruits, *Sci.*, **66**: 382–385.

———, 1930: The swelling of citrus fruits, *Am. J. Bot.*, **17**: 971–983.

RENNER, O., 1911: Experimentelle Beiträge zur Kenntnis der Wasserbewegung, *Flora*, **103**: 175–247.

———, 1911: Zur Physik der Transpiration. I, *Ber. bot. Ges.*, **29**: 125–132.

———, 1912: Zur Physik der Transpiration. II, *Ber. bot. Ges.*, **30**: 572–575.

———, 1912: Versuche zur Mechanik der Wasserversorgung. I, Der Druck in den Leitungsbahnen von Freilandpflanzen. II, Über Wurzeltätigkeit, *Ber. bot. Ges.*, **30**: 576–580, 642–648.

———, 1915: Theoretisches und Experimentelles zur Kohäsionstheorie der Wasserbewegung, *Jahrb. wiss. Bot.*, **56**: 617–667.

———, 1918: Versuche zur Mechanik der Wasserversorgung, *Ber. bot. Ges.*, **36**: 172–179.

REWALD, B., and W. RIEDE, 1933: Das Verhalten von Fett, Phosphatiden und Eiweiss während der Samenreifen, *Biochem. Z.*, **260**: 147–152; *Biol. Abst.*, **8**: 876.

RHINE, J. B., 1926: Translocation of fats as such in germinating fatty seeds, *Bot. Gaz.*, **82**: 154–170.

RIPPEL, A., 1921: Die Frage der Eiweisswanderung beim herbstlichen Vergilben der Laubblätter, *Biol. Zent.*, **41**: 508–523.

RIVETT, M., and F. RIVETT, 1920: The anatomy of *Rhododendron ponticum* and of *Ilex aquifolium* in reference to specific conductivity, *Ann. Bot.*, **34**: 525–550.

ROACH, W. A., 1934: III, Injection for the diagnosis and cure of physiological disorders of fruit trees, *Ann. App. Biol.*, **21**: 333–343.

ROBERTSON, J. H., 1933: Effect of frequent clipping on the development of certain grass seedlings, *Plant Physiol.*, **8**: 425–447.

ROGERS, C. F., 1929: Winter activity of the roots of perennial weeds, *Sci.*, **69**: 299–300.

ROHRBAUGH, P. W., 1934: Penetration and accumulation of petroleum spray oils in the leaves, twigs, and fruit of citrus trees, *Plant Physiol.*, **9**: 699–730.

ROSHARDT, P. A., 1910: Ueber die Beteiligung lebender Zellen am Saftsteigen bei Pflanzen von niedrigem Wuchs, *Beih. bot. Cent.*, I, **25**: 243–357.

RUHLAND, W., 1911: Untersuchungen über den Kohlenhydratstoffwechsel von *Beta vulgaris*, *Jahrb. wiss. Bot.*, **50**: 200–257.

RUMBOLD, C., 1915: Methods of injecting trees, *Phytopath.*, **5**: 225–229.

———, 1920: The injection of chemicals into chestnut trees, *Am. J. Bot.*, **7**: 1–21.

———, 1920: Effect on chestnuts of substances injected into their trunks, *Am. J. Bot.*, **7**: 45–57.

SACHS, J., 1862: Zur Keimungsgeschichte der Gräser, *Bot. Ztg.*, **20**: 145–150.

———, 1862: Microchemische Untersuchungen, *Flora*, **45**: 289–301, 313–320, 325–336.

———, 1878: Ein Beitrag zur Kenntniss des aufsteigenden Saftstroms in transpirirenden Pflanzen, *Arb. Bot. Inst. Würzburg*, **2**: 148–184.

———, 1879: Ueber die Porosität des Holzes, *Arb. Bot. Inst. Würzburg*, **2**: 291–332.

SALMON, S. C., C. O. SWANSON, and C. W. MCCAMPBELL, 1925: Experiments relating to the time of cutting of alfalfa, *Kans. Sta. Tech. Bull.* 15.

SAMPSON, A. W., and R. SAMISCH, 1935: Growth and seasonal changes in composition of oak leaves, *Plant Physiol.*, **10**: 739–751.

SAMPSON, H. C., 1918: Chemical changes accompanying abscission in *Coleus blumei*, *Bot. Gaz.*, **66**: 32–53.

SANFORD, F., 1914: An experiment on killing tree scale by poisoning the sap of the tree, *Sci.*, **40**: 519–520.

———, 1915: In regard to the poisoning of trees by potassic cyanide, *Sci.*, **41**: 213–214.

SAYRE, J. D., V. H. MORRIS, and F. D. RICHEY, 1931: The effect of preventing fruiting and of reducing the leaf area on the accumulation of sugars in the corn stem, *J. Am. Soc. Agron.*, **23**: 751–753.

SCHERER, C. M., 1927: Tree injection for control of fungous diseases and insect pests, abst., *Phytopath.*, **17**: 51.

SCHMIDT, R. H., 1891: Ueber Aufnahme und Verarbeitung von fetten Ölen durch Pflanzen, *Flora*, **74**: 300–370.

SCHNEIDER-ORELLI, O., 1909: Die Miniergänge von *Lyonetia clerkella* und die Stoffwanderung in Apfelblättern, *Cent. Bact.*, II, **24**: 158–181.

SCHROEDER, J., 1871: Die Frühjahrsperiode der Birke (*Betula alba*) und des Ahorns (*Acer platanoides*), *Landw. Versuchs-Stat.*, **14**: 118–146.

SCHULZE, E., 1900: Über den Umsatz der Eiweissstoffe in der lebenden Pflanze. II, *Z. physiol. Chem.*, **30**: 241–312.

——— and N. CASTORO, 1903: Beiträge zur Kenntnis der Zusammensetzung und des Stoffwechsels der Keimpflanzen, *Z. physiol. Chem.*, **38**: 199–258.

SCHUMACHER, W., 1930: Untersuchungen über die Lokalisation der Stoffwanderung in den Leitbündeln höherer Pflanzen, *Jahrb. wiss. Bot.*, **73**: 770–823.

———, 1932: Direct detection of material transport in sieve tubes of plants, *Naturwiss.*, **20**: 364–365; *Chem. Abst.*, **26**: 4628.

———, 1933: Untersuchungen über die Wanderung des Fluoreszenz in den Siebröhren, *Jahrb. wiss. Bot.*, **77**: 685–732.

SCHWENDENER, S., 1892, 1893: Zur Kritik der neuesten Untersuchungen über das Saftsteigen, *Sitzungsber. Akad. Wiss. Wien*, **44**: 911–946; *Bot. Cent.*, **53**: 290.

———, 1909: "Vorlesungen über mechanische Probleme der Botanik," W. Engelmann, Leipzig.

SEIDEN, R., 1926: Vergleichende Untersuchungen über den Einfluss verschiedener äusserer Faktoren insbesondere auf den Aschengehalt in den Pflanzen, *Landw. Versuchs-Stat.*, **104**: 1–50.

SEREX, P., 1917: Plant food materials in the leaves of forest trees, *J. Am. Chem. Soc.*, **39**: 1286–1296.

SHATTUCK, C. H., 1915: Effect of cyanide of potassium on trees, *Sci.*, **41**: 324.

SHAW, R. H., and P. A. WRIGHT, 1921: A comparative study of the composition of the sunflower and corn plant at different stages of growth, *J. Agr. Res.*, **20**: 787–793.

SHULL, C. A., 1923: Review of physiology of the ascent of sap by Bose, *Bot. Gaz.*, **76**: 316–317.

SHUTT, F. T., S. N. HAMILTON, and H. H. SELWYN, 1930: The protein content of grass, chiefly meadow foxtail (*Alopecurus pratensis*), *J. Agr. Sci.*, **20**: 126–135.

SIMON, W. J. M., 1906: Recherches sur l'alimentation artificielle des plantes, *J. Soc. Nat. Hort. France*, III, **7**: 678–686; *Gard. Chron.*, **3**: ser., **41**: 8 (1907).

SMITH, E. F., 1896: The path of the water current in cucumber plants, *Am. Nat.*, **30**: 372–378, 451–457, 554–562.

SMITH, F., R. B. DUSTMAN, and C. A. SHULL, 1931: Ascent of sap in plants, *Bot. Gaz.*, **91**: 395–410.

SMITH, H. F., 1933: The physiological relations between tillers of a wheat plant, *Australia Council Sci. and Ind. Res. J.*, **6**: 1; *Biol. Abst.*, **9**: 1716. Quoted by Bartel, Martin and Hawkins.

SMITH, O., 1931: A study of growth and development of the potato plant, *Proc. Am. Soc. Hort. Sci.*, **28**: 279–284.

SNOW, R., 1925: The transport of organic foodstuffs in plants, *Nature*, **116**: 360.

SNYDER, H., 1893: The draft of the wheat plant upon the soil in the different stages of its growth, *Minn. Sta. Bull.*, **29**: 152–160.

SPRAGUE, H. B., 1933: Root development of perennial grasses and its relation to soil conditions, *Soil Sci.*, **36**: 189–209.

STAMM, A. J., 1929: The capillary structure of softwoods, *J. Agr. Res.*, **38**: 23–67.

STARK, O. K., 1927: The protein metabolism of the soybean, *Am. J. Bot.*, **14**: 532–548.

STEWARD, F. C., 1928: A preliminary study of the diffusion of sugars through membranes of living plant tissue, abst., *Am. J. Bot.*, **15**: 629–630.

——— and J. H. PRIESTLEY, 1932: Movement of organic materials in plants: A note on a recently suggested mechanism, *Plant Physiol.*, **7**: 165–171.

STRASBURGER, E., 1891: Ueber den Bau und die Verrichtungen der Leitungsbahnen in den Pflanzen, *Histol. Beitr. Jena*, **3**: 609.

———, 1893: Ueber das Saftsteigen, *Histol. Beitr. Jena*, **5**: 1–94; abst. *Bot. Cent.*, **60**: 270–272.

STREETS, R. B., 1934: The treatment of deciduous fruit trees and nut trees infected by *Phymatotrichum omnivorum* with ammonium compounds, *Sci.*, **79**: 417–418.

STUART, N. W., and C. O. APPLEMAN, 1935: Nitrogenous metabolism in Irish potatoes during storage, *Md. Sta. Bull.* 372.

STURKIE, D. G., 1930: The influence of various top-cutting treatments on rootstocks of Johnson grass (*Sorghum halepense*), *J. Am. Soc. Agron.*, **22**: 82–93.

SULLIVAN, J. T., and H. R. KRAYBILL, 1930: Seasonal changes in the composition of Stayman apple trees. II, Forms of nitrogen, *Proc. Am. Soc. Hort. Sci.*, **27**: 220.

SURFACE, H. A., 1914: Cyanide of potassium in trees, *Sci.*, **40**: 852–853.

SUROZ, J. I., 1891: Oel als Reservestoff der Bäume, *Beih. bot. Cent.*, **1**: 342–343.

Swarbrick, T., 1927: Studies in the physiology of fruit trees. I, The seasonal starch content and cambial activity in one- to five-year-old apple branches, *J. Pom. and Hort. Sci.*, **6**: 137–156.

Swart, N., 1914: "Die Stoffwanderung in ablebenden Blättern," Jena, 118 pp.

Thomas, E. E., and A. R. C. **Haas**, 1928: Injection method as a means of improving chlorotic orange trees, *Bot. Gaz.*, **86**: 355–362.

Thomas, L. A., and W. A. **Roach**, 1934: Injection of fruit trees: Preliminary experiments with artificial manures, *J. Pom. and Hort. Sci.*, **12**: 151–166.

Thomas, M., 1933: The natural abscission of twigs in Canadian black poplar (*Populus serotina* Hartig), *Naturalist*, **1933**: 79–86.

Thomas, W., 1919: Soil investigations. Unpublished data.

———, 1932: Composition of current and previous season's branch growth in relation to vegetative and reproductive responses in *Pyrus malus*, *Plant Physiol.*, **7**: 391–445.

Thompson, T. G., 1915: The total amino nitrogen in the seedlings of the Alaska pea, *J. Am. Chem. Soc.*, **37**: 230–235.

Thornton, H. G., and H. **Nicol**, 1934: Some effects of clipping the tops upon the root development of lucerne, *Medicago sativa*, *J. Agr. Sci.*, **24**: 532–539.

Thut, H. F., 1928: Demonstration of the lifting power of evaporation, *Ohio J. Sci.*, **28**: 292–298.

———, 1932: Demonstrating the lifting power of transpiration, *Am. J. Bot.*, **19**: 358–364.

Tison, A., 1900: Recherches sur la chute des feuilles chez les Dicotyledonées, *Mém. Soc. Linn. Normandie*, **20**: 121–327. Reviewed by Sampson, 1918.

Toole, E. H., 1924: The transformations and course of development of germinating maize, *Am. J. Bot.*, **11**: 325–350.

Trandafirescu, E., 1934: La variation quantitative des matières minérales et organiques au cours du développement des feuilles de *Aesculus hippocastanum*, *Quercus robur*, *Juglans regia* et *Zea mays*, *Contr. Univ. Bucuresti Lab. Anyl. Chem.*

Traub, H. P., 1927: Regional and seasonal distribution of moisture, carbohydrates, nitrogen, and ash in 2–3 year portions of apple twigs, *Minn. Sta. Tech. Bull.* 53.

Trowbridge, P. F., L. D. **Haigh**, and C. R. **Moulton**, 1915: Studies of the timothy plant. II, The changes in the chemical composition of the timothy plant during growth and ripening with a comparative study of the wheat plant, *Mo. Sta. Tech. Bull.* 20.

Tucker, G. M., and B. **Tollens**, 1900: Über den Gehalt der Platenenblätter an Nahrstoffe und die Wanderung dieser Nahrstoffe beim Wachsen und Absterben der Blätter, *J. Landw.*, **48**: 39–63.

Tukey, H. B., and E. L. **Green**, 1934: Gradient composition of rose shoots from tip to base, *Plant Physiol.*, **9**: 157–163.

Unger, F., 1868: Weitere Untersuchungen über die Bewegung des Pflanzensaftes, *Sitzungsber. Akad. Wiss. Wien*, **58**: 392–418.

Ursprung, A., 1905: Untersuchungen über die Beteiligung lebender Zellen am Saftsteigen, *Beih. bot. Cent.*, I, **18**: 147–158.

———, 1906: Die Beteiligung lebender Zellen am Saftsteigen, *Jahrb. wiss. Bot.*, **42**: 503–544.

———, 1907: Ueber die Ursache des Welkens, *Beih. bot. Cent.*, I, **21**: 67–75.

———, 1907: Studien über die Wasserversorgung der Pflanzen, *Biol. Cent.*, **27**: 33–60.

———, 1907: Abtötungs- und Ringelversuche an einigen Holzpflanzen, *Jahrb. wiss. Bot.*, **44**: 287–349.

———, 1913: Zur Demonstration der Flüssigkeitskohäsion, *Ber. bot. Ges.*, **31**: 388–400.

———, 1913: Ueber die Bedeutung der Kohäsion für das Saftsteigen, *Ber. bot. Ges.* **31**: 401–412.

———, 1915: Ueber die Kohäsion des Wassers im Farnannulus, *Ber. bot. Ges.*, **33**: 153–163.

———, 1915: Zweiter Beitrag zur Demonstration der Flüssigkeitskohäsion, *Ber. bot. Ges.*, **33**: 253–265.

———, 1916: Dritter Beitrag zur Demonstration der Flüssigkeitskohäsion, *Ber. bot. Ges.*, **34**: 475–487.

———, 1925: Einige Resultate der neuesten Saugkraftstudien, *Flora*, **18-19**: 566–599.

VAN DEN HONERT, T. H., 1932: Mechanism of the transport of organic materials in plants, *Proc. Acad. Sci. Amsterdam*, **35**: 1104–1112; *Chem. Abst.*, **27**: 752.

VESQUE, J., 1884: Recherche sur le mouvement de la sève ascendante, *Ann. Sci. Nat. Bot.*, VI, **19**: 159–199.

VICKERY, H. B., G. W. PUCHER, C. S. LEAVENWORTH, and A. J. WAKEMAN, 1935: Chemical investigations of the tobacco plant, *Conn. Sta. Bull.* 374.

VINES, S. H., 1896: The suction force of transpiring branches, *Ann. Bot.*, **10**: 429–444.

VIRTANEN, A. I., and M. NORDLUND, 1934: Synthesis of sucrose in plant tissue, *Biochem. J.*, **28**: 1729–1732.

——— and M. NURMIA, 1936: Studies on the winter hardiness of clover: I, Effect of cutting on the carbohydrate reserves in red clover roots, *J. Agr. Sci.*, **26**: 288–295.

VLADESCU, I. D., 1934: La variation quantitative des substances minérales et organiques pendant le développement de la *Nicotiana tabacum*, *Contr. Univ. Bucaresti Lab. Anyl. Chem.*

VON MOHL, H., 1860: Über den Ablösungsprozess saftiger Pflanzenorgane, *Bot. Ztg.*, **18**: 273–274.

WALTER, H., 1931: "Die Hydratur der Pflanzen und ihre physiologischökologische Bedeutung," Jena. Quoted by James and Baker, 1934.

WANN, F. B., 1929: Experiments on the treatment of chlorosis in Utah, abst, *Am. J. Bot.*, **16**: 844.

WEEVERS, T., and J. WESTENBERG, 1931: Versuche zur Prüfung der Münchschen Theorie der Stoffbewegungen in der Pflanze, *Rec. trav. bot. Néerl.*, **36**: 1173.

WEINBERGER, J. H., and F. P. CULLINAN, 1932: Further studies on the relation between leaf area and size of fruit, chemical composition and fruit bud formation in Elberta peaches, *Proc. Am. Soc. Hort. Sci.*, **29**: 23–27.

——— and ———, 1934: Nitrogen intake and growth response in peach trees following fall and spring fertilizer applications, *Proc. Am. Soc. Hort. Sci.*, **32**: 63–69.

WELTON, F. A., V. H. MORRIS, and A. J. HARTZLER, 1930: Distribution of moisture, dry matter, and sugars in the maturing corn stem, *Plant Physiol.*, **5**: 555–564.

WERNER, H. O., 1935: The effect of temperature, photoperiod and nitrogen level upon tuberization in the potato, *Am. Potato J.*, **12**: 274–280.

WESTERMAIER, M., 1883: Zur Kenntniss der osmotischen Leistungen des lebenden Parenchyms, *Ber. bot. Ges.*, **1**: 371–383.

———, 1884: Bedeutung todter Röhren und lebender Zellen für die Wasserbewegung, *Sitzb. preuss. Akad. Wiss.*, **48**: 1105–1117.

WHITE, P. R., 1927: Studies of the physiological anatomy of the strawberry, *J. Agr. Res.*, **35**: 481–492.

———, 1938: "Root Pressure"—an unappreciated force in sap movement, *Am. J. Bot.*, **25**: 223–227.

WIESNER, J., 1871: Untersuchungen über die herbstliche Entlaubung der Holzgewächse, *Litz. Akad. Wiss. Wien*, **64**: 465–509.

———, 1904: Über Sommerlaubfall, Trieblaubfall, und Hitzelaubfall, *Ber. bot. Ges.*, **22**: 64–72, 316–323, 501–505.

———, 1905: Über Frostlaubfall nebst Bemerkungen über die Mechanik der Blattablösung, *Ber. bot. Ges.*, **23**: 49–60.

WILFARTH, H., H. RÖMER, and G. WIMMER, 1905: Über die Nährstoffaufnahme der Pflanzen in verschiedenen Zeiten ihres Wachstums, *Land. Versuchs-Stat.*, **63**: 1–70.

WILLARD, C. J., 1927: An experimental study of sweet clover, *Ohio Sta. Bull.*, **405**: 3–84.

———, 1930: Root reserves of alfalfa with special reference to the time of cutting and yield, *J. Am. Soc. Agron.*, **22**: 595–602.

WINKLER, A. J., 1929: The effect of dormant pruning on the carbohydrate metabolism of *Vitis vinifera*, *Hilgardia*, **4**: 153–173.

———, 1932: The lateral movement of elaborated foods in the grape vine, *Proc. Am. Soc. Hort. Sci.*, **29**: 335–338.

WOODHOUSE, E. D., 1933: Sap hydraulics, *Plant Physiol.*, **8**: 177–202.

WRIGHT, J. G., 1928: The pit closing membrane in the wood of the lower Gymnosperms, *Trans. Roy. Soc. Canada*, **22**: 63–94.

YOCUM, L. E., 1925: The translocation of the food materials of the wheat seedling, *J. Agr. Res.*, **31**: 727–744.

YOUNG, P. A., 1934: Penetration, distribution, and effect of petroleum oils in apple, *J. Agr. Res.*, **49**: 559–571.

———, 1935: Oil mass theory of petroleum-oil penetration into protoplasm, *Am. J. Bot.*, **22**: 1–8.

———, 1935: Distribution and effect of petroleum oils and kerosenes in potato, cucumber, turnip, barley, and onion, *J. Agr. Res.*, **51**: 925–934.

ZIMMERMAN, P. W., A. E. HITCHCOCK, and W. CROCKER, 1931: The movement of gases into and through plants, *Contr. Boyce Thompson Inst.*, **3**: 313–321.

——— and M. H. CONNARD, 1934: Reversal of direction of translocation of solutes in stems, *Contr. Boyce Thompson Inst.*, **6**: 297–302.

CHAPTER XIII

THE PROCESS OF RESPIRATION IN PLANTS

1. Energy Relations of the Plant.—The various processes that go on in a green plant involve the execution of work. Some of these processes are the formation of carbohydrates, fats, proteins, and other organic compounds from the simple materials obtained from the soil and the air; the evaporation of water from the stem and leaves; the penetration of the roots through the soil; the elevation of the stem into the air; the accumulation of solutes; and the movements and activities of the protoplasm not included in the processes mentioned above. To accomplish the various kinds of work, a supply of kinetic energy must be furnished continually to the different plant parts concerned in the performance of the various functions. Let us now inquire into the source of the energy of the plant and its distribution, transformance, and general availability in the various activities of plant life.

The sun is the source of the energy that is necessary for all life, and this energy is transmitted in the form of heat, light, and chemical rays. About 70 per cent of this energy which falls upon the plant is absorbed by it, while the remainder is either reflected or transmitted. It has been mentioned in Chap. VIII that slightly over 70 per cent of this absorbed energy is utilized in the evaporation of water in transpiration. From 1.5 to 4.5 per cent of it is utilized in the synthesis of carbohydrates, fats, proteins, and other organic compounds and is thus transformed into potential or stored energy. Another portion of the absorbed energy may be used directly in the various processes of the parts by which it is absorbed and in raising the temperature of the plant, while the remainder is lost by radiation into the surrounding air.

Although the rays of the sun are the original source of the energy supply of the plant, the structure of the plant and the conditions under which it grows prevent it from utilizing this energy directly in all its parts. The structure of the plant is such that its interior portions are not exposed to the rays of the sun, so that they cannot utilize directly the energy of the sunshine. Also the living cells must be furnished with a continuous supply of energy during darkness as well as light. It is evident, therefore, that some supply of energy must be available for immediate use in all the cells of the plant under conditions of both light and darkness. In every living plant cell there is present a supply of proteins, carbohydrates, fats, and other organic compounds in varying amounts. Every living plant cell thus has in the form of these compounds, a supply of potential energy, which, if changed into the kinetic form, would be available for the life processes of the cell. The living cell has the power so to transform this stored energy whenever the proper conditions exist. The processes associated with the transformation of potential energy in the living cell into the kinetic form are primarily chemical decompositions in which the stored foods and the protoplasm of the cell are involved. The most important process concerned in the liberation of stored energy in the cells is termed "respiration." Respiration may be defined as a process that goes on in every living cell whereby energy is released in a form that is available to the protoplasm. The term "energesis" has been suggested (Barnes, 1905) as a name for the processes involved in energy release, but this has never come into general use.

According to Stiles and Leach (1932) respiration has sometimes been called "dissimilation," in contrast to the process of assimilation. Malpighi (1679) reported that seeds require a supply of free oxygen to germinate. Jean Ingen-Housz in 1774 and De Saussure (1804) noted that plants in the light apparently produced oxygen, and, in the dark, carbon dioxide. The latter investigator noted the difference between the green and colorless parts of plants in this regard, but it remained for Sachs (1882) to explain the difference between the two basic processes concerned in these gaseous exchanges.

2. Importance of Respiration.—The energy necessary for the performance of any function by a living organism or part of an organism is obtained by the process of respiration in the protoplasm. Two conditions, at least, are necessary for cells to perform the functions of life: (*a*) there must be substances present that can readily be oxidized, and from the oxidation of which the available energy can be obtained; and (*b*) there must be oxygen present with which to oxidize these materials. Most organisms can perform some work in the absence of free oxygen, but usually it is only a limited amount and for only a limited time (Meldrum, 1933).

3. Nature of Respiration.—In the process of respiration under ordinary conditions, oxygen is absorbed, organic compounds, especially carbohydrates and fats, disappear, and carbon dioxide and water are formed as end products. The materials used and the waste products formed in the process are well defined, but the changes that occur in the cell whereby the organic compounds are utilized and the waste products produced are not clearly understood. We can thus observe the beginning and end of the process but know little or nothing of the intermediate changes that occur.

a. Direct Oxidation of Foods.—Respiration is generally considered to be an oxidation process because the final products—carbon dioxide, water, and kinetic energy—are the same as those derived from the combustion of organic substances in the air. The oxidation of organic substances in respiration, however, is striking in that it occurs at ordinary temperatures generally below 35°C., while oxidation of the same substances outside the living organism is accomplished only by the aid of very high temperatures or of powerful reagents. The oxidation of organic substances in respiration apparently occurs in the protoplasm itself or in close proximity to it, and it is considered that the oxidation is accomplished by the action of oxidizing enzymes which are secreted by the protoplasm. Herzfeld and Klinger (1919) stressed the possibility that certain molecular structures may weaken the union of the oxygen atoms in the oxygen molecule so that "activation" (increased oxidizing power) results, whereby loose chemical combinations are made with H_2O, with OH ions, and with O_2 absorbing surfaces. Such organic compounds as the lower fatty acids or their salts might thus be oxidized to carbon dioxide and water, fulfilling the conditions for oxidation processes in the living organism, namely active O_2 and easily oxidizable simpler metabolic products, thereby rendering the assumption of certain oxidases unnecessary.

According to Stiles and Leach (1932), Blackman (1928) presented a scheme of the respiratory process involving a chain of four consecutive processes as follows:

1. Hydrolysis of starch and disaccharoses to hexoses.
2. Activation: This includes the steps leading to the formation of heterohexoses with the less stable type of internal ring structure. This involves an isomeric change of the hexose molecule.
3. Glycolysis: This stage, involving the action of the zymase complex, includes the degradation of activated hexose to methyl glyoxal, pyruvic acid, acetaldehyde, and probably lactic acid.
4. Aerobic Respiration: This is concerned with the production of carbon dioxide and water from the intermediate substances produced in glycolysis.

The carbon dioxide formed in the last process may account for only a fraction of the carbon that appears to be involved in glycolysis. The conclusion is thus drawn that a part of the products of glycolysis are in some manner reconstructed into the system. To this process the name "oxidative anabolism" is given. In this scheme oxygen plays a double role in respiration. It not only produces an oxidation of the products of glycolysis, but it also affects the preglycolytic stages in that it brings about an alteration in the rate of production of the direct substrate for glycolysis.

In 1897 Büchner obtained zymase from yeast. It is now apparent that the substance originally termed zymase is a mixture of enzymes, and it is now usually called the "zymase complex." The following are some of the enzymes present or accompanying the substance termed "zymase": (1) hexosephosphatase, concerned in the formation of hexosephosphate; (2) phosphatase, which hydrolyzes hexosephosphate into hexose and phosphate; (3) glycolase, which acts on hexose to form pyruvic aldehyde or methylglyoxal; (4) carboxylase, which acts on α-ketonic acids, *e.g.*, pyruvic acid, to give acetaldehyde and carbon dioxide.

Allison, Hoover, and Burk (1933) claimed that they obtained, by extracting commercial sucrose with alcohol, a factor that was specific and essential for respiration. This factor is soluble in water and in absolute alcohol, but insoluble in the ordinary fat solvents. These characteristics classify it as a coenzyme.

According to Stiles and Leach (1932) it is misleading to speak of respiration as combustion. There are only two points of resemblance between them. The substrate and the end products may be alike; and, when this is true, the amount of energy released will be the same in respiration and in combustion. It should, however, be remembered that in no case has it been definitely proved that the carbohydrates, fats, or proteins are directly attacked outside the protoplasm, and it will also be shown in the discussion of the oxidizing enzymes that they are incapable, as prepared by present methods, of oxidizing under laboratory conditions any of the carbohydrates or fats that are apparently used in respiration.

b. Decomposition of Protoplasm.—It is believed by some that the facts of respiration may be explained more satisfactorily by assuming that the process is concerned in the decomposition of the living matter itself. Barnes (1905) assumed that in the labile protein components of the protoplasm there occurs a shifting of atomic groups within the molecule as a result of the addition of hydroxyl groups from the water present. Dissociation of these components follows this procedure, the rate depending upon the prevailing conditions. The groups or radicals of the protoplasm that have thus been dissociated may contain structural components that have been derived from sugars, fats, or proteins. The portion of the protoplasm that has been dissociated is then rebuilt at the expense of the sugars, fats, or proteins present in the cell, and as a consequence they disappear during the process of respiration. An attempt was made by Palladin (1896) to determine the relationship of the amount of protoplasm to the rate of carbon dioxide production. He determined the amount of carbon dioxide given off by wheat seedlings in the dark, and in parallel experiments he determined the amount of protein present which was indigestible in gastric juice. It was assumed that such proteins represented those that were components of the protoplasm. Palladin found that with an adequate supply of carbohydrate, the elimination of carbon dioxide was proportional to the amounts of indigestible proteins present in the seedlings. Although this method is far from exact, it nevertheless gives some indication of the relationship between the protoplasm and the process of respiration. The part of the protoplasm, however, that is the most actively concerned in respiration has received but little attention. The leaf cells of the Indian pipe (*Monotropa uniflora*) contain a colorless chromogen that oxidizes very easily if the proper conditions are provided. Osterhout (1917) determined that the oxidation of this chromogen in the cell is much more rapid in the nucleus than in the cytoplasm. From that he inferred that respiration is much more rapid in the nucleus than in the other parts of the protoplasm. It is not known, however, whether the nuclei of other plant cells show a similar behavior.

4. Organic Materials Used in Respiration.—One of the striking results of respiration is the loss in weight of the organism, which, however, is not ordinarily noticed in plants on account of the manufactured food equaling or exceeding the amount lost. If food is withheld from an animal, or if a plant is prevented from manufacturing food, the loss in weight due to respiration is very apparent. The most striking examples of the loss in weight from respiration are shown in the case of germinating seeds. The seedling plant always weighs less in dry matter than the seed from which it originated. Thus corn seedlings, including the residues of the seeds 2 weeks after planting, contain from 5 to 10 per cent less dry matter than the original seeds. Miller (1910) noted in sunflower seedlings 10 days old that the dry weight of 100 seedlings was 5.7 g. as compared with 7.3 g., which was the dry weight of an equal number of seeds. This was a loss in weight of 1.6 g. or approximately 20 per cent of the original dry weight of the seeds. Thus, one-fifth of the foods stored in the seeds was consumed in respiration in order to generate sufficient energy to push the plant roots into the soil, to raise the stem or cotyledons above ground, and to carry on the various life activities of the seedling.

Haller, Harding, Lutz, and Rose (1931) noted in strawberries that at 80°F., the loss per day of citric acid amounted to 25 per cent of the amount present at the time of packing. Bolas and Melville (1933) found in the tomato plant that the loss in weight due to respiration during a 17-hr. period of darkness amounted to 4.6 per cent of the dry weight of the plant.

a. Fats, Oils, and Carbohydrates.—The materials which disappear in the process of respiration are, for the most part, the carbohydrates, fats and oils. It is well established that the carbohydrates furnish the greater portion of the materials that are used in respiration. Even when fats and oils are the only source of oxidizable material, it is very probable that they are transformed into sugars before they are ultimately utilized in respiration.

The relationship of the supply of carbohydrates, especially of the sugars, to the intensity of respiration has received considerable consideration. Palladin (1896) and Palladin and Komleff (1902) found that etiolated bean leaves floated on a solution of sucrose or with their petioles immersed therein respired much more rapidly when transferred to distilled water than did the controls. In one case, 100 g. of leaves low in carbohydrate gave off at room temperature 89.6 mg. of carbon dioxide per hour for a given period. After the same leaves were floated upon a solution of cane sugar in darkness for 2 days they gave off carbon dioxide at the rate of 147.8 mg. per hour. Maige and Nicolas (1910) showed that, in general, respiration increased as the concentration of the sugar in the medium increased until the concentrations were sufficient to produce plasmolysis when the intensity of respiration declined. In ripening bananas, Gore (1914) found that the graph of the rate of respiration paralleled that of the rate of starch hydrolysis in the fruit, while there were only slight changes in the amounts of fats and proteins. His results were substantiated by Olney (1926), who found that the value of the respiratory quotient during the ripening of this fruit was approximately unity, which indicated that the oxidation in this case consisted solely of the combustion of carbohydrates. It was noted by Knudson (1916) that there was an increased respiration in plants when sugars were added to the nutrient solutions in which they were growing. According to Hopkins (1927), the sugar content of wounded potatoes increased from 53 to 68 per cent of the original sugar present. The rate of respiration was also increased after the wounding, and he assumed that the basis of this augmentation of the intensity of respiration was the increase in the sugar content of the cells. The curves of the sugar content and of the rate of respiration, however, did not exactly parallel.

The hexoses are apparently the sugars utilized in respiration. Thus Hasselbring and Hawkins (1915) could observe no correlation between the total sugars in the sweet potato and respiratory activity. A simultaneous decrease, however, in the reducing sugars and the respiratory intensity was observed, which indicated that they were the immediate source of the materials used in respiration. In plants *d*-glucose is apparently the most easily oxidizable sugar. The amount of *d*-levulose has been found to be in excess of *d*-glucose in the leaves, which is taken to indicate that the former is the more stable and is thus used to a lesser degree in respiration. Spoehr and McGee (1922) found that the intensity of respiration in the excised leaves of *Helianthus annuus* was not increased when levulose was fed to them, while the presentation of *d*-glucose, sucrose, and mannose stimulated respiration. As will be mentioned later, the addition of amino compounds with these sugars stimulated the rate of respiration in all cases except with levulose.

When the carbohydrate content of plants is low, the rate of the emission of carbon dioxide increases with the increasing supply of sugars. This may be conceived of as a matter of mass action, in that as the concentration of the available material to be oxidized becomes greater, the rate of reaction is increased. This is true, however, only within the limits of the capacity of the protoplasm present to carry on respiration. Thus, the intensity of respiration in etiolated leaves of the bean was greatly accelerated by floating them for a 2-day period upon a sugar solution. A further increase in respiration, however, could not be obtained by extending the period of sugar feeding to 40 days, although the concentration of the sugar in the leaf cells was greatly increased. During a 2-day period upon the sugar solution, sufficient sugar was absorbed for the maximum rate of respiration for the quantity of protoplasm present, so that the further addition of sugar was without influence upon the process. The total quantity of carbohydrate in the cell will thus not influence the rate of respiration except to determine the length of time that the process may continue therein.

b. Proteins.—It was formerly considered that proteins were also directly utilized in respiration, but experiments have shown that under normal conditions they are not utilized to any appreciable extent in this way. In the absence of an adequate supply of carbohydrates, however, the proteins are apparently utilized in the process. Thus, Palladin (1888) found that if wheat plants were left in the dark until they were deprived of their carbohydrate reserves, they lost a portion of their protein. Deleano (1912) found that excised leaves of the grape during the first 100 hr. used only carbohydrates in respiration, while the protein remained intact. After the carbohydrate supply was depleted, however, the process of respiration was strikingly altered and the protein was split up into soluble products. Meyer (1918) made the same observation with leaves and noted that the protein supplied only the carbon chains for respiration and that no loss of nitrogen from the leaves could be observed.

Yemm (1935) considered that during the later stages of starvation it is highly probable that the breakdown of protein is one of the important reactions concerned with carbon dioxide production. According to Stiles and Leach (1932) the minimum constant rate of respiration of starved leaves has been termed "protoplasmic respiration," in contrast to "floating respiration" in which carbohydrates and oily reserves are utilized. This protoplasmic respiration may result from the utilization of protein as a respiratory substance after the other reserve materials have been exhausted.

It is evident that the amount of energy which could be obtained by the decomposition of proteins under normal conditions would be very small, for this decomposition is only a cleavage into the amino acids. Furthermore, the plant is capable of re-forming these amino acids again into proteins without the direct utilization of energy save that which is derived from the oxidation of carbohydrates. Spoehr and McGee (1923) believed, therefore, that the total gain in energy for the plant from protein decomposition would be nothing and that unless the proteins yield energy of a certain form or at a particular rate, which is necessary for the proper functioning of certain processes, it is difficult to see what advantage would result or what the mode would be of deriving energy by continual decomposition and subsequent resynthesis of the proteins. It may safely be assumed that the rate of energy release from protein decomposition is much slower than that from carbohydrates, but of the possible different forms of energy liberated in the decomposition of these two types of compounds little or nothing is known.

Since it is known that amino acids have a marked influence upon the action of certain enzymes, Spoehr and McGee suggested that the decomposition of proteins may provide materials essential to the proper functioning of these agents. As the amino acids are amphoteric electrolytes, they may also function in maintaining the hydrogen-ion concentration of the cell contents within definite limits. Spoehr and McGee investigated the effects of glycocoll and asparagine upon the intensity of respiration in excised leaves of *Helianthus annuus* and found that the leaves which were fed glycocoll respired more actively than those which had not thus been treated. The rate of respiration was increased by feeding the leaves with *d*-glucose, but this rate was further increased by feeding in addition glycocoll or asparagine. When the leaves were given sucrose there was an initial drop in the rate of carbon dioxide emission, which probably represented the time required for the sugar to migrate into the leaves, after which there was an irregular rise that became prominent after 35 hr. When glycocoll was added in addition to the sucrose, there was a marked rise in the rate of respiration immediately after the initial drop. Hafenrichter (1928), however, in his study of the germination of the soybean concluded that there was no evidence under the conditions of his experiments to support the theory that amino acids stimulate respiration.

c. Effect of Food Type.—Hafenrichter (1928) pointed out that since the stored food materials of seeds undergo a series of changes during the germination and development of the seedlings, there would be qualitative and quantitative differences in the materials available for respiration. The manner in which these qualitative and quantitative differences may affect respiration under various conditions has not, however, been investigated to any extent, and it is not known whether or not the required amount of energy under any set of conditions is always supplied by a definite metabolic process. It has been demonstrated for animals that it is possible to maintain the same energy relations by substituting foods that are qualitatively different provided that the caloric values remain constant. It has been observed, as has been previously mentioned, that plants apparently prefer one of two closely related carbohydrates for respiration and it has also been found that some plants may utilize different amounts of one food in the presence of different quantities of another. This indicates that external as well as internal conditions determine the kind of materials that may be utilized in respiration. From a study of the respiration of the seedlings of two varieties of soybean, Hafenrichter concluded that the difference in the respiration of different plants cannot be explained on the basis of quantitative differences in the reserve food materials. He obtained evidence, however, which suggested that plants exhibit a selection of organic compounds in their metabolism and that this is reflected in their respiration.

5. Aerobic and Anaerobic Respiration.—Under the conditions of a normal supply of free oxygen, the process of respiration is marked by an absorption of oxygen, by a consumption of organic compounds, especially of carbohydrates, and by the appearance of water and carbon dioxide as end products. When respiration proceeds after this manner, it is termed "normal," "aerobic," or "oxygen" respiration. If carbohydrates are the compounds concerned, the chemical procedure in the process is represented by the equation

$$\underset{\text{Glucose}}{C_6H_{12}O_6} + 6O_2 \longrightarrow 6CO_2 + 6H_2O$$

If ethyl alcohol is the compound concerned, the equation is written:

$$\underset{\text{Ethyl alcohol}}{C_2H_5OH} + 3O_2 \longrightarrow 2CO_2 + 3H_2O$$

In some cases, carbon monoxide appears to be a product of respiration. Thus Langdon and Gailey (1920) reported from 1 to 12 per cent of this gas but no carbon dioxide in the hollow pneumatocysts of *Nereocystis luetkeana.*

When plants are placed in an atmosphere devoid of free oxygen, the process of normal respiration is greatly modified so that the end products are carbon dioxide and a number of more or less complex compounds including ethyl alcohol, acetic, formic, oxalic, propionic, other organic acids, and probably molecular hydrogen instead of simply carbon dioxide and water. Under this condition, however, the evolution of carbon dioxide, as a rule, is at a much lower rate than during aerobic respiration. When respiration proceeds after this fashion it is termed "intramolecular" or "anaerobic" respiration, or "fermentation." Under the conditions of anaerobic respiration, the decomposition of the organic compounds may be accomplished in some instances by the rearrangement of the atoms within the molecule without the presence of any oxygen supply. Alcoholic fermentation is one of the best examples of this type of procedure. The process is represented by the equation

$$\underset{\text{Glucose}}{C_6H_{12}O_6} \longrightarrow \underset{\text{Ethyl alcohol}}{2C_2H_5OH} + \underset{\text{Carbon dioxide}}{2CO_2}$$

Fermentation and anaerobic respiration are either the same process, or are very similar. According to Stiles and Leach (1932) this assumption is substantiated by the following facts: (*a*) Hexose sugars are apparently utilized in both processes. (*b*) In both processes carbon dioxide is produced. (*c*) The presence of ethyl alcohol has been demonstrated in many cases of anaerobic respiration. Generally, it is found in smaller amounts than in fermentation, but some of it may be utilized in secondary reactions. (*d*) Zymase, which is concerned in alcoholic fermentation, has been found in various parts of the higher plants. (*e*) Acetaldehyde is probably formed as an intermediate product in both cases. (*f*) Phosphates increase the rate of both reactions.

The action of zymase on sugars, in the production of ethyl alcohol and carbon dioxide, is frequently termed "zymasis."

The partial oxidation of organic compounds under anaerobic conditions may also occur under the influence of oxygen that has been derived from the breaking down of other organic compounds. It is therefore conceivable that the oxygen thus derived from one part of a complex molecule may oxidize even another part of the same molecule. In no case, however, in anaerobic respiration is the oxidation of complex compounds entirely completed to carbon dioxide and water.

6. Relationship of Anaerobic and Aerobic Respiration.—According to Kidd (1917) Rollo as early as 1798 noted that plant tissues, as a vital process, produced carbon dioxide in the absence of free oxygen, but his statement was not accepted for many years afterward. Pfeffer (1881 to 1885) termed this anaerobic production of carbon dioxide "intramolecular respiration," as he believed that it was a normal process in respiration and genetically connected with a subsequent oxidation by the oxygen of the atmosphere under ordinary conditions. An impetus was given to this manner of considering anaerobic respiration when Stocklasa and Czerny (1903) reported the extraction of an enzyme from beet roots, potatoes, peas, and other plant tissues and also from animal tissues, which appeared very similar in its action to the enzyme zymase from yeast, which has the power of converting sugar into ethyl alcohol and carbon dioxide. It was through the extensive studies of Palladin and Kostytschew, however, from 1888 to 1920, and through the suggestions of Barnes (1905) and Peirce (1909, 1912) that the theory of the direct relationship between anaerobic and aerobic respiration came to be generally accepted. The student is referred to Kostytschew (1927) for a detailed review of the literature pertaining to this subject.

It is now generally believed that anaerobic and aerobic respiration are intimately connected in the vital activities of the plant. The present theories of respiration generally assume that anaerobic respiration constitutes a preliminary step in the process. This initial decomposition of the organic materials in the process of respiration is believed to be independent of free oxygen. This may be brought about by a rearrangement of the atoms within the molecule or by the intervention of oxygen derived from other compounds. The end products derived from this initial decomposition are believed to be ethyl alcohol (Palladin and Kostytschew, 1906), organic acids, and carbon dioxide. It is believed by some that ethyl alcohol and carbon dioxide are the principal compounds formed in this initial step (Genevois, 1927; Blackman, 1928). Under normal conditions, however, plants and animals utilize free oxygen to oxidize to carbon dioxide and water, the products produced in anaerobic respiration. If this is true, then anaerobic respiration is proceeding under the conditions of aerobic respiration, but it is not discernible, because its end products are almost immediately oxidized.

Most organisms, however, can live for only a short time by the process of anaerobic respiration and must be supplied with free oxygen if they continue to function normally. The inability to survive may be due to one or both of the following causes: In the first place, the products of anaerobic respiration may be toxic to the protoplasm, so that unless they are oxidized by the free oxygen their accumulation will render the protoplasm completely inactive. The production of blackheart in Irish potatoes is a good example of the detrimental effects that may be initiated by anaerobic conditions (Bennett and Bartholomew, 1924).

Thomas (1929) found that Newton Wonder apples, under anaerobic conditions, sooner or later suffered from "invasive alcohol poisoning," which was associated with, although not necessarily caused by, an accumulation of ethyl alcohol produced in zymasis. He reported later (1931) that apples stored in air at low temperatures above 0°C. are likely to show "low temperature, internal breakdown," and also to develop "soft and deep scald." After these pathologic conditions develop, ethyl alcohol and acetaldehyde progressively accumulate, but there is no evidence that their formation precedes the incidence of these diseases. Zymasis apparently follows many types of tissue injury. Thomas and Fidler (1933) noted that increasing concentrations of free oxygen progressively retarded zymasis. Fidler (1933) stated that senescence in apples was accompanied by a change in the respiratory metabolism of the cells of the fruits. This change caused an accumulation in them of considerable amounts of ethyl alcohol and acetaldehyde. In 1934, he described methods for estimating these two compounds in apples.

In the second place, if sufficient free oxygen is not present, the decomposition processes in the anaerobic stage may reach an equilibrium due to the accumulation of the products thus formed. There may be needed in the process some substance to disturb constantly in one direction or another the equilibrium that tends to be reached, and oxygen appears to be the agent in this regard.

The initial decomposition of the compounds that occur in anaerobic respiration is considered by some to be the process that liberates the energy necessary for the vital activities of the cell. The energy derived from the oxidation of the end products of anaerobic respiration is assumed to be greatly in excess of what is actually needed and is lost by radiation, conduction, and convection into the surrounding medium (Peirce, 1909 to 1912). The amount of energy liberated, however, in anaerobic respiration is small compared with that released in aerobic respiration. Thus in the transformation of a gram molecule of glucose to alcohol and carbon dioxide, 57 kg.-cal. of heat would be liberated. The oxidation of the alcohol thus produced to water and carbon dioxide would liberate 652 kg.-cal. of heat. The complete oxidation of a gram molecule of glucose would consequently give 709 kg.-cal. In order to liberate a given amount of energy, more than twelve times as much glucose must be used in fermentation or anaerobic respiration as would be required in aerobic respiration (see also McGinnis and Taylor, 1923).

The following observations are generally presented as evidence to substantiate the assumption that anaerobic respiration is a normal process of the organism under conditions of a free oxygen supply, and that it is intimately connected with oxygen respiration:

a. All the plants that have thus far been examined are able to produce carbon dioxide in the absence of free oxygen. This behavior is believed to be so general that it cannot be considered simply a biological adaptation to abnormal conditions but must be classed as a vital process of all cells.

b. When plants are transferred from an anaerobic condition to one that has a normal supply of free oxygen, the production of carbon dioxide is markedly increased over that which normally prevails under aerobic conditions. This excess carbon dioxide production is assumed to be due to the oxidation of the accumulated end products of anaerobic respiration.

This phenomenon has been observed by Gustafson (1930) for tomato fruits of all ages, and by Leach and Kent (1934), and Leach (1936) for various seeds. The latter investigators noted that when the seeds were changed from the anaerobic to the aerobic condition, the absorption of oxygen was very rapid. They suggested that under anaerobic conditions some product is formed which has a strong affinity for oxygen, and that it thus becomes oxidized rapidly as soon as free oxygen is supplied.

c. Any changes that are made in the external conditions, excepting the oxygen supply, do not modify the value of the ratio of the carbon dioxide output of anaerobic respiration to that of aerobic respiration. This indicates that the intensity of both types of respiration are influenced by the same factors and to the same extent. Thus Palladin (1894) showed that etiolated bean leaves which were artificially supplied with carbohydrates increased the formation of carbon dioxide in the absence of oxygen in the same proportion as did leaves that were similarly treated and placed under conditions of a normal supply of oxygen. Karlson (1925) found that the effects of the vapors of ether, benzene, and alcohol upon aerobic and anaerobic production of carbon dioxide by wheat seedlings were closely similar.

d. The oxidizing enzymes of plants are unable to attack sugar directly, but according to Kostytschew (1927) they are able to oxidize to carbon dioxide and water the substances contained in fermented solutions.

e. It has been demonstrated by Zaleski (1915), Kostytschew (1908, 1927), and others that green plants can oxidize ethyl alcohol. This is assumed to account for the failure to detect alcohol in the tissues under normal conditions, a fact which has been used as an argument against the theory that anaerobic respiration occurs normally.

It was also found by Kostytschew (1908) that sugar solutions fermented by yeast stimulated the aerobic respiration in plants.

According to Thomas (1925) it is now generally accepted that in the usual respiration of glucose by higher plants, a zymase system cooperates with an oxidation enzyme system to produce carbon dioxide. As long as the oxidation system maintains its normal activity, the zymase activity of cells is not demonstrable. When, however, the oxidation system is inhibited under anaerobic conditions, zymase activity becomes evident by the accumulation of ethyl alcohol. It is considered that acetaldehyde is the precursor of ethyl alcohol in alcoholic fermentation. Gustafson (1934) found acetaldehyde in tomato fruits under all conditions of treatment, but the amount did not increase as the time of anaerobosis lengthened. Ethyl alcohol, however, was found in all tomato fruits under all conditions of treatment, and the amount increased with the length of time the anaerobic conditions continued.

Some investigators, however, including Maquenne and Demoussy (1921), believed that anaerobic and aerobic respiration do not have a common origin but that they proceed from different causes and should both be regarded as autonomous functions as truly in their intimate mechanism as in the influence they exert upon the life of the green plant. A direct proof that anaerobic respiration is the initial step in normal respiration and is thus intimately associated with aerobic respiration is difficult and could be supplied only if it were possible to demonstrate which products of anaerobic metabolism are oxidized in the presence of oxygen. In the oxidation of organic compounds in respiration there are a great many steps between the original compound and the formation of carbon dioxide and water, and the most important products and reactions in the process cannot be measured or detected by the chemical methods at present available.

A few facts should be mentioned here in regard to the relative intensity of anaerobic and aerobic respiration as measured by the production of carbon dioxide. As a general rule, the production of carbon dioxide in anaerobic respiration is less than that under aerobic conditions ranging from 18 to 85 per cent of the latter. The production of carbon dioxide in anaerobic respiration is greater, however, than that which would be produced by the formation of ethyl alcohol and carbon dioxide by fermentation alone, so that some of the carbon dioxide evolved must have its origin from other sources than from the activity of zymase. Some cases have been reported in which the intensity of anaerobic respiration as measured by the evolution of carbon dioxide equaled or exceeded its production in aerobic respiration. Thus Hill (1913) found that the production of carbon dioxide by ripe cherries, blackberries, and grapes for a considerable period of time was as rapid as the aerobic production. The fruits that showed this, however, had apparently finished their growth and were ripe. He found, however, that green peaches and germinating wheat respired more than twice as intensely under aerobic as under anaerobic conditions.

Gustafson (1932) observed that certain of the cacti often produce nearly as much carbon dioxide in the absence of free oxygen as they do in the air.

Edwards (1933) reported that seeds of rice and arum germinate in the almost complete absence of oxygen. The coleoptile may elongate two or three times its original length, owing apparently to the lengthening of cells already formed. The seedlings of rice and arum possess either a type of respiratory metabolism that produces nontoxic intermediary products when complete oxidation is prevented or a greater tolerance to toxic products than other plants.

Emerson (1927) demonstrated the presence of two types of respiration in *Chlorella* and other green algae. The first is the typical respiration during the course of which a photosynthetic product is consumed. The rate of this type of respiration is slightly increased by cyanide. The second type is apparent when glucose is added to the suspending medium and the rate of respiration doubles. The mechanism responsible for this increase differs from that concerned with typical respiration in that it may be inhibited by cyanide.

7. Aerobic Respiration and Photosynthesis Contrasted.—Jean Ingen-Housz (1774) noted that green plants in darkness rendered the air impure just as did animals and that in daylight the fruits, flowers, and roots acted in the same manner. He was thus apparently the first to define in this general manner the conditions under which plants carry on respiration, although he did not recognize the significance of his observations. It remained for De Saussure (1804) to show that this absorption of oxygen and the formation of carbon dioxide were stages in the respiration of plants, and that the respiration of plants was identical with the respiration of animals. He also recognized the formation of water in the process and was one of the first to explain the relation of respiration in growth and the other vital activities of the plant.

For a long time, however, the processes of respiration and photosynthesis were not clearly defined, and it was considered that they did not occur concurrently in the same cells. It was argued that it was impossible that two processes which were direct opposite could occur in the cells at the same time. This erroneous idea still persists, and it is frequently stated that one of the marked differences between green plants and animals is that the plants take up carbon dioxide and give off oxygen while animals absorb oxygen and eliminate carbon dioxide. This statement is only partially correct and leaves the general impression that green plants do not carry on respiration. The best way in which to differentiate clearly between photosynthesis and respiration is to contrast them in regard to the time and place of their occurrence, the compounds used, the end products formed, and the relation of these two processes to the energy supply of the plant. These points may be contrasted after the following manner:

Photosynthesis	Respiration
1. Photosynthesis occurs only in the chlorophyllous cells of plants.	1. Respiration occurs in every living cell regardless of its color and regardless of whether it is of plant or animal nature.
2. Photosynthesis occurs only in the sunlight or under artificial illumination.	2. Respiration occurs all the time. It occurs in darkness as well as in light.
3. In the process of photosynthesis water and carbon dioxide are used.	3. In the process of respiration water and carbon dioxide are the end or waste products.
4. In the process of photosynthesis oxygen is given off as a waste product.	4. In the process of respiration oxygen is used.
5. Food is built up in the process of photosynthesis.	5. Food is destroyed in the process of respiration.

Photosynthesis	Respiration
6. Photosynthesis increases the weight of a plant.	6. Respiration decreases the weight of a plant.
7. Energy is stored in the process of photosynthesis.	7. Energy is released in the process of respiration.
8. The chemical equation of photosynthesis is written:	8. The chemical equation of respiration is written:
$6CO_2 + 6H_2O \rightarrow C_6H_{12}O_6 + 6O_2$	$C_6H_{12}O_6 + 6O_2 \rightarrow 6CO_2 + 6H_2O$

8. Respiration and Breathing.—In the preceding topics, the process of respiration in plants has been discussed in regard to the place of its occurrence, the materials used, and the waste products produced. It is the intention here to describe the means by which the oxygen is supplied to the cells for use in respiration and the manner in which the waste products are eliminated from the plant body. In discussing these processes, it is necessary to distinguish between the terms "respiration" and "breathing." These two terms are commonly used synonymously, but from a biological standpoint each has a distinct meaning and the difference between them should be clearly fixed. Breathing is the forcible inhaling and exhaling of air by the animal body, while respiration is an energy-releasing process that occurs in every living cell. In the higher animals there are special organs—the nose, windpipe, lungs, and certain muscles—that are employed to draw the air into the body and to force it out again. The air entering the body contains the oxygen supply needed in respiration, while the air leaving the body contains the carbon dioxide and water that are the waste products of the process. After the air has entered the lungs, a portion of its oxygen is taken up by the blood stream and carried to the cells of the body. This oxygen is taken up by the cells while the carbon dioxide and water formed in the cells as a result of respiration diffuse into the blood stream and are carried primarily to the lungs, where they are eliminated with the exhaled air.

Plants, in contrast with animals, do not have any special organs for drawing air into the body and forcing it out again. The air simply diffuses in and out through the stomata, lenticels, and cuticle of the younger plant parts. After entering the plant, it circulates through the intercellular spaces in the plant and possibly to a considerable degree in solution in the water movements of the plant. As the air thus comes in contact with the cells of the plant, they absorb the oxygen therefrom and give up the carbon dioxide thereto that has been formed in respiration. The entrance and exit of these gases from the cells are due primarily to the difference in gradient that exists between the respective gases in the cells and in the surrounding air. The oxygen exterior to the plant diffuses inwardly through the lenticels, stomata, and intercellular spaces to restore the oxygen equilibrium that has been disturbed by the absorption of oxygen by the cells. Likewise, the carbon dioxide diffuses outwardly through these openings, since its concentration within the intercellular spaces is higher than that of the surrounding air. Plants thus differ markedly from animals in regard to the manner in which they obtain their air supply and by which they eliminate the end products of respiration into the external atmosphere. The general statement can thus be made that animals breathe but plants do not. The student should keep clearly in mind that respiration is a chemical process which goes on continuously in every living cell whether of plant or animal nature, while breathing is a mechanical process that supplies air to and eliminates it from the animal body but does not occur in plants at all.

9. Methods of Measuring the Intensity of Aerobic Respiration.—The intensity of aerobic respiration in plants may be measured by the loss in weight, the amount of oxygen absorbed, or the amount of carbon dioxide produced by the plant or plant part concerned. A detailed description of the various methods is beyond the scope of this work, and the student is referred to Harrington and Crocker (1923) and Kostytschew (1927) for references and more complete information until 1927. More recently, methods for studying respiration have been reported by Hottes and Hafenrichter (1928); Harding, Maney, and Plagge (1929); Maney, Harding, and Plagge (1929); Emerson (1930); Stiles and Leach (1931); Bennet-Clark (1932); Tang (1932); Wynd (1932, 1935); Leach (1932); Balch (1933); Jeffrey and Cruess (1933), and Mitchell (1935).

It is the intention here to describe in detail only two of these methods and to describe only the general procedure to be followed in others. Most of the methods that have been devised for determining the rate of respiration are based on the measurement of the carbon dioxide that is produced. These methods may be grouped as follows according to the principles involved in their manipulation:

a. One of the oldest and most generally used types of procedure is to allow the carbon dioxide that is respired to diffuse into a stream of carbon dioxide free air, which is constantly being drawn through the apparatus containing the respiring material. This air then passes through a solution of alkali which depletes it of the carbon dioxide that it thus acquires. The amount of carbon dioxide may then be determined by titration, by the increased weight of the absorbent, or by the weight of the precipitate, provided an insoluble carbonate is formed. Apparatus of this type have been devised or modified by Pfeffer (1885), Kostytschew (1908), Gurjar (1917), Ginsburg (1925), and Beaumont, Willaman, and DeLong (1927). For relatively large amounts of actively respiring materials, methods of this type are excellent for measuring the carbon dioxide produced. They have the advantage in that the air is constantly renewed over the respiring material so the disturbing effects of the changes in the partial pressures of carbon dioxide and oxygen that occur in closed containers are avoided. The methods of this type, however, neglect entirely the question of the absorption of oxygen, which is not always proportional to the amount of carbon dioxide given off. They are not adapted for use with limited quantities of small or inactive material, as the experimental error is relatively large, so the amount of carbon dioxide given off in a reasonable time may fall within the limit of error.

One of the most common methods of the above type which can easily be used by students in the laboratory is conducted after the following manner: The apparatus, which consists of a connected series of closed containers, U-shaped calcium chloride tubes, and potash tubes, is attached at one end to a filter pump, by means of which a continuous stream of air is drawn through the apparatus. As the air enters, it is drawn through a concentrated solution of sodium or potassium hydroxide and thence through a solution of barium hydroxide, which will indicate whether or not the entering air has been deprived of its carbon dioxide. The carbon dioxide-free air now passes over the respiring material, which is in a closed container, and thence through calcium chloride tubes, which deplete it of moisture. The moisture-free air is thence drawn into weighed potash bulbs containing a concentrated solution of sodium or potassium hydroxide where the carbon dioxide is absorbed. The air that has absorbed moisture from the alkali solution next passes through weighed calcium chloride tubes where this moisture is again absorbed. After the experiment has proceeded the required time, the alkali tubes and the last two calcium chloride tubes are weighed. The initial weight of the alkali tubes is subtracted from their final weight plus the increased weight of the two calcium chloride tubes. This difference in weight represents the weight of the carbon dioxide that has been eliminated by the respiring material during the period of the experiment.

b. Numerous types of microrespirometers have been devised for the detection of very slight changes in the composition of the gases under consideration. Instruments of this type have been devised and used by Tashiro (1913), Osterhout and Haas (1917), Lund (1919), and others. Some of these methods are too complicated for general use and are limited in regard to the quantity of material that may be used. None of them provides for the simultaneous determination of both sides of the respiratory exchange in the same material.

c. An indicator method has been perfected by Haas (1916) and Osterhout (1918) for the detection of small quantities of carbon dioxide respired. The method has the advantage of sufficient sensitiveness so that the experimental period can be very short. The method is also useful for the rapid and simultaneous determination in several different experiments. By a special arrangement the apparatus can be used to determine approximately the amounts of oxygen absorbed. Indicator methods for the determination of the oxygen absorbed by respiring material have been described by Osterhout (1918). The use of the indicator method requires care to avoid errors due to buffer effects or to the production of acid-reacting substances other than carbon dioxide or of alkaline-reacting substances (Harrington and Crocker, 1923).

d. A method that has been used to a considerable extent is to place the respiring material in a closed container and occasionally sample the air for analysis in some form of microapparatus. This method is described in detail by Thoday (1913) by whom it was used with success.

e. The absorption by a solution of alkali of the carbon dioxide respired in a closed container with manometric measurement of the oxygen absorbed has been used by Stich (1891) and others. This type of apparatus gives both the oxygen consumption and the carbon dioxide produced. The method, however, has a disadvantage, since the carbon dioxide is absorbed as rapidly as it is produced and thus does not compensate for the oxygen absorbed by the respiring material. This fact causes relatively large changes in pressure within the apparatus, which might affect the behavior of the plant in the experiment.

An apparatus for the measurement of the gaseous exchange in the respiration of seeds and small parts of plants, devised by Harrington and Crocker (1923) and later modified by Davis (1925), is shown in Fig. 35. This instrument, based upon the principle of the respirometer of Ganong (1908), is one of the most accurate and easy to operate of the many respirometers that have been devised. With this apparatus, the oxygen consumption and the carbon dioxide evolved are determined in the same apparatus and for the same period of time, using the whole volume of air instead of a sample. This apparatus also has an advantage in that the carbon dioxde given off by the respiring material accumulates and tends to compensate for the oxygen absorbed so no large changes in pressure occur during the experimental period.

This apparatus consists of a cylindrical glass tube *c* in the form of a bottle with an extension *c′*, which serves as a receptacle for the seeds or material to be studied. Into the mouth of this cylinder is fitted a ground-glass stopper *s*, upon which are mounted a short open manometer *m* with a 2-mm. bore and an inlet tube provided with a stop cock *sc* and a small chamber *ch*. The stopper fits deep into the neck of the cylinder so that a mercury seal *mc* can be made. The apparatus has been used in three sizes of about 20, 40, and 80 cc. capacity. Before conducting an experiment, the combined capacity of the corked cylinder *c* including the seed chamber *c′* and the pores of the manometer to the level of the mercury and of the intake tube to the stop cock *sc* must be determined. This is found by measuring the volume of water necessary to fill these parts of the apparatus.

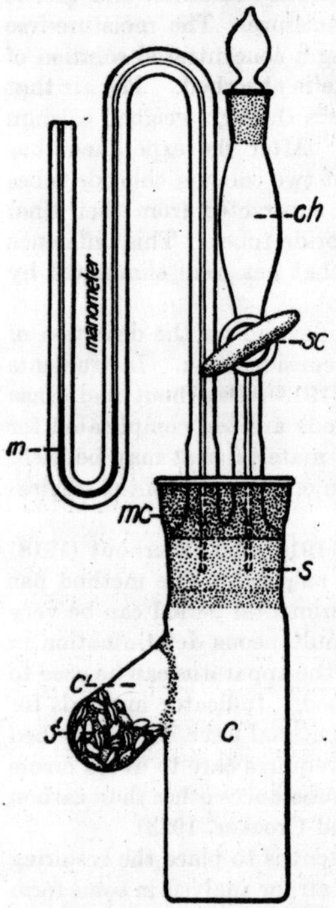

Fig. 35.—Diagram of respirometer. *c*, chamber of respirometer. *c′*, seed chamber. *s′*, seeds. *s*, ground-glass stopper with attached manometer and supply tube provided with a small chamber *ch* and a stop cock, *sc*. *mc*, mercury seal. *m*, mercury. (*After Davis*, 1925.)

After the volume of the instrument has been determined, the soaked seeds *s′* whose volume and weight have been obtained are placed in the seed chamber *c′*. The glass stopper is then inserted in the cylinder, the stop cock *sc* and cork of the intake tube being left open to equalize the pressure, and the mercury seal made. The apparatus is then placed in any convenient thermostat, preferably a water bath, and as soon as the instrument has come to the desired temperature the stop cock *sc* in the intake tube is closed and the chamber *ch* stoppered. Care must be taken that the respirometer stands vertically and that the mercury in the manometer *m* is at the same level in the two arms. The experiment is now ready to proceed, and the temperature of the medium surrounding the respirometer is taken, and the barometric pressure is recorded.

After the experiment has proceeded for the desired period, which is generally from 24 to 48 hr., from 1 to 3 cc., as required, of 20 per cent sodium hydroxide is measured into the chamber *ch* of the intake tube, which is then left unstoppered. The manometer reading is then taken by holding a millimeter scale behind the arm of the manometer. The stop cock *sc* is now opened and the alkali carefully admitted into the respiratory cylinder *c*, care being taken that no air escapes from the cylinder. The temperature of the medium surrounding the respirometer and the reading of the barometer are recorded. After absorption is completed, the manometer reading is again determined.

From the manometer and barometric readings, the temperature, and the volume of the apparatus corrected for the volumes of the respiratory material and of the alkali admitted, the three significant volumes of air under standard conditions (0°C., 760 mm.) can be determined. From these three volumes, the volume of oxygen absorbed and the volume of carbon dioxide given off may be calculated after the following manner (Sherman, 1921; Harrington and Crocker, 1923):

V_r = volume of the respirometer
V_s = volume of the imbibed seeds
V_a = volume of the absorbent alkali
T_1 = initial absolute temperature
T_2 = final absolute temperature
B_1 = initial barometric pressure
B_2 = final barometric pressure
m_1 = initial manometer reading
m_2 = final manometer reading

$$A.\ (V_r - V_s)\frac{273}{T_1} \cdot \frac{B_1}{760}$$

$$B.\ (V_r - V_s)\frac{273}{T_2} \cdot \frac{B_2 \pm m_1}{760}$$

$$C.\ (V_r - V_s - V_a)\frac{273}{T_2} \cdot \frac{B_2 \pm m_2}{760}$$

$B - C$ = the volume of carbon dioxide eliminated
$A - C$ = the volume of oxygen absorbed

From this the respiratory quotient CO_2/O_2 is easily calculated. The number of milligrams of carbon dioxide eliminated and of oxygen absorbed per gram of imbibed weight for any given period can also be determined; 1 cc. of carbon dioxide weighs 1.96 mg. and 1 cc. of oxygen weighs 1.428 mg. at 0°C. and 760 mm. barometric pressure.

Stiles and Leach (1931), and Leach (1932) described an instrument known as a "katharometer," which they developed for the measurement of small amounts of respiration. The principles involved in the instrument are: A wire becomes heated when an electric current is passed through it. The temperature of this wire is determined by the thermal conductivity of the medium surrounding it. The thermal conductivities of different gases are different, so that if in the respiration chambers oxygen is replaced by carbon dioxide there is a change in the thermal conductivity of the gases in the vessel, and hence in the temperature and electrical resistance of the wire. The output of carbon dioxide is thus followed by determining the changes in electrical resistance of a heated wire in the respiratory chamber. This method is sufficiently sensitive to detect a change of 0.001 per cent in volume of carbon dioxide. However, it is only applicable when the concentrations of oxygen and carbon dioxide are changing.

10. The Respiratory Quotient.—The ratio of the volume of carbon dioxide given off to the volume of oxygen absorbed during respiration is termed the "respiratory quotient" and is written CO_2/O_2. Great diversity of opinion, however, exists as to the importance to be attached to the value of this ratio as an index to the reactions occurring in respiration.

a. Factors Influencing.—A discussion of the factors that may cause the fluctuation of this ratio will serve to show some of the conditions prevailing in the tissues that may be indicated by its changes in value.

1. *Kind of Material.*—The types of materials consumed in respiration will influence the value of the respiratory quotient. Thus under the ordinary aerobic conditions of respiration, if carbohydrates are the

compounds utilized in the process and if all the oxygen absorbed by the plant is consumed in their combustion, the volume of oxygen absorbed would be equal to the volume of carbon dioxide given off according to the equation:

$$\underset{\text{Glucose}}{C_6H_{12}O_6} + 6O_2 \longrightarrow 6CO_2 + 6H_2O$$

Thus the respiratory quotient would be unity. If organic acids that are relatively rich in oxygen were the materials utilized, the respiratory quotient would be greater than unity. Thus the respiratory quotient for the complete oxidation of tartaric acid would be 1.6, while that of oxalic acid would be 4.0. If fats are the materials utilized in respiration, the respiratory quotient is less than unity, since most of these compounds are relatively low in oxygen. The fats are relatively low in oxygen and generally absorb oxygen to satisfy their unsaturated condition before their disintegration is initiated in the process of respiration. If the fats are converted into sugar before they are ultimately used in respiration, this transformation requires a quantity of oxygen which is not indicated by the quantity of carbon dioxide produced when the sugar thus formed is oxidized. This accounts for the low respiratory quotient of plant parts which contain fats as reserve foods. During the ripening of oily seeds the respiratory quotient is also greater than 1. As mentioned in Chap. X, the fats in oily seeds are formed at the expense of carbohydrates. In the transformation of carbohydrates to fats a certain amount of oxygen is set free. The oxygen that has thus been freed is utilized directly in respiration, and consequently a smaller amount is absorbed from the exterior than would otherwise occur. Consequently the volume of carbon dioxide produced is greater than the volume of oxygen absorbed, and the respiratory quotient is greater than unity.

2. *Temperature.*—The respiratory quotient may also vary according to the temperature at which the observations are made. This has been explained by the fact that both the absorption of oxygen and the exhalation of carbon dioxide vary with differences of temperature, but they vary in different degrees. Harrington (1923) noted in respiratory studies of dormant apple seeds that the respiratory quotient increased with an increase in temperature. He believed that this was due to an impoverishment in easily oxidizable substances and might also be indicative of an oxygen deficiency in the respiring tissues. The respiratory quotient decreased with a decreasing temperature. This indicated a storage of oxygen that became very considerable at 10 and 0°C. In the respiration of these seeds the temperature coefficient was affected by the previous treatment. It was higher after treatments that tended toward after-ripening and lower after treatments that induced deeper dormancy. The influence of external and internal conditions upon the respiratory quotient

is, however, but little understood, and no definite statements can be made in that regard.

Tang (1931, 1932) reported that the temperature characteristics for the production of carbon dioxide by *Lupinus albus* are different from those for the consumption of oxygen, so that the respiratory quotient changes accordingly with temperature.

3. *Organic Acids.*—A portion of the oxygen that is absorbed by the plant may be utilized for purposes other than that of respiration. This extra oxygen may be used in the formation of organic acids. The depression of the value of CO_2/O_2 due to this factor is noticed in the storage of organic acids in ripening fruits and in succulents (Richards, 1915).

According to Bennet-Clark (1933) it has long been known that some members of the *Crassulaceae* possess the peculiarity of producing a large quantity of an acid, now known to be malic acid, when they are placed in darkness. Since the quantities of carbon dioxide that they produce are very small in comparison with the quantity of malic acid formed, the view became established that this formation of acid was an abnormal type of respiration—a respiration in which a product is formed that is less highly oxidized than carbon dioxide, the normal end product of complete respiration. Bennet-Clark considered that no more than 1 molecule of malic acid is formed from each molecule of sugar, so that 2 or 3 atoms of carbon are converted into other substances that appear to be polysaccharoses. He believed that the carbon dioxide which is liberated must originate from some intermediate product that arises in the conversion of sugar into malic acid.

When succulent plants are placed in darkness, the amount of oxygen absorbed in the process of respiration is in excess of the carbon dioxide evolved, and in extreme cases the respiratory quotient may approach zero (Aubert, 1892). After the plants have been in darkness for a time, the accumulation of organic acids decreases, and the rate of evolution of carbon dioxide gradually increases with a corresponding increase in the value of the respiratory quotient, which, however, does not reach unity. When these succulent plants are exposed to sunlight the acids are decomposed and carbon dioxide is freed. Some of this carbon dioxide is used in photosynthesis. This procedure is considered to be an advantage to fleshy plants because they are handicapped by the relatively slow diffusion of gases through their tissues. The respiratory quotient under these conditions greatly exceeds unity, and, according to Bennet-Clark (1932), maximal values of 1.39 and 2.05 have been observed in *Sedum praealtum.* He believed that these high values are inconsistent with the view that the disappearance of malic acid is due to oxidation, and that this procedure strengthens the view that malic acid forms a link in the carbon cycle in succulent plants other than that of the oxidation of carbon dioxide.

4. *Other Factors.*—Carbon dioxide may be formed without the absorption of oxygen. This occurs under anaerobic conditions, as has been mentioned previously. The same effects may occur, but to a lesser degree if the supply of oxygen is rather limited. Under these conditions the respiratory quotient is greater than unity.

Plants with leaves colored red by anthocyanin behave similarly to succulents in respiration. The respiratory quotient is lower in red than in green leaves, and this is apparently due to a more rapid absorption of oxygen by the red leaves. This difference in the value of the quotients can be related to a greater accumulation of organic acids in the leaves containing anthocyanin than in those from which this pigment is absent.

b. Germination of Seeds.—The respiratory quotient of a large number of different types of germinating seeds has been studied by Godlewski (1882), Bonnier and Mangin (1884), Gerber (1900), Sherman (1921), Harrington (1923), Ermakoff and Iwanoff (1931), Stiles and Leach (1932, 1933), and Dastur and Desai (1935). If carbohydrates are the only reserves present and the supply of free oxygen is unlimited, the respiratory quotient will approximate unity. If fats and oils are the only foods present in the seed and a supply of free oxygen is readily available, the respiratory quotient will be considerably less than unity until these foods have been converted into sugars. In practically all seeds that have been examined, however, the respiratory quotient during the first few hours of germination is unity or greater than unity. As germination proceeds, the value of the quotient falls. In seeds containing fats and oils the fall is very marked and continues for a considerable time, after which it rises and approaches unity.

These fluctuations in the value of the respiratory quotient may be explained for the most part upon the basis of the nature of the food available for respiration and upon the structure of the seed coats of the seed. Let us consider first the influence of the types of food. Practically all seeds, whether of a starchy or an oily nature, contain a small reserve of sugars that serve the seedling in the initial stage of germination, so that while this is being utilized the respiratory quotient would be approximately unity. Even starchy seeds contain more or less oil, which is generally concentrated in the embryo. If, as would be expected, this oil is drawn upon during very early germination, the large use of oxygen in its conversion to sugar would give a temporarily low value to the respiratory quotient of such seeds. In the germination of oily seeds, the value of the respiratory quotient will be near that for complete oxidation (approximately 0.70), provided that the stored fats are easily oxidized. If these fats, however, are difficult of oxidation, the respiratory quotient will be low and near to that for the conversion of fats to sugar.

After the fats and oil in seed have been converted into carbohydrates, the value of the respiratory quotient approaches unity.

The structure of the seed coats may influence the respiratory quotient after the following manner: In the initial phase of germination, the presence of the unbroken testa around the seed prevents the free diffusion of gases, so that a relatively small amount of oxygen enters, and anaerobic respiration occurs. The rupture of the seed coats causes a quick rise in the absorption of free oxygen and a less rapid rise in the output of carbon dioxide, so that the value of the respiratory quotient falls.

Some observations have been made on the value of the respiratory quotient of other plant parts. Thus Cerighelli (1924) reported that the respiratory quotient of the roots of wheat, sunflower, corn, peas, buckwheat, rhubarb, and others at the beginning of development was always less than unity. In annuals low in stored foods, it remained less than unity during growth, while in those which accumulated reserve foods it became greater than unity.

According to Maquenne and Demoussy (1913), the respiratory quotients of the leaves examined by them were begonia, 1.11; lilac, lily, maize, and pea, 1.07; pear, 1.10; tobacco and wheat, 1.03; and rhubarb, 1.02.

11. Respiration and the Production of Heat.—It is a familiar fact that respiration produces heat in the animal body, but that this occurs in plants is not known so generally. The heat produced by respiration in plants is quickly dissipated into the surrounding air, primarily because they have no means of controlling its loss and because they have a large surface in proportion to their mass. Fleshy fruits, bulbs, and tubers may, however, retain sufficient heat to raise their temperature slightly above their surroundings. Thus ripening bananas maintain themselves at temperatures distinctly above that of their surroundings, while Richards (1897) stated that the temperature of the Irish potato was 0.17°C. and that of the onion 0.5°C. higher than the surrounding air. Wright and Whiteman (1931) noted that the internal temperature of cases of holly stored at 50°F. was from 2 to 6°F. higher than the surrounding air. The production of heat by respiration in plants, however, cannot readily be observed unless the proper precautions are taken to prevent its loss.

The method that is now the most commonly used for demonstrations in laboratory studies was developed and perfected independently by Ganong (1908) and Peirce (1908). The general procedure is to place the plant parts to be studied in Dewar flasks or thermos bottles, insert a thermometer, carefully plug with cotton, insulate the containers as thoroughly as possible, and note the changes in temperature as indicated by the thermometer. This experiment, however, is not so easily performed as would at first appear, since, to obtain reliable results, certain conditions, of which some are difficult to obtain, must be satisfied. According to Potter (1917), some of these conditions are (*a*) a sufficiently large amount of material, (*b*) a supply of oxygen, (*c*) a means for the disposal of the carbon dioxide produced, and (*d*) efficient control for the prevention of putrefaction and fungal growth. This last condition is the most difficult to obtain in this experiment. The heat equivalent of the material under consideration, of the water used, and of the container must also be known and taken into consideration if accurate results are to be obtained.

The results obtained by Peirce (1912) with peas are commonly cited as illustrations of the liberation of heat by plants in respiration. He placed 75 g. of air-dry peas that had been rinsed with a solution of mercuric chloride in a Dewar flask, added 100 cc. of sterilized distilled water, and noted the changes in temperature. The temperature of the flask rose from 18°C. at the beginning of the experiment to 38°C. at the end of 5 days, after which it remained constant. In the 7 days this quantity of peas liberated 2,593 cal. or an average of 370 cal. per day. In similar experiments Darsie, Elliott, and Peirce (1914) determined the average daily increase in temperature of several kinds of seeds due to respiration, when infection from bacteria and fungi was prevented. They found that the increase in temperature to be expected from 10 g. of viable seeds was 1.82°C. for hemp, 0.80° for barley, 0.75° for clover, 0.73° for wheat, 0.55° for oats, and 0.49° for corn. Their experiments indicated that the quantity of heat liberated in respiration when they were placed under conditions favorable for germination gave some indication of the quality of the seeds, of their viability, and of the vigor of their growth immediately following germination.

The question arises as to the value of the heat liberated in respiration to the plant. Peirce (1909, 1912) was of the opinion that although the essential product of respiration is energy, it is the energy that is immediately converted into work which is utilized by the plant in its vital activities. If the aerobic phase of respiration is merely considered a process of purification whereby useless or injurious substances are converted into forms which can be eliminated, then the heat thus liberated may be considered as of no special benefit to the plant. It may be considered as a waste product of respiration just as carbon dioxide and water.

The heating of moist grain when placed in storage is due primarily to the heat generated in the respiration of the cells of the seeds. A high temperature and moisture content are two factors that especially stimulate respiration, as will be discussed in subsequent topics. In this self-heating of grains, respiration may raise the temperature to a point where other chemical reactions may occur (Gore, 1911). Bacterial action may also play a part under these conditions. When grain with a high moisture content is stored, the heat generated from respiration cannot readily be dissipated into the atmosphere, so that it accumulates and raises the temperature to a point where it is injurious to the seed and where it may even cause the disintegration of the tissues. Hot weather accelerates the heating of stored moist grain in two ways. It affects it biochemically by increasing the rate of enzyme action and physically by reducing the rate of conduction of heat into the atmosphere. Bailey (1918) found that all the normally plump spring wheat that heated contained over 14.3 per cent of moisture. He observed that sound, plump, hard wheat containing less than 14.5 per cent of moisture would keep without heating in storage in a temperate climate. A lower moisture content must be employed, however, in storing shriveled or frosted wheat and possibly also with plump wheat in a mild climate.

It was observed by Gilman and Barron (1930) that the rise in temperature due to the germination of oats, wheat, and barley was 6.8, 2.6, and 3.6°C., respectively. The presence of *Aspergillus flavus* on these grains, however, raised their temperature (at 18 per cent moisture) to 26, 7.4, and 16°C., respectively, while *Aspergillus niger* had approximately the same effect. It was found by Larmour, Clayton, and Wrenshall (1935) with red spring wheat that when the action of fungi is prevented by the vapor of toluene or carbon tetrachloride, the production of carbon dioxide occurs at a slow rate, and no heating occurs even when the moisture content is raised to 25 per cent.

Bakke and Noecker (1933) found with oats that there is a general rise in the consumption of oxygen as the moisture content increased. The optimum moisture content for heat production is between 24 and 40 per cent, depending on aeration. Molds and microorganisms are almost always associated with grain in storage. When the water content is below 15 per cent, under conditions of farm storage, respiration is so low that the little amount of heat produced is rapidly radiated. The optimum temperature for heat production by moist seeds is between 30 and 35°C. Swanson (1934) observed that wheat which was stored at 60°F. produced very little growth of mold when the moisture amounted to 20 per cent and none when it was 18 per cent or lower. However, if the grain was placed at 95°F. some growth of mold was observed when the moisture content was as low as 14 per cent. Mold growth may be inhibited by the exclusion of air.

The question of the spontaneous heating and ignition of hay has long been discussed and many conflicting hypotheses have been proposed to account for these reactions. Hoffman (1935) stated it is commonly accepted that the production of heat in a mass of uncured or wet hay at temperatures up to 70°C., or higher, is due mainly to the process of respiration of the living cells of the plant and of the microorganisms present. Chemical reactions are considered to account for increases in temperature above which microorganisms are inactive. He found in the heating haymow that along with the operation of biological agencies there occurs a purely chemical oxidation as evidenced by a loss of oxygen considerably in excess of the carbon dioxide formed. This chemical oxidation is more marked beyond the temperature range usually ascribed to the activity of microorganisms.

12. Factors Affecting Respiration. *a. Temperature.*—It is apparently well established that the rate of respiration increases with the temperature, until the latter begins to injure the tissues. Clausen (1890) found the temperature coefficient for the respiration of germinating wheat to be 2.86 between 0 and 10°C., and 1.09 between 30 and 40°C., with an average value of 2.71 between 0 and 40°C.

Gerhart (1930) stated that the temperature coefficient for respiration in the fruit of the strawberry was 2.5 for temperatures below 25°C. Bělehrádek and Bělehrádková (1929) found that the temperature coefficient of respiration first increased and later decreased with the age of the tissues. They attributed this to the change with age of the colloidal nature of the protoplasm. This modifies the rate of diffusion in the reacting protoplasmic phase. Matthaei (1905) observed that at 5.8°C. 2 g. of leaves respired 0.1 mg. of carbon dioxide per hour, while at 33.1°C. the rate was 1.35 mg. per hour. Gore (1911) obtained an average temperature coefficient of 2.38 in 49 respiration experiments with 40 different fruits. Olney (1926) noted that the rate of respiration in ripening bananas at 12°C. was only one-third of that at 20°C. Bigelow, Gore, and Howard (1905) noted that the respiration of apples was twice as high in cellar storage as in cold storage. Morse (1908) studied the intensity of respiration of apples at different temperatures by determining the amount of carbon dioxide evolved. Some of his results are shown in the following table:

MILLIGRAMS OF CARBON DIOXIDE EXHALED PER KILO OF APPLES PER HOUR

Experiment	0°C.	5° to 10°C.	18 to 25°C.	Experiment	0°C.	5 to 10°C.	18 to 25°C.
1	2.3	8.0	16.4	4	2.8	8.8	18.2
2	2.6	7.3	18.7	5	2.2	7.9	18.0
3	3.8	8.7	12.6	6	5.2	13.2	21.9

The effect of the temperature upon the rate of respiration of Wagener apples directly after picking was determined by Burroughs (1922). At

0°C. the carbon dioxide eliminated amounted to only 3.1 mg. per kilogram of apples per hour as compared with 7.5, 12, and 24.4 mg. per kilogram per hour, respectively, for temperatures of 9, 20, and 30°C. Hopkins (1924) noted that the carbon dioxide eliminated by Cobbler potatoes at 0, 4.4, 7.2, and 10.0°C. was, respectively, 5.0, 5.1, 6.0, and 8.0 mg. per kilogram of weight per hour. It was observed by Drain (1926) that apples of different varieties showed marked differences in their response in the respiration rate to temperature changes. Thus 6 days after picking, the carbon dioxide produced by Maiden Blush apples was 280 mg. per kilogram per hour at 25°C. as compared with 46.8 mg. produced by those at 0°C. Apples of the Oldenburg variety, however, produced 55.8 mg. of carbon dioxide per kilogram of weight at 25°C. as compared with 39.3 mg. at 0°C. The localization of the oxidase activity in the different varieties is of interest in regard to the intensity of respiration. The Maiden Blush variety shows this localization in the periphery, which suggests that the localization may be influenced by gas exchange. Many apple varieties show greater oxidase activity near the core, which suggests that they have a gas exchange through the calyx tube and core.

Bushnell (1925) found that the rate of respiration in the aerial parts of the potato increased with temperature during the night. The reduction in the amount of carbohydrate that resulted from increased respiration at higher temperatures roughly corresponded to the reduction in tuber growth under these conditions. Kertesz (1933) showed in green peas that the inactivation of respiration with water at 80, 90, and 100°C. was slow after 90 per cent inactivation had been reached. A complete inactivation was obtained by a treatment at 100°C. for 2 min., while Mack (1930) found in germinating wheat that the rate of carbon dioxide production increased with a rise in temperature. Harding (1935) noted that temperature was the outstanding factor influencing the intensity of respiration in stored Grimes Golden apples. Werner (1935) gives a complete review of the literature on the physiological reactions of potatoes in storage.

The composition of the air contained in the intercellular spaces of apples, potatoes, and carrots stored at different temperatures was determined by Magness (1920). Some of his results are shown in the table on page 957.

These figures show that marked variations may occur in the composition of the gaseous content in tissues under varying conditions of temperature. It is to be noted that at the lower temperatures the sum of the percentages of oxygen and carbon dioxide closely approximates their total percentage in normal air. As the temperature increases, however, the sum of these gases increases, which indicates that at higher temperatures more than 1 molecule of carbon dioxide is liberated for each molecule of oxygen absorbed. At the higher temperatures it is probable that some anaerobic respiration occurred, owing to the relatively small amount of oxygen present. In the tissues of ripening Bartlett pears, Magness and Ballard (1926) also found that the

Storage temperature, degrees Centigrade	Percentage of		
	Carbon dioxide	Oxygen	Nitrogen
Apples			
2	6.7	14.2	79.1
6	8.4	12.9	78.7
11	12.2	10.7	77.1
20	17.2	5.5	77.3
30	21.4	3.2	75.4
Potatoes			
11	19.6	10.9	
22	34.4	5.7	
Carrots			
11	12.2	13.1	
24	28.6	5.2	

concentration of carbon dioxide was higher and that of oxygen lower in the fruit that was held at higher temperatures than in that held in cold storage. The high concentration of carbon dioxide and the correspondingly low oxygen content in this fruit ripening at a temperature of about 15°C. appear to be associated with the development of high aroma and flavor. The main factors determining the amount of carbon dioxide and oxygen in the intercellular spaces are (1) the rate of oxidation or the rate at which oxygen is taken up from and carbon dioxide given off into the intercellular spaces, (2) the permeability of the skin or epidermal coverings to carbon dioxide and oxygen, and (3) the difference in the pressure of carbon dioxide and oxygen within the tissue, which determines the rate of gaseous exchange when the permeability factor is a constant.

Rigg and Henry (1935) stated that there are three possible sources of the gases contained in the internal cavities of marine algae: (1) atmospheric air, (2) gases dissolved in sea water, and (3) the respiratory metabolism of the plant.

It was stated by Willaman and Brown (1930) that the amount of carbon dioxide in apple twigs varied directly with temperature. This amount may be as high as 360 mg. per kilogram of twigs at −22°C., and from 150 to 250 mg. per the same weight at 0°C. Dowd (1933) noted that the percentage of carbon dioxide in the internal atmosphere of apples tended to decrease during the growing period. Kertesz (1934) observed that the content of carbon dioxide of peas remained at approximately 1.6 per cent under general conditions. However, when the pods were sealed with paraffin and stored at room temperature, the content of this gas increased to 43 to 50 per cent. This indicates that ordinarily there is a very vigorous exchange of gases through the intact pods.

Besides the direct effect of temperature upon respiration, there is evidence to indicate that there is a stimulating effect upon the process caused by temperature changes. Thus Müller-Thurgau (1882) noted that respiration was much more intense in potatoes which had been kept at 0°C. for some time and then returned to the original temperature than in those tubers which had not been so treated. Similar results were observed by Palladin (1899) with bean seedlings and by Burroughs (1922) for freshly picked apples. The latter investigator noted that this stimulation in respiration was more marked for immaturely than for maturely picked fruit.

Kimbrough (1924) determined the respiration rate of potato tubers that had been stored at 32, 40, and 72°F. and found that the lower the temperature of storage the higher was the initial intensity of respiration when they were exposed to warmer temperatures. It was also observed that the initial respiration rate was higher the longer the tubers had been in storage up to 3 weeks, after which longer storage did not further increase the rate. It was noted by Hopkins (1924) in studying the respiration of potatoes at low temperatures that there was a marked acceleration in the rate at 0°C., so that for a considerable period the intensity was greater than at 4.5°C. The curve of respiration for temperatures from 0.8 to 11.5°C. showed a minimum point at 3.0°C., and as the temperature increased it showed an increase at 0°C. and then a decline. Carrick (1928) found that Winesap apples frozen from 3 to 6 hr. at temperatures of from −7.5 to −8.5°C. increased their respiration at 0°C. as much as 85 per cent above their previously normal performance at that temperature. This acceleration lasted for several days, gradually declining so that slight effects were to be noted 1 month after the treatment. When the respiration rate was measured at 20°C., the treated fruit showed an increase in the rate, although generally with a smaller gain than when measured at 0°C. When the fruit was frozen sufficiently to kill 80 per cent of the cells, the excretion of carbon dioxide at 0°C. was only one-third that which was normally produced at that temperature. In 1930, he found that the respiratory rate of Emperor grapes decreased 26 per cent between 1.75 and 0°C., and 40 per cent between 0 and −3°C. DeLong, Beaumont, and Willaman (1930) reported that if apple twigs were transferred from a lower to a higher temperature, a peak of evolution of carbon dioxide occurred for several hours, following which a constant level was attained. The lower the previous temperature the higher was the peak of the evolution of carbon dioxide. They suggested that this behavior indicated that the carbon dioxide is less soluble at the higher temperature and thus diffuses outward more readily. Dexter (1934) found that the rate of respiration of cabbage and winter wheat continuously decreased during storage at 2°C. regardless of the simultaneous increase in the percentage of sugar.

These changes in the rate of respiration can apparently be explained, in some cases at least, by the changes in the sugar content that accompany the changes in temperature. Thus it was noted by Müller-Thurgau (1882), Appleman (1911), and Hopkins (1924) that sugar accumulated in the tubers of the potato at 0°C. Appleman also found that after 2 to 3 weeks at this temperature the diastatic activity had greatly increased over that of the tubers stored at room temperatures.

Wright (1932) stored potatoes at 32, 36, 40, 50, 60, and 70°F. from Nov. 5 to Mar. 24, and found that at 32°F. the sugar in the tubers increased steadily during the experiment. This accumulation was less at 36°F. than at 32°F., while at 40°F. there was only a slight increase. At 50, 60, and 70°F. the sugar content decreased during the period of storage in direct proportion to the increase in temperature. The causes of this accumulation of sugar are not definitely known. It has been suggested that the lower temperatures are unfavorable for respiration, while the activity of diastase is not affected to so great a degree, so that sugar is thus formed in excess of its consumption by respiration and as a consequence accumulates. It is thus assumed that the increased respiration of tubers and fruits following their exposure to low temperatures is caused by the increased sugar supply available for the process (Butler, 1913). This assumption has been strengthened by the observations of Hopkins (1924), who noted, as previously mentioned, that the respiration of potatoes at lower temperatures showed a marked acceleration at 0°C. over that occurring at

4.5°C. It was found that sugar accumulated in these tubers at 0°C. but began to disappear rapidly when stored at 4.5°C. The respiration curve and that of the change in sugar content thus approximately paralleled each other.

Barker (1936) considered there is strong evidence that sucrose is closely related to the supply of substrate for respiration, and that neither glucose nor fructose is directly associated with this function.

Barker (1933) found that the sugar-respiration relation was influenced by a temperature effect that consisted of an enduring depression of the rate of respiration by exposure to low temperature. This lag persisted for several weeks after a return to higher temperatures. The factors causing this depression are apparently the result of the accumulation under conditions of low temperature of an inhibiting substance and its subsequent activation by higher temperatures. The accumulation of inhibiting substance is greater the lower the temperature. It is greatest at −1°C., and declines to zero at 8°C. The developmental effect of warmth is very rapid at 10°C., and falls to a negligible rate at −1°C. Newton and Anderson (1931) reported that the rate of respiration for different wheat varieties at −7°C. was in the inverse order of hardiness. This probably accounts for the observed fact that hardy varieties maintain their sugar reserves better than nonhardy plants during the winter.

b. Moisture.—The amount of moisture in the tissues, within certain limits, is one of the determining factors in the intensity of respiration. This fact has been especially observed in the case of seeds and grains. In air-dry seeds respiration is proceeding at a relatively low rate, but directly following an increase in their moisture content the intensity of respiration increases. Thus Kolkwitz (1901) found that barley kernels containing 10 to 11 per cent of moisture liberated only 0.33 to 1.5 mg. of carbon dioxide per kilogram per hour, while after they had absorbed 33 per cent of moisture, 2,000 mg. of this gas were liberated in a similar period. White (1909) reported that the respiration rate of air-dry seeds of wheat with a moisture content of 12 per cent was 0.03 part of what it was in active seedlings, while the intensity in air-dry seeds of *Acacia melanoxylon* amounted to only 0.0001 of that from the germinating seeds. The coefficient of increase in the respiration rate due to changes in the moisture content was observed in wheat by Bailey and Gurjar (1918) to be as follows:

Increase in moisture, per cent	Acceleration in respiration, per cent
12 to 13	0.16
13 to 14	0.17
15 to 16	1.41
16 to 17	3.02

In corn seed, Bailey (1921) observed that respiration became marked when the moisture content exceeded 13 per cent, and that when the moisture content was raised to 15 to 16 per cent an increase in the respiration rate of 400 per cent was obtained. Coleman, Rothgeb, and Fellows (1928) in a study of the respiration rate of the seeds of the sorghums noted the following responses in yellow milo and feterita seeds to changes in moisture content at a temperature of 37.8°C.

Percentage of moisture in the grain	Milligrams of carbon dioxide evolved per 24 hr. per 100 g. of dry matter	
	Yellow milo	Feterita
11	0.53	1.00
12	0.70	1.10
13	1.00	1.58
14	1.56	3.00
15	3.27	5.80
16	7.58	11.80

Observations similar to those which have been mentioned in regard to the relation of the moisture content and respiration have been made by Duvel (1904), Babcock (1912), and others.

The length of time the excess moisture is present in seeds bears a relation to the rate of respiration, as may be observed from data mentioned by Bailey and Gurjar (1918) for wheat. The quantity of carbon dioxide respired by a sample of wheat which contained 15.2 per cent moisture and which was stored 55 days after it was dampened was approximately four times as great as that respired by freshly dampened wheat of the same moisture content. The carbon dioxide evolved from a sample which contained 15.7 per cent of moisture and which was stored for 108 days was about eight times as great as the wheat of the same moisture content that had just been moistened. It is considered that the dampened grain is favorable for the formation of sugar, which tends to accumulate as time goes on so that there is a supply sufficient for a relatively high rate of respiration.

It is worthy of note here that the humidity of the atmosphere may affect the rate of respiration in stored grain by increasing its water content. Thus Bailey (1921) noted that at 35 per cent humidity the moisture content of corn was 8.5 per cent, while at 85 per cent humidity it was 17.8 per cent.

The effect of moisture upon the rate of respiration varies to a considerable degree for different plants or plant parts. Thus Jacquot and Mayer (1926) in the case of the seeds of corn and the peanut found that the rate of carbon dioxide production per unit of moist weight increased with increased water content up to 30 to 35 per cent, when the rate began to fall off, while in the bean the respiration rate increased up to 60 per cent before it began to decline. When the carbon dioxide production, however, was determined per unit of dry matter, it was found to increase up to the complete imbibition of the bean seeds (117 per cent) and up to 58 to 60 per cent in peanuts. The respiratory quotient was greater than unity in all the seeds at low water content and as complete saturation was approached. Fraymouth (1928) studied the relationship of respiration to the moisture content of certain terrestrial algae and other lower plants. In the algae the respiration rate increased rapidly on the addition of water to the air-dry material until the amount of water had reached 100 per cent. There was then a gradual rise in rate until a water content of 600 per cent was reached, when gaseous exchange was apparently seriously impaired. After prolonged drought, respiration in these algae was practically zero, but immediately after the addition of water, it was found to be almost as intense as for fresh material of a similar water content. In the lichen (*Parmelia*) the respiration rate showed no great increase until a water content of 60 per cent was reached, after which it rapidly increased until a water content of 200 per cent was attained. In *Hypnum sp.* there was a gradual rise in the respiratory rate with increasing amounts of associated water until 250 per cent was reached, after which the tissues were saturated with water and gaseous exchange was apparently impaired. The effects of desiccation upon respiration and certain organic constituents of the cacti have been reported by MacDougal, Long, and Brown (1915) and Long (1917).

It was reported by Singh and Varadpande (1930) that a deficiency of water in leaves lowered their rate of respiration, and Smyth (1934) found that the respiration rate of certain lichens increased with a rise in water content to saturation. Livingston (1934) estimated that approximately 8 per cent of the water incorporated in growth was derived from respiration. The quantitative importance of this water may vary according to environmental conditions.

Several experiments have been reported, however, that show some striking variations from the majority of the observations in regard to the relationship between the moisture content of the tissues and respiration. Smith (1915) by means of a vacuum desiccator diminished the water content of the leaves of the snowdrop and the stem tips of nasturtium and young asparagus and found that the respiratory rate increased with a diminishing water content until about 30 per cent of the water had been withdrawn. The rate of respiration then remained about the same until the water was diminished to about 50 per cent of that originally present. Below this limit to complete dryness, respiration decreased proportionately to the amount of water lost. Palladin and Sheloumova (1918) used potato tubers and exposed a part of the material for drying in the air at room temperature, with the result that during the first few days, in spite of a great loss of water, the intensity of respiration was gradually increased, but after further loss of water there was a distinct lowering of the respiratory rate. When the same tubers were immersed in water, the respiratory intensity was again stimulated for a few days, after which it fell to the normal rate. They thus could not establish any definite correlation between the amount of water lost and the intensity of respiration. Luthra (1924) in experiments with pears kept in moist and in dry air noted that both transpiration and respiration of some varieties were increased in the dry air.

The question naturally arises as to why respiration is accelerated by an increased moisture content. Bailey and Gurjar (1918) assumed in the case of grains that the colloids of the cells imbibe water and form elastic gels and that if sufficient water is supplied the gels become more and more dilute. Diffusion through the stronger gels is relatively slow, while through the dilute gels it is relatively rapid. Increasing the moisture content thus results in progressively less viscous gels and a corresponding increase in the rate of diffusion. It is assumed that the rate of respiration depends upon the rate of diffusion between the various cells of the tissues, so that the less viscous the gels of these cells the more rapid the rate of respiration.

c. *Injury*.—It has been known for a considerable time that wounded plant parts respire more intensely than those which are intact. One of the first to observe this fact was Boehm (1887), who called attention to the abnormal production of carbon dioxide by potatoes that had been injured by cuttings. Richards (1896) by using potatoes and carrots cut in four pieces, slit leaves, split hypocotyls, and decapitated roots of seedlings demonstrated increases of as much as 100 per cent in the intensity of respiration when measured by the output of carbon dioxide. In the case of the potatoes and carrots he noted a very marked and sudden increase during the first 2 hr., which again fell rapidly, although to a point still considerably above the normal respiration, and then rose more slowly later to the maximum production of carbon dioxide. The respiratory activity reached its maximum 24 to 30 hr. after the injury had been inflicted. Lutman (1926) found an increased respiration in potatoes during the second and third days after cutting, which was followed by a gradual fall until the normal rate of the uninjured tuber was attained. If the cutting was repeated in 5 to 7 days, the respiration curve did not rise so high as after the first cutting, but a similar intensity could be attained by increasing the cut surface. Repeated cuttings lowered the respiration curve until finally the tubers would not respond. By lengthening the intervals between cuttings to 11 to 12 days, the respiration curve more nearly approximated the one obtained after the first cutting. Johnstone (1925) by means of a cork borer removed cylindrical plugs from sweet potatoes and noted that from one lot of roots with an injured area of 7,385 sq. mm. the evolution of carbon dioxide increased from 64.3 mg. per kilogram per hour before wounding to 126.8 mg. during the same time after wounding. It was noted by Coleman, Rothgeb, and Fellows (1928) that cracked grains of the sorghums showed a higher intensity of respiration than did the intact grains.

It was noted by Richards (1897) that the temperature of injured parts of plants was materially higher than that of the surrounding tissues. In the potato no increase in temperature was observed until 24 hr. after injury or at the time of maximum respiratory activity due to injury. This increase in temperature was only 0.4 per cent and was very local, since 2 cm. back of the wound no perceptible differences in temperature could be noted. Similar reactions were also observed in the radish and carrot. Onion bulbs showed the most marked changes in temperature due to injury of any of the material used. The rise in temperature due to injury was as much as 3.3°C. and in contrast with the potato the temperature effect extended a considerable distance into the surrounding tissue.

Harris (1929, 1930) reported that pruning or any injury to the roots of fruit trees resulted in a temporary increase in their rate of respiration. Lall (1934) noted that the cutting of apples produced an increase in the rate of respiration which disappeared at a temperature of 12°C. in about 7 days. Audus (1935) and Godwin (1935) showed that handling starved laurel leaves greatly increased their rate of respiration. Barker (1935) could find no increase in respiration due to handling firm potatoes but obtained a marked increase by handling those that had softened from desiccation. The stimulus that causes this increased respiratory rate arises apparently from a deformation of the cells by bending or rubbing.

The probable causes of this increased respiratory activity of tissues following their injury will now be considered. One cause is apparently the facilitation of the gaseous exchange. Thus when Stich (1891) sealed together the cut ends of a potato with neutral gelatin, no carbon dioxide was evolved. Johnstone (1925) noted in injured sweet potatoes that the rate of respiration increased 97.3 per cent when the injured area was exposed, while it increased only 17.1 per cent when this area was not directly exposed to the air. The sudden and marked increase in the output of carbon dioxide directly following injury is probably due to the escape of the gas that has accumulated in the intercellular spaces of the tissues. This assumption is strengthened by the fact that only bulky tissues show this marked increase in the evolution of carbon dioxide directly following injury. The absorption of oxygen is also rendered more favorable by the freer access of air across the tissues due to injury. This facilitation of gaseous exchange is indicated by the composition of the air taken from the intercellular spaces of uninjured and injured apples, as shown in the following table (Magness, 1920):

State of apple	Storage, temperature degrees centigrade	Percentage of	
		Carbon dioxide	Oxygen
Whole apple.........	1	6.6	14.6
Ends peeled.........	1	1.7	15.8
Whole apple.........	20	17.8	7.0
Ends peeled.........	20	7.4	9.9
Whole apple.........	30	23.9	1.8
Ends peeled.........	30	12.6	8.9

It has been noted by Drain (1926) for apples and Lutman (1926) for potatoes that an injury which is not accompanied by the rupture of the skin does not increase the respiratory rate. This indicates also that the increase in respiration rate due to injury is due primarily to an acceleration of the gaseous exchange between the tissues and the surrounding air.

Lutman (1926) believed that the increased respiration due to injury in the potato may be regarded, in the main, as an irritable response of the protoplasm to the stimulus of the contact with the oxygen of the air of the tuber cells, which under ordinary conditions are protected from such a contact by the external layer of cork cells. He was able to remove a large part of the effect of injury on respiration by washing the cut surfaces in running water for 1 to 3 hr. The activators of respiration or the substances upon which they act are apparently soluble in water.

The chemical changes that occur in the injured regions of tissues have also been observed to a limited degree. Thus Grüss (1907) found in the cells around wounded potatoes an accumulation of sugar in the subphellogen and an increase in the oxidizing enzymes and in the diastatic activity. Friedrich (1908) in potatoes noted in the cells bordering the cut surfaces an increase in the total nitrogen and acidity and a decrease in total carbohydrates but an increase in reducing sugars. He considered that the sugars increased the intensity of respiration, while the acids were waste products of this process. Lutman (1926), however, did not consider this increase in sugar content to have any effect upon the respiratory rate. It was observed by Hopkins (1927) that the sugar content of wounded potatoes increased from 53 to 68 per cent of the original amount present. This sugar content began to increase after injury, reached a maximum after several days, and then fell off. He believed that this increase in sugar was brought about by activities leading to callus formation.

d. Kind and Age of Tissues.—The intensity of respiration varies markedly with the kind and age of the tissues. This variation of the rate of respiration of different tissues may be illustrated by the observations of Benoy (1929), who determined the rate of respiration of a number of different vegetables at 30°C. for a 30-hr. period following their harvesting. With respect to the amount of carbon dioxide evolved for equal weights of dry matter, they stood in the following order: asparagus, lettuce, green bean, okra, green onion, carrot, tomato, beet, green mango, and red pimento. It should be noted here that the amount of glucose which accumulated per 100 g. of material was the highest (13.68 g.) in asparagus and the lowest (1.29 g.) in the red pimento. Beaumont and Willaman (1924) found that the output of carbon dioxide from the buds and twigs of a hardy variety of apple, as Charlamoff, at temperatures of 5.3 and 12.4°C. was greater than that from the Delicious variety, which is considered nonhardy. This difference in carbon dioxide evolution held on the basis of the number of milligrams per unit of length of twig, weight, volume, or number of buds. This difference in the intensity of respiration was not due to any differences in the extent of the rest period or dormancy.

The difference in the rate of respiration of different tissues has been especially studied in seeds and seedlings by determining the relative respiratory intensity of the embryo and endosperm. In this regard, Burlakow (1898) stated that the respiratory activity in the embryo of wheat was twenty times that of the endosperm. Kolkwitz (1901) found that when wheat kernels were divided transversely into equal portions, the embryo end respired three times as much carbon dioxide as the opposite end when equal weights of material were compared, while Karchevski (1903) reported the aerobic respiration of the embryo of wheat to be twelve times as great as that of the endosperm. The embryo of the rice seed, according to Bailey and Gurjar (1920), also shows a higher rate of respiration than does the endosperm. The same authors also noted (1918) that shriveled wheat respired at a rate two to three times greater than did the plump grains. They attributed this to the higher ratio of embryo weight to endosperm in the shriveled grain as compared with that of the plump wheat. In many of the studies of the relative rate of respiration in the embryo and endosperm, the embryos have been removed. As the tissues are more or less injured by this operation, this method is open to criticism. The production of embryoless seeds of wheat in certain instances furnished Lyon (1928) an opportunity to compare the respiration rate of these seeds with the normal ones without inflicting any injury upon their tissues. It was found that when these two types of seeds were placed under conditions favorable to germination, the seeds devoid of embryos respired 22.1 mg. of carbon dioxide per gram of seed in 6 days, while the normal seeds gave off 26.5 mg. for the same unit weight. As the embryo made up only 3 per cent of the entire seed by weight, its rate of respiration was decidedly greater than that of the endosperm.

The age of the tissues has a marked influence upon the intensity of respiration. Thus Nicolas (1918) noted that the respiratory intensity of young leaves and stems was from three to seven times greater than that of corresponding fully developed organs taken from the older parts of the plant. Bezagu (1919) observed that respiration was relatively weak in young leaves and increased to a maximum at the time the leaf reached its full development, after which it decreased as the leaf became older. Bergman (1925) and Wakabayashi (1925) found the rate of respiration of the flowers of cranberries much less than that of buds. The shoots of the current year's growth respired two to three times as rapidly as those of the previous year and at about the same rate as the flowers. The young fruits for a period of 1 to 3 days after the petals fell respired very rapidly, the rate exceeding that of the buds, but decreased rapidly with the development of the fruit. Hover and Gustafson (1926) in observations on the respiratory activity of the leaves of corn, oats, sorghum, and wheat noted a decreased activity with increasing age until middle age and thereafter a gradual increase as determined per unit of dry weight. They assumed that this increase was due to the decrease of the dry matter in the leaves after middle age. Maige and Nicolas (1910) considered that both the respiratory intensity and the respiratory quotient increased with the increased turgidity of the cells. An increase in turgidity, however, may be only an indicator of favorable conditions for growth and vital activity and thus not a direct causative factor in increasing respiration.

Kidd, West, and Briggs (1921) determined the respiratory index for *Helianthus annuus* at different stages of growth. They defined the respiratory index as the milligrams of carbon dioxide liberated per gram of dry weight per hour at 10°C. when the amount of respiratory material is not limited and when the external concentration of the oxygen is that of the atmosphere. This index is consequently a measure of the effective amount of respiring cell matter per gram of dry weight. The observed respiratory indexes for the sunflower plant and its parts at different ages are shown in the following table:

RESPIRATORY INDEXES OF SUNFLOWER PLANTS AND PLANT PARTS AT DIFFERENT STAGES OF GROWTH

Days from germination	Number of plants used	Dry weight of a single plant, grams	Respiratory index			
			Entire plant	Stem	Leaves	Stem apex
1	30	0.0225	2.90			
2	25	0.0223	3.00			
4	25	0.0242	3.00			
13	10	0.1009	2.80			
22	8	0.630	3.00			
29	2	4.065	2.30			
36	1	12.850	1.21	0.81	1.56	
43	1	22.050	1.03	0.69	1.38	
50	1	45.150	0.94	0.46	1.52	2.56
59	1	93.200	0.66	0.33	1.32	1.72
89	1	294.700	0.48	0.31	0.90	1.13
99	1	377.400	0.37	0.25	0.45	0.89
112	1	818.300	0.26	0.098	0.38	0.75

It is to be observed from this table that the respiratory index of the entire plant falls off continuously from 3 to about one-tenth of this value at the end of the life cycle. The respiratory index of the stem, of individual leaves, and of the flowers, respectively, decreased with the age of the organ. The initial respiratory index of successive leaves decreases with the age of the plant, indicating that the respiratory index of the meristematic tissue decreases with age. The fall in the respiratory index of the meristematic tissue and young leaves indicated that the fall in the respiratory index of the whole plant was not due entirely to the increase in proportion of mechanical tissue. The fall in the value of the respiratory index with age followed closely that of the value of the relative growth rate, indicating a close connection between the "internal" factor for respiration and the "internal" factor for growth.

Gustafson (1929) found a decrease in the production of carbon dioxide during growth in the John Baer variety of tomato fruits until a minimum point was reached at about the time increase in size stopped. This was followed by an increase in carbon dioxide production, which reached its maximum when the fruits were orange red in color, after which there was a final decrease in respiration. Gustafson attributed the increase in respiration to a lowering of the hydrogen-ion concentration in the cell sap, while the final decrease was thought to be due to a slowing down of all life activities as the fruit approached its maturity. Hafenrichter (1928) found, however, in the germination of the soybean that if the growth rate of the entire plant was considered instead of its parts, the maximum rate of respiration was reached earlier in the development of these plants than was the maximum rate of growth.

It was observed by Kidd and West (1929) that the rate of respiration during the senescence of the apple rises at first and then falls. Bergman (1929) noted that the rate of respiration in developing blueberries was highest when the fruits had reached the pink-red stage, and that it then decreased until the berries had reached the blue-ripe stage. Harley (1929) found that the accumulation of acetaldehyde was more rapid in the late-picked pears than in those which had been picked earlier. Overholser, Hardy, and Locklin (1931) reported that the greater firmness of the flesh of strawberries is apparently not directly correlated with a low intensity of respiration in either immature or mature fruit.

Smith (1929) believed that the reason more mature potato tubers respire less than younger ones is that an increase in the development of the periderm prevents the exchange of gases.

It was observed by Harding (1929) in Grimes Golden apples, Bonazzi (1931) in the sugar cane, and Luthra and Chima (1931) in grape that the rate of respiration was more intense in the younger than in the older tissues. Johansson (1933) noted that the maximal respiration values in the bark occurred at the time of greatest growth.

e. Anesthetics.—It was observed by Irving (1911) that very small doses of chloroform applied as vapor to leaves increased their respiration. This increased rate could be maintained if the chloroform was applied continuously, but when it was withdrawn the respiration rate dropped to normal. When medium doses were applied, an initial outburst of carbon dioxide resulted, which was followed by a decline much below normal, the rate and extent of this decline increasing with the amount of chloroform used. When strong doses were applied, there was no initial increase of carbon dioxide, and its production rapidly dropped to zero. Thoday (1913) substantiated the observations of Irving in regard to the production of carbon dioxide and noted further that, in the stimulation of respiration due to small doses of chloroform, the absorption of oxygen and carbon dioxide increased in like proportion, but that, when the doses of chloroform were large, the absorption of oxygen was no longer closely correlated with the production of carbon dioxide. The behavior of different leaves in this regard is of interest. Thus in the leaves of *Tropaeolum*, which contained no tannin, the absorption of oxygen was depressed more than the production of carbon dioxide. In leaves of cherry laurel and *Helianthus*, however, which contained tannins and the oxidation of which imparted a brown or black color to the disorganized leaves, the absorption of oxygen was very rapid for a short time and, though quickly falling, remained at a much higher level than the production of carbon dioxide.

With wheat seeds soaked in 7.3 and 3.6 per cent ether solutions, Thomas (1918) obtained an increase in the rate of respiration, which was followed by a decrease. Smith (1921) treated the seedlings of wheat when the roots were 1 to 1.5 in. in length with 1, 3.65, and 7.3 per cent of ether in water and found that the first effect was an increase in the rate of respiration, which in turn was followed by a decrease. With all these concentrations the rate of respiration was ultimately reduced to approximately the same level, and the stronger the concentration the more rapid the fall. When respiration was reduced below normal, recovery was possible on the removal of the ether and appeared to be complete if sufficient time was allowed. If respiration was too far depressed, however, no recovery was possible. Haas (1919) exposed *Laminaria* to 0.3 per cent chloroform, 0.8 per cent formaldehyde, and 10 per cent ethyl alcohol and observed that the rate of respiration was respectively 1.4 to 2.8, 2.7, and 6.2 times that of the controls. Upon the exposure of the alga to these compounds, the initial effect was an increase in respiration, which was followed by a decrease if they were sufficiently toxic. No decrease in respiration, however, was observed when the concentration was too low to be toxic.

The probable action of anesthetics upon the oxidative mechanism of the cells is indicated by the observations of Ray (1923). He found that when *Ulva* was killed in such a manner as to destroy the oxidizing enzymes no carbon dioxide was produced unless hydrogen peroxide and ferric sulphate, which acts like a peroxidase, were present. If the concentration of the iron was low, there was an increase of the carbon dioxide production when chloroform was added. If the concentration of iron was high, the rate of carbon dioxide production appeared to decrease from the start. It was also observed that if oleic, tannic, succinic, or malic acid were treated with hydrogen peroxide and ferric sulphate, carbon dioxide was produced at a rate that could be measured by the indicator method. In the case of acids with a double bond, the rate of production of carbon dioxide could be varied by the addition of an anesthetic. The changes in the rate of production of carbon dioxide under the influence of an anesthetic, such as chloroform, showed a striking resemblance to the reaction of an organism. Ray considered that there is a definite relation between the rate of production of carbon dioxide and the ability of the compounds to absorb iodine. An analysis of the effects of chloroform on the production of carbon dioxide by the living and killed cells of *Ulva* and unsaturated organic acids indicates that the same process is taking place in all three cases. The conclusion of Ray was that the action of chloroform on the oxidative mechanism of the cell was chemical in nature, and that it acted either by catalysis or by the formation of a loose compound with some portion of the system.

f. Afterripening Agents.—The effects of ethylene, which is now extensively used in the ripening of fruits, upon respiration might well be noted under this topic. Denny (1924) found that concentrations of this gas of 1 part in 1,000, 1 part in 10,000, 1 part in 100,000, and even 1 p.p.m. of air increased the respiration of green lemons. The increase in carbon dioxide output ranged from 100 to 250 per cent and appeared to be greatest at the intermediate concentrations. After the discontinuance of the application the rate of respiration decreased. Davis and Church (1931) also have reported that the treatment greatly stimulated the rate of respiration in Japanese persimmons. Regeimbal, Vacha, and Harvey (1927) treated ripening bananas with 1 part ethylene to 1,000 parts air for 15 to 20 min. and then determined their rate of respiration. In all cases the rate of respiration expressed in milligrams of carbon dioxide per hour was doubled or trebled within a few minutes, after which the rate fell off to a value lower than normal. It was assumed that the high initial increase might have been due either to the increase of oxidation or to the increase in the permeability of the membranes, allowing the diffusion outward of the carbon dioxide already in the cells. The rise in the respiratory rate after a second dose of the ethylene seemed to indicate that the oxidation rate rather than the permeability rate was increased. This stimulation wears off in less than an hour. It should be noted in this regard that the treated bananas had one-fifth to one-fourth more sugar in them than the untreated ones, while the starch content was proportionately decreased. The activity of diastatic enzymes as well as respiratory enzymes was apparently increased by the treatment with ethylene. It is not known, however, whether permeability changes indicated by the gas allow the enzyme and substrate to come together, activate the enzymes, or actually increase the amount of enzymes present.

It was reported by Guthrie (1931) that the expressed juice of potatoes that had been treated with ethylene chlorhydrin, ethyl alcohol, and acetaldehyde showed an increase in pH value and the power to reduce iodine in solution. The treatment with thiourea and potassium thiocyanate produced only a small change in the hydrogen-ion concentration and reducing properties. Hanes and Barker (1931) reported that the rate of respiration of potato tubers at 15°C. was increased when exposed to an atmosphere containing a low concentration of hydrogen cyanide. Shill (1931) stated that citrus plants treated with hydrocyanic acid increased their rate of respiration activity about 75 per cent, and that after 35 hr. the rate of respiration gradually decreased to normal. Guthrie, Denny, and Miller (1932) found that treatment with ethylene chlorhydrin produced an increase in the peroxidase, catalase, and sulphydryl content and in the pH value in both the dormant and nondormant corms of gladiolus.

Miller (1932, 1933, 1934, 1935) reported large increases in the rate of respiration when the whole tubers of the potato are subjected to the vapors of such chemicals as ethylene chlorhydrin, ethylene mercaptan, ethylene bromohydrin, hydrogen sulphide, ethyl disulphide, acetaldehyde, and halogenated derivatives. These increases occur very soon after the beginning of the 24-hr. period of treatment and reach their maximum the second or third day. The rate of respiration then declines approximately to that of the untreated tubers. There is neither a close correlation between the effect of these chemicals on the rate of respiration and their efficiency in breaking dormancy, nor between the increase in sucrose and the breaking of dormancy. Miller, Guthrie, and Denny (1934) stated that the decrease in citric acid and the increase in pH value begin approximately at the same time as the increase in the rate of respiration. They believed that the acid is used in respiration, and that the pH value increases as a consequence of the utilization of the acid. The large increase of glutathione that follows the treatment by ethylene chlorhydrin is not the cause of the increased respiratory rate, since the increment of this compound occurs 40 to 60 hr. after the beginning of the treatment, which is much later than the increase in the rate of respiration. Tubers treated with ethyl alcohol always show a decrease in the rate of respiration.

It should be noted here also that Haas (1919) found that when *Laminaria* was killed by alcohol, acetone, formaldehyde, or ethyl bromide, as well as by drying, the rate of respiration was even greater than under normal conditions. This rate, however, was not long maintained but gradually fell off and eventually became negligible.

g. Concentration of Oxygen and Carbon Dioxide.—The effects of the concentration of oxygen upon the rate of respiration will depend not only upon the plant or plant part under consideration but also upon its condition when the observation is made. When the supply of oxygen is sufficient for the normal respiration of the plant, the presence of a surplus produces in most cases no perceptible effects. The intensity of respiration in many plants is not markedly modified when the oxygen supply is reduced to one-half that normally present in the air or when it is increased to five to ten times that which normally occurs. According to Stich (1891), the absorption of oxygen becomes insufficient for most plants when the supply in the air falls below 5 to 8 per cent. If a plant is confined in a container, it can generally absorb the last trace of oxygen from the medium. When the oxygen supply is increased, the plant may continue to respire normally for a period of time even when the partial pressure of oxygen has been increased to twenty or thirty times the normal amount. In a few hours or days under such conditions, respiration declines and death ultimately ensues.

The variation of plants in their oxygen needs under different conditions of their development is illustrated by the embryo of *Alisma*, as reported by Crocker and Davis (1914). They found that the hypocotyl was capable of elongating 1.2 times its original length in the total absence of free oxygen. For greening and branching, however, it required an air pressure of at least 5 mm. to supply the minimum amount of oxygen, while for the development of the primary root an even greater amount was required. The amount of oxygen that must be supplied to produce a certain intensity of respiration or rate of growth will depend in many cases upon the permeability of the membranes that it must penetrate in reaching the seat of its utilization. This is illustrated by the coats of seeds through which oxygen diffuses with difficulty. When such coats are present, a higher supply of oxygen is required to produce a certain rate of respiration or growth than when they are removed. Thus Shull (1914) found with *Xanthium* seeds that when the oxygen supply was increased, it brought about, in some manner, an immediate and rapid increase in the rate of oxygen absorption, resulting in an immediate germination of the seeds. Shull considered that it was difficult to say what function or functions were affected by this treatment. It is certain that oxygen acted as a limiting factor on some function, but whether it was by limiting the process of respiration, by limiting enzyme formation, or by limiting the action of oxygen carriers could not be stated. Becker (1912) suggested that in some instances oxygen may act as a stimulus for certain functions without being directly concerned in respiration.

Blackman (1930) stated that the respiration rate of apples decreased with diminishing oxygen supply until a minimum concentration of 5 to 10 per cent of this gas was reached. As the concentration of oxygen was increased above this point, the rate of respiration continued to increase until this gas composed approximately 100 per cent of the gaseous medium. Zimmerman (1930) reported that, with cuttings of *Salvia* in 2 in. of water, roots were formed at the base of the stem; in 5 in. they had a tendency to root along the water surface; while, in 7 in. of water, roots were only sparsely produced, if at all. When the water was aerated, roots were produced equally at all three depths. Willow cuttings, however, produced roots when the oxygen supply in the water was as low as 1 p.p.m.

Loehwing (1934) noted that the aerated plants of sunflower and soybeans grown in sand and loam soil differed from the unaerated controls by being taller, heavier, and possessing a more fibrous and a more highly branched root system. The aerated plants had a higher absolute amount of ash, calcium, potassium, phosphorus, crude fiber, starch, sugars, and total nitrogen. The expressed sap of the aerated plants was more alkaline.

Jones (1933) considered that a deficiency of oxygen is probably the principal factor which causes a lack of germination in rice seeds continuously submerged in water. Earlier seeding is preferable because the temperatures of the air and water are lower early in the spring. Since more oxygen is dissolved in water at low temperatures, there is more of this gas available to the germinating seeds earlier in the season.

Soil preparation	Average yield in bushels per acre	
	1933	1934
Check (no tile)	311.6	330.6
Over plain tile	362.2	363.5
Over perforated tile	381.2	400.9

Bushnell (1935) in Ohio laid 4-in. ordinary drain tile and tile perforated with 36 holes per section at a depth of 10 in. in plowed soil. Potatoes were planted over these tile, and the yields were obtained as shown on page 970.

These data indicate that the potato plant is sensitive to soil aeration, and that in certain soils the lack of sufficient oxygen may be a limiting factor in crop yields.

The quantity of oxygen in the intercellular spaces of fruits and other plant parts due primarily to poor gaseous exchange is frequently considerably below that of the air and may thus in some cases be a limiting factor in the respiration of these parts. Injury to plants due to an inadequate supply of oxygen is shown by the effects upon cranberry plants that have been submerged by flooding (Bergman, 1921, 1925). This injury was more pronounced on cloudy than on clear days, owing to the difference in the amount of oxygen released in photosynthesis. Shading the flooded plants had the same effect as clouds in this regard. Low temperatures operated to reduce the amount of injury by lowering the respiration rate and thus decreasing the need for oxygen while increasing the capacity of the water to absorb this gas. The flowers and tips of the shoots were the parts of the plants most seriously affected by flooding, due to their high respiration rate (Bergman, 1930).

The amount of carbon dioxide in the atmosphere surrounding the plant may have some effect upon the respiration rate. Kidd (1917) noted that the rate of respiration was lowered by the accumulation of carbon dioxide, and that this effect was more marked upon the process of aerobic respiration. These depressing effects were evidently not due to permanent disorganization, as they always passed away as soon as the depressing concentrations of carbon dioxide were removed. In their study of the respiration of apple twigs at 0°C., potato tubers at 22°C., and wheat at 40°C., Willaman and Beaumont (1928) found that if carbon dioxide was allowed to accumulate in the respiratory chamber, the rate of carbon dioxide production continuously decreased with time. When aspiration of the atmosphere surrounding the tissues was commenced after 30 or 40 hr. of accumulating carbon dioxide, the respiration rate as measured by the output of carbon dioxide immediately assumed a much higher value than it had during the accumulation period. The work of Spoehr and McGee (1924) indicates that under such conditions, the establishment of an equilibrium between the carbon dioxide in the air surrounding the tissues and that which is dissolved in them is the result that is being observed. Thus they found that when the carbon dioxide content of the air surrounding a leaf was changed from a lower to a higher concentration, the leaf showed a reduced rate of carbon dioxide emission for a period following the change and then finally again attained the same rate as before the change in carbon dioxide content was made. When the carbon dioxide of the air surrounding the leaf was changed from a higher to a lower concentration, the leaf showed a primary increased rate of carbon dioxide emission and a subsequent decrease to the original rate.

The accumulation of carbon dioxide in some cases, however, apparently has no marked effect upon certain activities. Thus Newcombe and Bowerman (1918) in observations upon 2,000 seedlings of 12 different species reported that ventilation had no effect in producing better seedlings in small or large chambers in the dark and that ventilation had no visible effect on the sensitive reactions of these plants to gravity and light.

Schulz (1926) and Michaels (1932) with potato, and Clements (1935) with apple, studied the relationship between the number of lenticels per unit of surface and the rate of outward diffusion of carbon dioxide. They concluded that the carbon dioxide within these plant parts escapes with equal speed regardless of the number of lenticels. Gerhardt and Ezell (1934) found that carbon dioxide was absorbed to a much greater extent by Bosc pears than by Jonathan apples, when these fruits were stored in this gas. Thornton (1931, 1932, 1933, 1935) found that, in the presence of 20 to 70 per cent of carbon dioxide and 20 per cent of oxygen, the rate of uptake of the latter gas by different tissues differs greatly, so that they may show an increase, decrease, or no change in the rate of absorption. Thus under the preceding conditions the absorption of oxygen was increased 83 per cent for dormant potatoes, 239 per cent for nondormant ones, 117 per cent for those undergoing sprouting, 89 per cent for onion bulbs, 11 per cent for tulip buds, and 95 per cent for beet roots. On the other hand, the amount of absorption was decreased 33 per cent for asparagus, 17 per cent for bananas and strawberry fruit, and 34 per cent for shelled lima beans, while there was no variation for the roots of carrots.

h. Various Salt Solutions.—The effects of sodium chloride and calcium chloride upon respiration have been studied by Lyon (1921) and Inman (1921). It was observed in the case of *Elodea* that solutions of sodium chloride caused an increase in respiration that was followed by a decrease, while solutions of calcium chloride caused only a decrease. In a mixture containing 99.65 mols of sodium chloride and 0.35 mol of calcium chloride, the rate remained normal, while a mixture of 98.6 mols of sodium chloride and 1.38 mols of calcium chloride caused a great increase in respiration. In highly hypertonic solutions of sea water, the rate of respiration of *Laminaria agardhii* was rapidly reduced, while in hypotonic solutions it was reduced somewhat less rapidly. Hypertonic solutions of sodium chloride, calcium chloride, and mixtures of sodium chloride and calcium chloride in the proportion of 50 to 1 all caused a decrease in the respiration of wheat seedlings.

The stimulating action of phosphates upon oxidation reactions has been reported by Spoehr (1924) and Lyon (1924, 1927). The former investigator noted that glucose and other hexoses were oxidized by air, in solutions containing disodium phosphate and methylene blue, with the formation of carbon dioxide. This reaction was greatly accelerated by small quantities of iron salt. Sodium ferropyrophosphate was a more active catalyst than methylene blue in causing the oxidation of carbohydrates with air. Lyon (1924, 1927) reported that phosphates exerted a promoter action upon potato oxidase so that carbon dioxide was produced by an oxidation of some component in a solution of glucose. He also stated that phosphate catalyzes the slow oxidation of pyrogallol by atmospheric oxygen. He assumed that phosphate increases the rate of production of carbon dioxide by anaerobic processes because of its role in the early stages of alcoholic fermentation, and that it effects an increase in the production of carbon dioxide by the aerobic phase of respiration through its action as a catalyst toward oxidases. In *Elodea canadensis* and wheat seedlings Lyon was able by the application of phosphate to obtain an increase in the production of carbon dioxide amounting to 55 per cent for the former and 35 per cent for the latter plant. Lyon also observed that arsenates exerted a catalytic effect on the oxidation of pyrogallol by atmospheric oxygen and by metallic iron and on the production of carbon dioxide by *Elodea*. Warburg (1921, 1925, 1928) believed that iron is the oxygen-carrying component of the respiration ferment. He stated that if respiration is considered as an oxidation, the respiration ferment is the sum of all the catalytically active iron compounds that occur in the cell. The respiration ferment is thus considered a substance that takes up the atmospheric oxygen and transfers it to the organic substance and in its chemical constitution is related to the hemoglobin or red pigment of the blood.

It was observed by Smith (1921) that the respiratory rate of wheat roots which had attained a length of about 2 in. was greatly accelerated and then depressed after they were placed in a 0.0025 *M* solution of lactic acid. As the concentration of the acid was increased, this preliminary rise in respiratory intensity became less marked until a concentration was reached where the rate began to fall at once. Recovery was complete even when the respiratory rate was reduced to 25 per cent of normal by a 2 *M* solution of the acid. These results were apparently not due to acidity or osmotic pressure but to some specific action of the lactic acid. They also indicate that, in the case of wheat, lactic acid is not an important intermediate product of the respiratory process.

Hutchinson (1922) reported that, when radish seedlings were placed in 0.003 and 0.002 per cent solutions of adrenalin, they showed a falling off of 20 per cent in carbon dioxide production at the end of 15 min. and then a gradual rise. More dilute solutions of 0.0005 and 0.0002 per cent gave a much smaller decrease, with a final rise to normal in about 1 hr.

Thomas (1931) found that the rate of respiration in the tissue of carrots in buffered solutions increased with increasing hydrogen-ion concentration. The relation of the respiratory rate to the absorption of salts should be considered here. Briggs and Petrie (1931) established the view that a general parallelism exists between the rate of respiration in the tissues and the rate of conductivity of the external solution. They considered that the respiratory rate is an important factor in determining the concentration of electrolytes in the water in which pieces of plant tissue have been placed. Steward, Wright, and Berry (1932) believed that the accumulation of salt in a tissue is dependent upon respiratory phenomena that presumably supply the energy necessary for it. Becker (1936) considered, however, that the total energy available from respiration is far in excess of the presumed energy expenditure for accumulation alone, and that if the fraction of the total energy utilized in accumulation were to vary, the relation between the processes would doubtless differ from the one here described.

i. Light and Electricity.—Spoehr (1915) found respiration in the sunlight higher than in the dark. He believed that the air may possess higher oxidative power during the hours of illumination than during darkness, owing to the ionizing power of the sun's rays. He assumed that deionization of the air has its influence on oxidation. In wheat seedlings the ratio of the day rate of respiration to the rate at night in ordinary air during two experiments was 1.042 and 1.091, while in deionized air these ratios were 1.010 and 1.014, respectively. The rate of respiration in deionized air thus was approximately the same during the day and night. For the autoxidative process the dissociation or liberating of free valences in the oxygen molecule is necessary, and by this is explained the accelerating influence of light and heat upon oxidative processes.

Middleton (1927) found that when the air passing over barley seedlings was artificially ionized by means of polonium, the rate of respiration was increased by as much as 29.1 ± 5.6 per cent. In the case of the leaves of *Pelargonium zonale* with air ionized to a degree varying from 7.28×10^2 to 3.64×10^2 that of normal air, Whimster (1927) observed a percentage increase of 85.7 ± 7.1 in the respiration rate. After the removal of the polonium there was a definite aftereffect, the succeeding 2 hr. showing a percentage increase of 28.0 ± 7.3 over the control period. This effect of ionized air on respiration is apparently due to the ions themselves and not to the gaseous products produced by the action of the ionizing agent.

Portsmouth (1934) found that the respiration rate in barley seedlings in ionized air increased about 2 per cent over the control. Ionized air apparently had little or no effect on the rate of respiration in the leaves of the geranium. Sapozhnikova (1928) reported that ionization of the air decreased the rate of respiration in the plants that he observed. DeBoer (1930) grew *Phycomyces blakesleeanus* and *Polyporus destructor* in ionized air in which the degree of ionization varied from a few hundred to several million times that of normal air. He, however, could observe no effects of ionization on the rate of respiration.

Tang (1932) reported that the consumption of oxygen by the germinating seeds of *Lupinus albus* can be reversibly inhibited by carbon monoxide to a maximum of 36 per cent with a mixture of 24 per cent oxygen and 76 per cent carbon monoxide at 18°C. in darkness. This inhibition, however, is completely removed when the seeds are illuminated. Parija and Saran (1934) found that red light had no effect on the rate of respiration of starved leaves, while blue and violet light increased it after the manner of white light. They considered that light might cause this increase by the activation of the enzyme system or by changing the permeability of the cells. Groner (1936) found that if albino corn were suddenly exposed to light after a period in darkness, an increase in the output of carbon dioxide resulted. Wynd and Reynolds (1935) give a thorough review of the effects of ultraviolet light on rate of respiration and respiratory enzymes.

In 1914 Knight and Priestley reported some observations on the effects of electricity on respiration. Direct currents of a density 10^{-6} to 10^{-4} amp. had no effect on the respiration of peas other than that due to accompanying changes of temperature. The proportion of the currents actually traversing the peas, however, was small, the greater portion being taken by the water films of the seeds. Overhead discharges producing a current of density less than 3×10^{-6} amp. had no effect upon respiration. When higher currents were employed, a definite increase of the carbon dioxide output was observed. This increase, however, could be wholly attributed to the rise of temperature caused by the discharge. In the field where the currents are too small to produce any appreciable rise in temperature, electrification will have no effect on respiration, and explanations of the acceleration of growth due to electricity must be sought in other functions of the plant.

Marsh (1935) derived a quantitative relation between the velocity of respiratory rate and the inherent e.m.f. of the cells considered. Rosene and Lund (1935), and Rosene (1935) believed that a quantitative relation exists between cell oxidation and the continuously maintained e.m.f. in the roots of the onion.

j. Other Factors.—Other factors that have been observed to influence the rate of respiration are as follows:

1. *Sprays.*—The effects of the application of oil sprays on the rate of respiration have been reported by several investigators. Kelley (1930) noted that all oil sprays applied to the dormant twigs of apple increased their rate of respiration but retarded it after the unfolding of the first leaves. Green and Johnson (1931) found that the respiratory rate of bean leaves was increased 7.5 per cent when sprayed with dark petroleum oils containing more than 16 per cent sulphonatable residues. The rate, however, was decreased 5.0 per cent when the leaves were sprayed with light-colored oils containing less than 16 per cent of sulphonatable residue. Green (1936) stated that the rate of respiration was increased more markedly by oil sprays that were poorly refined than by those of a better quality. The average gain in the respiratory rate of various plants due to the application of oil sprays was 4.2 per cent. According to Hoffman (1935) the application of lime-sulphur sprays to leaves slightly increased their rate of respiration.

2. *Materials.*—It was observed by Gregory and Richards (1929) that the respiratory rate of barley plants was subnormal when nitrogen and potassium were deficient but normal when there was a phosphorus deficiency. Benoy and Webster (1930) in a study on fresh vegetables could find no relationship between the rate of respiration and the composition of the parts concerned. When different sugars were injected into starved leaves, Singh (1930) found that the intensity of the respiratory rate at first increased and then decreased. Pal and Chatterji (1936) found that the injection of a mixture of 40 units of insulin in 100 cc. of water into the leaves of *Hispage*

madablota and *Allium tuberosum* increased the rate of respiration as compared to those which had been injected with water. LeVan (1930) found that a stimulation of the production of carbon dioxide occurred in seedlings of *Lupinus albus* when placed in solutions of aluminum chloride, sodium chloride, magnesium sulphate, and ferrous nitrate. These increases ranged from 15 to 50 per cent. Overholser and Claypool (1931) reported that the average intensity of the respiratory rate in strawberries from plots that had received nitrogen applications was practically 10 per cent greater than that in berries from plots to which no nitrogen had been applied. Kertesz (1933) found that the sugar which disappears in green peas after shelling is utilized largely in respiration and is not transformed into starch as has commonly been assumed.

3. *Diseases.*—It was reported by Richter, Dvoretskaia, and Grechushnikov (1929) that there was a decreased respiratory rate in badly diseased plants. They attributed this decrease to a prolonged deficiency of foodstuffs. Whitehead (1934) stated that, except for the short period extending from the end of tuber dormancy to the unfolding of the first leaves, potato plants infected with potato leaf roll respire at a much higher rate than do healthy plants.

13. Substances or Enzymes Concerned in or Associated with Oxidation in Plants.—It was demonstrated by Schönbein as early as 1856 that there are present in plants and animal juices substances that act as oxidizing agents for certain compounds. Traube (1877) termed these substances "oxidizing ferments," and Bertrand (1896) proposed the term "oxydases" as a group or class name. This term has come into general use, although in England and America it is now commonly written "oxidases." As the behavior of these substances was more thoroughly studied it was found that they varied widely in their composition and in the nature of their reactions. As a result, numerous terms have been introduced into the literature, so that the nomenclature in regard to substances concerned in or accompanying oxidation by plant and animal tissues has become rather confusing. Before beginning a detailed discussion of the nature and action of the oxidizing substances of the plant it would perhaps be best, for clearness, to define four terms that have been the most generally used. These are "oxidases," "oxygenases," "peroxidases," and "catalases."

An *oxidase* is a complex of substances which has the power at ordinary temperatures to bring about the oxidation of certain compounds in the presence of gaseous or dissolved free oxygen.

An *oxygenase* is a substance that is capable of causing the formation of organic peroxides. By some authors an oxygenase is defined as a preformed organic peroxide resulting from auto-oxidation. Some consider that the oxygenase itself takes up oxygen and thus becomes a peroxide, while others believe that it acts only as a catalyzer in the production of peroxides. An oxygenase is sometimes called a "peroxide-forming enzyme."

A *peroxidase* is a substance that is capable of rendering active the oxygen of peroxides. The oxygen is freed from the peroxides in the atomic or active form and is thus capable of causing the oxidation of certain compounds.

A *catalase* is a substance that is capable of decomposing hydrogen peroxide into water and molecular oxygen without being able apparently to activate the oxygen thus liberated toward oxidizable substances.

a. The Oxidases.—When the juice or watery extract of a plant tissue causes the oxidation of compounds, which cannot be brought about by oxygen alone to any appreciable extent at ordinary temperatures, it is said to contain a direct-oxidizing enzyme or oxidase. Some of these oxidation reactions are manifested by the bluing of guaiacum, the conversion of hydroquinone into quinone, the liberation of iodine from potassium iodide, and the formation of phenolphthalein from phenolphthalin by the liberation of 2 atoms of hydrogen. The action of oxidases is manifested by the blackening of cut potatoes, the staining of lumber (Bailey, 1910), and the blackening of tea leaves upon curing. Since the oxidases are in most cases destroyed by heat, mineral acids, and poisons, and since they apparently act in a catalytic manner, they have been termed "oxidizing enzymes," although they possess certain characteristics that are different from those of enzymes. Thus in most cases they do not behave as true catalysts, as they are unable to accomplish the oxidation of unlimited amounts of oxidizable materials.

It was stated by Bach (1897), Kastle and Lovenhart (1901), and Bach and Chodat (1903) that the action of the oxidases indicated that they have the power to produce peroxides, and they believed that an oxidase is composed of two substances—an oxygenase and a peroxidase. They suggested that the oxygenase has the power of fixing the atmospheric oxygen in such a manner as to produce a peroxide, but the manner in which this occurs is not clear. Some suggest that the oxygenase catalyzes the formation of organic peroxides from auto-oxidizable substances in the cells, while others believe that the oxygenase itself combines with the atmospheric oxygen and thus becomes a substituted organic peroxide of a very unstable nature (Moore and Whitley, 1909).

It was stated by Kertesz (1933, 1934) that oxygenase catalyzes the oxidation of certain compounds with the *o*-dihydroxy-grouping characteristics of the catechol tannins. The oxidation is accompanied by the formation of the dark-colored *o*-quinone, the presence of which causes the darkening of peaches.

Gallagher (1923) found evidence that the production of peroxide in the potato is intimately associated with the lecithin of the tuber. He believed that the substance from which the peroxide is derived may either be the lecithin itself or a compound intimately associated with it. Gallagher isolated from fresh potatoes a substance that bore a close relation to the lipins. In contact with air or oxygen a solution of this substance acquired the property of causing the immediate oxidation of guaiacum in the presence of peroxidase. He thought this indicated that the so-called "oxygenase" of the potato is really an auto-oxidizable lecithin-like substance. He believed this evidence substantiates the view that there is a relationship between lecithins and plant respiration. Onslow (1919) believed that the peroxides are formed under the influence of oxygenase from aromatic substances having the dihydroxy grouping of catechol. The peroxidase is considered to catalyze the decomposition of the peroxide that is formed by the oxygenase, with the liberation of active oxygen. This active oxygen then combines with the oxidizable substances with which it comes in contact. This theory of the nature and reaction of oxidases has been generally accepted and has largely dominated the field of thought in that regard.

The two constituents of the oxidases have been separated by Bach and Chodat (1903) and Reed (1916) in a fairly satisfactory manner. Reed found that the filtrate from finely ground potato peel in 55 per cent alcohol contained a peroxidase, while the residue after washing with alcohol gave a water extraction containing oxygenase. When these two preparations were mixed, the oxidation increased in the case of pyrogallol about six times over that where oxygenase alone was used.

1. *Classification of Oxidases.*—The oxidases of plants that have been studied have been grouped into two main classes according to the substances upon which they act. These are the laccases or phenolases and the tyrosinases.

(*a*) *The Laccases or Phenolases.*—It has long been known that a blue coloring matter is produced when a tincture of gum guaiacum is treated with certain oxidizing agents, and the first knowledge of oxidases is closely associated with this fact (Kastle, 1910). The laccases or phenolases include those oxidases which are capable of reacting directly, in the presence of free oxygen, with a fresh alcoholic solution of gum guaiacum with the production of its characteristic blue oxidation product (Gallagher, 1923). Each of the oxidases in this group is capable of oxidizing one or more of the phenol compounds. Some of the compounds that can be oxidized under the influence of the phenolases are pyrogallol, benzidine, α-naphthol, phloroglucin, pyrocatechol, hydroquinone, phlorizin, resorcin, guaiacol, *o*-cresol, *m*-cresol, *p*-cresol, *o*-toluidine, *m*-toluidine, *p*-toluidine, phenolphthalin, tannins, guaiac tincture, and others.

In 1883, Yoshida reported that the hardening of the milk sap of the lac tree (*Rhus vermicifera*) was apparently an oxidation process that was catalyzed under the influence of a nitrogenous substance having the characteristics of an enzyme. In 1894 and 1895 Bertrand confirmed these results and called the oxidizable substances present in the sap "laccol" and the compound that catalyzed the oxidation "laccase." On account of the resemblance of "laccol" to the phenols, Bertrand studied the effects of laccase upon a number of them and found that they were oxidized under its influence. He found guaiacum to be a useful reagent for laccase and detected its presence by this means in many plants and plant parts. Laccase has been found to be not a specific enzyme in the narrow sense, because it will bring about the oxidation of many phenols and phenol derivatives. It is not able, however, to oxidize tyrosine or any of its derivatives. It thus apparently acts only upon substances containing a certain grouping in their structure.

Substances that give the same or similar reactions as laccase toward the phenol compounds are widely distributed in the plant kingdom and have been named collectively the "laccases," the "phenol oxidases," or the "phenolases." Most of the oxidases that have been studied in plants belong to this group, but practically no progress has been made toward their further classification. Some idea of the widespread distribution of the laccases may be had from the work of Onslow (1920, 1921), who examined 320 species of angiosperms representing more than 180 natural orders and 309 genera and found that 62 per cent of the orders examined contained oxidases of the laccase type. She found the laccases or phenolases in the fruit of the apple, pear, quince, plum, cherry, peach, apricot, strawberry, grape, fig, and mulberry. They have also been found in apple bark and leaves, in alfalfa, clover, turnips, potatoes, peas, cabbage, silks of green corn, asparagus, and in many other plants and plant parts. Bunzel (1913, 1914) noted that the seeds of the beet had the highest phenolase activity of any of the plant parts, while the leaves stood next in order. In the normally developing leaves of the potato plant the phenolase activity was greatest in the early stages of development. It fell off with the growth of the plants and again rose when growth was completed. Bunzel (1916) noted in the case of the onion, potato plant, and sugar beet that the relative activity of the phenolases, as manifested toward the reagents which he used, varied only in intensity for the different parts of the same plant but found that the enzyme from each plant was markedly different from those of the other plants.

Drain (1923) in tests made at various periods during the time of fruit development, at harvest, and late in the storage season of apples, found that the majority of them fell in that class where the greatest phenolase activity was noted in the vicinity of the core and core line. He believed that oxygen may enter this fruit largely through the fibrovascular bundles that connect the calyx tube and the core cavity.

(*b*) *The Tyrosinases.*—It has long been noted that when certain fungi were exposed to the air they turned pink or red and finally black, while other species became blue due, as we now know, to the activities of the laccases. Bertrand (1896) proved that the former colorations were due to the oxidation of tyrosine by the absorption of oxygen under the influence of a specific oxidizing enzyme which he named "tyrosinase." He proved that this enzyme also occurs in the roots of the beet and in the tubers of the potato and dahlia, and it has since been demonstrated to be rather widely distributed in the plant and animal kingdoms. It is especially abundant in wheat bran. Miller (1929) considered that the insoluble tyrosinase of the seed coat of the velvet bean might be a peroxide.

Tyrosinase exhibits two general reactions. It oxidizes tyrosine, with the production finally of the black pigment melanin and the oxidation of *p*-cresol with the formation of a deep orange color. The former is the most common reaction shown by this enzyme. When tyrosinase is added to a solution containing tyrosine, the solution passes through a series of color changes ranging from pink to rose, to red, to violet, to blue black, and finally to the formation of a black precipitate. The tyrosinases are nondialyzable and are destroyed by heat. They are apparently more or less specific, since they act upon a group of compounds that are closely associated in structure.

2. *Methods of Determining Activity.*—The presence and activity of oxidases in plant extracts and tissues have been studied by color reactions and by the determination of the quantity of oxygen absorbed.

(*a*) *Color Methods.*—Much of the knowledge of the oxidases has been obtained from the study of their action upon compounds that undergo a change of color upon oxidation. The compounds generally employed for this purpose are those which do not oxidize spontaneously, or do so only very slowly, when exposed to the air in very dilute solutions, but which when brought into contact with living tissues or their extracts show a change in color.

The color reactions have been especially useful in determining the distribution of oxidases, and their behavior toward various indicators has also shown that these enzymes vary in the type of reactions that they catalyze. The color methods, however, give no quantitative indication of the amount of oxidation that has taken place, although it has been so assumed in many experiments. The amount of oxidation necessary to produce the colored appearance is, in most cases, very small, so that after the color appears, the oxidation may continue without necessarily changing the intensity of the color. There is also no definite means of determining the amount of oxidation that a particular color reaction represents. It should also be noted in this connection that the amount of oxidation necessary to produce a color in the various reagents varies over a wide range (Reed, 1916). Some of the most common preparations used in the color determinations of oxidase activity and their method of application are given below.

(1) *Guaiacum Reaction.*—Two grams of gum guaiacum are dissolved in 100 cc. of 95 per cent ethyl alcohol. To 5 cc. of plant extract is added 10 drops of this solution. If oxidases of the laccase type are present, a blue color develops.

(2) *Benzidine Reaction.*—One gram of benzidine is dissolved in 100 cc. of 50 per cent ethyl alcohol. To 5 cc. of plant extract is added 10 drops of this solution. This is allowed to stand for ½ hr., and if oxidases are present, a blue color develops.

(3) *α-Naphthol Reaction.*—A 1 per cent solution of α-naphthol in 50 per cent ethyl alcohol is prepared. When a few drops of this reagent is added to plant extracts, a lilac or lavender color results if oxidation occurs.

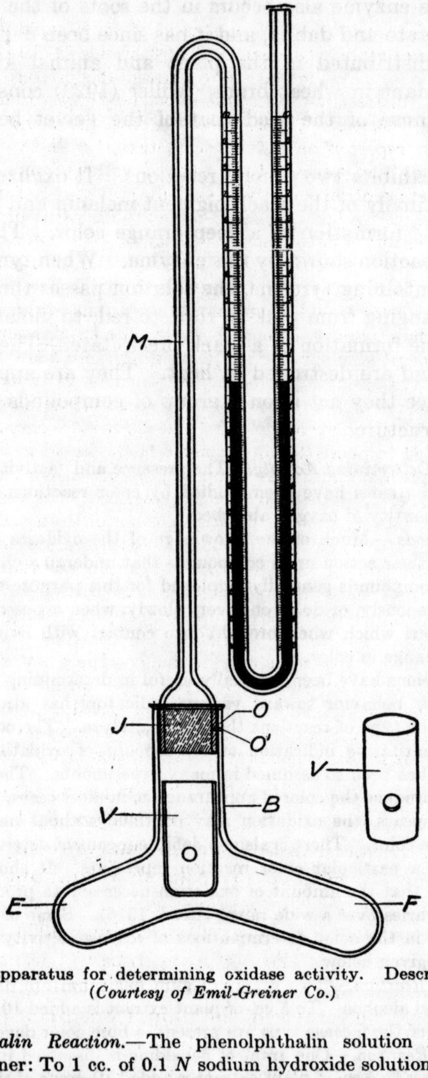

Fig. 36.—Bunzel apparatus for determining oxidase activity. Description in the text. (*Courtesy of Emil-Greiner Co.*)

(4) *Phenolphthalin Reaction.*—The phenolphthalin solution is prepared after the following manner: To 1 cc. of 0.1 *N* sodium hydroxide solution is added 0.1 g. of phenolphthalin. This is dissolved as thoroughly as possible, after which it is diluted with 25 cc. of water, filtered, and made up to 100 cc.

To 10 cc. of plant extract is added 5 cc. of the phenolphthalin solution and the mixture allowed to stand for 15 min., after which it is made alkaline with 0.05 *N* sodium hydroxide. If oxidase activity has occurred, the solution acquires a pink or red color due to the formation of phenolphthalein by the oxidation of the colorless phenolphthalin.

(5) *Röhmann-Spitzer Reaction.*—Equal parts of the α-naphthol solution and of a 1 per cent solution of *p*-phenylenediamine hydrochloride in distilled water are mixed just before the oxidase test is made. Five drops of this mixture is added to 2 cc. of plant extract. If oxidases are present, the colors developed within 15 min. vary from pink to deep reddish purple. This is one of the most delicate of the oxidation tests.

Guthrie (1930) reported an iodimetric method for determining the action of oxidase, while Fong and Cruess (1929) studied various indicators as a means of determining the extent of the activity of oxidases occurring in fruits.

(*b*) *Measurement of Oxygen Absorbed.*—The measurement of the oxygen used in oxidation offers a means of determining quantitatively the intensity of oxidase activity. Apparatus for this purpose has been devised by Bunzel (1912, 1914) and Harvey (1920). The simplified apparatus of Bunzel is shown in Fig. 36. This consists of two parts—a manometer *M* and a basal part *B*, which has two compartments *E* and *F*. The two parts of the apparatus are connected by a ground-glass junction *J*. The total volume of the apparatus minus the vial *V* up to the zero point on the manometer is 76.1 cc. In preparing for an experiment the manometer *M* is detached at the ground joint *J*, and a measured quantity of the plant juice is run into the compartment *E*, and the solution containing the oxidizable substance is run into compartment *F*. Enough water is then added to make the total volume of the liquid in the apparatus 18 cc. Rose (1915) used the materials in the following proportions: 3 cc. of plant extract, 3 cc. of water, and 12 cc. of a 1 per cent solution of pyrogallol. The vial *V* with 1 cc. of concentrated alkali is next inserted in the basal portion of the apparatus. The volume of this vial and contents is 3.5 cc. The manometer is replaced in such a manner that the two openings *O* and *O′* in the ground joint coincide, and the apparatus is brought to the temperature at which the experiment is to be carried out. The apparatus is then closed by turning the manometer in the ground joint through an angle of 90 deg. The plant extract and the oxidizable material are mixed by shaking the apparatus. The basal portion of the apparatus may be immersed in a water bath of the desired temperature, and continuous shaking by mechanical means may be carried on for the duration of the experiment. After 18 cc. of liquid and the vial and contents are added, the volume of air in the apparatus is 54.1 cc., so that every change in pressure of 1 cm. of mercury corresponds to the absorption of 0.71 cc. of oxygen.

3. *Factors Influencing Activity.*—The oxidases, as a rule, increase in oxidizing power with an increase in temperature up to 42 to 45°C., but above that they become less and less active and are completely inactivated at 100°C. They are very sensitive to acids and apparently are most active in a neutral or slightly alkaline medium. Reed (1916) found that potato oxidase was completely inactivated by 0.0005 to 0.007 *N* acids. Bunzel (1916) found that the inhibiting range of hydrogen-ion concentration for oxidase activity in the case of the potato tuber was pH 3.55 to 3.70. In the buds and leaves of the tulip it was pH 2.30 to 2.80, and for the buds and leaves of magnolia pH 2.45 to 3.05.

Cruess, Jeffrey, and Pancoast (1932) noted that the oxidase activity of the juice expressed from peaches, apricots, olives, avocados, asparagus, and spinach, was greatly affected by pH values. The action of these oxidases almost ceased at pH values below 2.5. Cruess and Fong (1929) stated that in any comparison of the activity of fruit oxidase the pH value of the surrounding medium must be carefully controlled. According to Haber (1928) the soil reaction had little or no effect on the oxidase activity of the tomato. Rose, Kraybill, and Rose (1920) studied the effects of various salts upon the rate of oxidation of pyrogallol by apple bark. They found that tenth-normal solutions of the chlorides of potassium, sodium, lithium, cesium, ammonium, calcium, manganese and ferric iron decreased oxidation, while it was slightly increased by solutions of the same concentration of the sulphates of these cations. The nitrates of potassium, sodium, and magnesium in tenth-normal concentration had no effect on oxidation, while the nitrates of carbon, barium, manganese, and ferric iron decreased it. Oxidation was increased in the presence of the tartrates, oxalates, citrates, acetates, and carbonates.

In their experiments the increased oxidation seemed to be due, in part at least, to the low acidity of the mixture of bark, pyrogallol, and salt, but marked decreases in oxidase activity were not necessarily accompanied by high acidity of the mixture. They believed that ions other than hydrogen and hydroxyl may be important in influencing oxidase activity so that in neutralizing these ions in the study of oxidases the possible effects of the salts formed should be considered.

The relative oxidase activity of healthy and diseased tissues has received considerable attention. Woods (1902) found a greater oxidase activity in the spotted leaves of tobacco affected with mosaic than in the normal ones. Bunzel (1913) noted that the juice from the leaves of beets affected with curly top had an oxidase content two to three times greater than the healthy and normally developing ones. An abnormally high oxidase activity was also shown in those plants whose growth had been retarded. Bunzel (1914) also found that the juice of the tubers and foliage of potato plants affected with curly dwarf showed a greater oxidase activity than the juice from these parts of healthy plants. The oxidase activity of tumor tissue in the beet caused either by freezing or by *Bacterium tumefaciens* was observed by Harvey (1920) to be greater than in the adjacent healthy tissue. Rose (1915) found that the extracts of apple-tree bark affected with Illinois canker caused greater and more rapid oxidation of pyrogallol than did the extract of healthy bark. The diseased bark extract was less acid than the healthy bark extract, hence he concluded that oxidase activity was in approximately inverse ratio to the acidity of the extract in the range of concentrations used by him. His conclusions were substantiated by the fact that the addition of acid to the solution in the apparatus decreased oxidation, while the addition of alkali increased it. Rose (1919) found in later work that apple bark attacked by the blister canker caused about twice as much oxidation with pyrocatechol, guaiacol, and benzidine as did the healthy bark. He considered that the lower tannin content of the diseased bark might have some influence on the oxidase activity.

Guthrie (1931) noted that the treatment of dormant potato tubers or gladiolus corms with ethylene chlorohydrin caused an increase in the capacity of their extracted juices to reduce sulphur to hydrogen sulphide. This action was increased by boiling the juice and was decreased by the addition of unboiled to boiled juice. This action is explained by assuming the presence of oxidase in unboiled juice. Christensen and Samisch (1934) stated that the exposure of plant material to high-frequency sound waves decreased considerably the oxidase activity, but no complete inactivation was obtained after an exposure of 12 hr.

4. *The Relation to Physiological Functions.*—The phenomena of oxidation in plants are good examples of the ability of living organisms to carry out changes at ordinary temperatures without the presence of chemical agents. Thus the living plant is continually oxidizing materials in the process of respiration under conditions under which oxidation cannot be accomplished in the laboratory. In an attempt to explain the mechanism of respiration, the oxidases have generally been assigned a role in the process, although their relationship to it is almost entirely hypothetical. The activity of oxidases has been studied, for the most part, by the action of plant juices upon phenol compounds under the conditions of the laboratory. In one instance, however, Kastle and Buckner (1917) injected phenolphthalin into the stalks of corn and obtained evidence that the phenolphthalin had been oxidized to phenolphthalein in the cells into which it had penetrated. They thus considered that oxidation similar to that obtained by oxidases under laboratory conditions prevails in the plant.

It is, however, difficult to assign a function to the oxidases in respiration, since, by the present known methods of measuring their activity, they have no direct action at all upon the ordinary substances consumed in respiration. The decomposition products of respiration are also incapable of being oxidized directly by the oxidases under the laboratory conditions that have so far been tried. Appleman (1916) found no correlation between oxidase activity and the rate of respiration in potato tubers. He stated that although oxidases may play some role in the process they are certainly not the controlling factor in regulating the rate of respiration.

Navez (1929) believed that some respiratory oxidative process controlled the onset of geotropic curvature. Bonazzi (1931) reported that there was more oxidase in the younger than in the older tissues of the sugar cane. Dexter (1934) found that oxidase activity was unchanged in the plants of cabbage and winter wheat when stored at 2°C. He considered, therefore, that the amount or the activity of oxidase is not correlated with the rate of respiration, with an increase in sugar content, or with a change in cold hardiness. Tyson (1930) noted in the sugar beet that oxidase activity was greatest in those tissues where growth was inhibited. Samisch (1935) found that the oxidase activity from the juice of ripe apricots was greatest in the epidermis and in the cells surrounding the fibrovascular bundles.

Crocker and Harrington (1918) observed in the case of the seeds of Johnson grass and *Amaranthus* that the intensity of oxidase activity decreased markedly with age. One-year-old seeds of Johnson grass showed only two-thirds the oxidase activity of fresh seeds, while 3-year-old seed had only one-third the activity of 1-year-old seeds. Two-year-old *Amaranthus* seeds were only about one-half as active as fresh seeds. The oxidase activity of the dry seeds of Johnson grass showed no difference in intensity from that of germinating seeds. It was also observed that the oxidase activity of the nonliving bracts that enclose the mature caryopses of Sudan and Johnson grass was as high as or higher than that of the caryopses.

Davis (1931) reported that phenolase activity was greater in young seeds with a high percentage of germination than in old seeds with a low percentage of germination. In wheat and cucumber, the phenolase and catalase activities parallel.

It was reported by True and Stockberger (1916) that the intensity of oxidase activity roughly paralleled the distribution of the latex in the poppy (*Papaver somniferum*). With the exception of the roots, the oxidase activity ran roughly parallel with the alkaloid content. The alkaloids appear not to exist as such in the poppy plant but appear as products of the action of the oxidases on constituents present in the latex in the presence of oxygen. Annett (1922) noted that the latex of the Indian poppy has a powerful oxidizing action on guaiacum tincture, pyrogallol, benzidine, and tyrosine in the absence of hydrogen peroxide. He found that opium stored for 3 years possessed an oxidizing enzyme that acted on benzidine and suggested that the loss of morphine from opium on storage might be due to the action of oxidizing enzymes.

There appears to be some connection between the occurrence of oxidases and certain sap pigments in the stems, leaves, and flowers. A vast amount of work has been reported in this regard by Wheldale (1910); Keeble, Armstrong, and Jones (1912); Keeble and Jones (1913); Combes (1913); Everest (1914); Atkins (1915); Robinson (1924); and many others. A fine review of the literature in this field is given by Atkins (1916). The discussion of this relationship of oxidases, however, is beyond the scope of this work.

b. The Peroxidases.—When the juices or watery extracts of tissues produce oxidation only after the addition of hydrogen peroxide, they are said to contain an indirect oxidizing enzyme, which was called "peroxidase" by Linossier (1898). The usual test for a peroxidase is to note if the juice or extract gives a blue color with a solution of guaiacum after the addition of hydrogen peroxide. Guthrie (1930) described a method for estimating peroxidase activity. It is based on the formation of indophenol from a mixture of *p*-phenylenediamine and α-naphthol in a citrate-buffered solution with a pH value of 4.5. A peroxidase is a peroxide-activating enzyme activating both hydrogen peroxide and organic peroxides (Bach and Chodat, 1903). A peroxidase apparently combines with the oxygen from the peroxides to form an intermediate compound which is a more energetic oxidizing agent than the original source of the oxygen. The final stage of oxidation is then effected by the intermediate compound.

The peroxidases are probably more widely distributed in plants than any other enzyme with the exception of catalase. Of the 320 species examined by Onslow (1921), 95 per cent gave the peroxidase reaction with hydrogen peroxide and guaiacum. Clark (1911) reported that 75 per cent of the plants examined by him contained peroxidase. It is especially abundant in the root of the horse-radish, in the tuber of the potato, and in the beet. Onslow (1920) found peroxidase in the blackberry, red and black currants, gooseberry, grape fruit, pineapple, melon, and tomato. McHargue (1920) and Coupin (1925) found that peroxidase predominates in the seeds of most species of plants.

Peroxidase is apparently the most stable of the enzymes associated with oxidation in plant tissues.

Apparently peroxidases from different sources behave differently. Thus Elliot (1932) found that the peroxidase from milk would oxidize both nitrate and tryptophane in the presence of H_2O_2, but that from horse-radish would not.

1. *Preparation.*—Methods for the preparation of peroxidase have been reported by Onslow (1919), Getchell and Walton (1931), and Elliot (1932). According to Onslow it is obtained from the pear or potato as follows:

Thin slices of the material are rapidly pounded in a mortar after sufficient 95 per cent alcohol has been added to prevent exposure to the air. After filtering off the alcohol with a filter pump, the residue is again ground with more alcohol and again filtered. This process should be repeated two or three times, and the grinding should be very thorough. The insoluble residue or white powder that is finally obtained upon drying contains the peroxidase. A cold-water extract is then made of this residue and filtered. This filtered extract does not darken on exposure to air, and in the presence of hydrogen peroxide it oxidizes guaiacum, guaiacol, benzidine, and other phenol compounds. In the absence of hydrogen peroxide it is without action on these substances.

Elliot (1932) gives in detail the following method for the preparation of peroxidase from horse-radish: Wash, chop, and pass through a meat mincer about 1,500 g. of horse-radish. Extract this for 2 hr. with 1 l. of water and squeeze the liquid portion through a cloth. Repeat this process until three or four extractions are made. The combined extracts are then centrifuged or filtered on fluted filter paper, and the translucent liquid that is thus obtained is saturated with ammonium sulphate. This completely precipitates the enzyme with certain impurities. The precipitate is then obtained by centrifuging or by filtering on a Büchner funnel, and lixiviated with 250 cc. of water. The turbid solution is then dialyzed for 3 days until salt free, and the liquid, which doubles in volume during the dialyses, is filtered until clear; twice its volume of denatured alcohol is added. The resulting precipitate is collected by centrifuging and dried in a vacuum desiccator. The product obtained is not readily soluble and should be ground to a paste with a little water before diluting. The entire time necessary for this preparation of peroxidase approximates 7 days.

2. *Factors Influencing Activity.*—It has been observed that SO_2 prevents the darkening of apple tissue. Overholser and Cruess (1923) considered that this inhibitory action is due to the effect of this radical upon the organic peroxide in the tissue and not to the influence upon either the oxidases or the chromogen from which the dark compound is formed by oxidation. Balls and Hale (1934) found that glutathione or cysteine salts applied to sliced apples prevent their discoloration upon drying. They considered this effect to be due to the sulphydryl component upon the peroxidase. Denny (1935) found that treating apple tissue with a solution of thiourea of a concentration as low as 0.1 per cent prevented the browning of the cut surface upon drying in the air. Favorable results were also obtained with pear, banana, and egg plant, but the darkening of potatoes was only slightly inhibited. Balls and Hale (1935) stated that the darkening of the freshly cut surfaces of apple is a reaction that is catalyzed by peroxidase, and that the formation of hydrogen peroxide by a respiratory enzyme which uses molecular oxygen is a necessary preliminary step.

It was noted by Fong and Cruess (1929) and Fong (1931) that the temperature required for the inactivation of the peroxidase of apricots, pears, prunes, figs, tomatoes, bananas, and dates varied with the pH value. The resistance to heat was greatest between pH 5 to 7, decreasing with both increased and decreased values beyond this range until at pH 2.0 and 12.0 the peroxidase was inactivated at room temperatures.

3. *Relation to Physiological Functions.*—It was noted by Flemion (1931) that the activity of peroxidase increased during the afterripening of *Sorbus aucuparia* until germination, and then decreased. Garner, Bacon, Bowling, and Brown (1934), in a study of the nitrogen nutrition of tobacco, noted that the peroxidase activity stood in direct relation to the vegetative activity in the leaf. Josyln and Marsh (1933) considered that peroxidase is not the chief causative agent involved in the deterioration of plant tissue following freezing.

c. The Catalases.—It was noted by Schoenbein (1863) that all the animal and plant tissues with which he worked had the power to decompose hydrogen peroxide with the evolution of molecular oxygen. A further study of this phenomenon by him and others led to the general conclusion that it was a property possessed by all the various enzymes in plant and animal tissues. In 1901, however, Loew isolated from tobacco leaves a body capable of decomposing hydrogen peroxide into molecular oxygen but possessing no other enzymic properties. This body he called "catalase," and with but few exceptions (Reed, 1916; Harvey, 1924) it has been found in the cells of all living tissues that have been examined and is thus apparently the most widely distributed of any of the enzymes. Catalase is classified as an enzyme upon its sensitiveness toward heat, acids, and various poisons, but, as will be discussed later, it exhibits certain behaviors that are not characteristic of enzymes.

1. *Preparation.*—Any description of a method for the preparation and study of catalase must necessarily be general, since the details will depend upon the type of material used, upon the amount of catalase present, and upon other conditions under which the experiment is conducted. For details concerning the preparation and study of various plant tissues, the experiments of Appleman (1910 to 1916), Heinicke (1923, 1924, 1933) Rhine (1924), Davis (1925), Lantz (1927), Knott (1927), Overholser (1928), Bunzel (1930), and Dexter (1934) should be consulted.

The material to be studied is generally ground or macerated in a small quantity of water with an excess of calcium carbonate, since catalase is very sensitive to acids. Quartz sand is sometimes used in the grinding in order to pulverize the tissue more thoroughly. The pulverizing process is usually limited to 2 to 5 min., since in some cases it has been found that the activity of catalase is sensitive to agitation. In some cases the pulverized material is forced through bolting cloth or a sieve to obtain a uniform sample. The fineness to which the material must be reduced to give the best results must be determined by actual experiment, as it has been observed in some cases that the activity of the enzyme may be decreased by pulverizing the tissue too finely. In the case of the seeds of Johnson grass, Crocker and Harrington (1918) found that the maximal activity was obtained when the material was forced through a 70- to 80-mesh screen. They considered that the best results were obtained when the average diameter of the sieve mesh was several times the diameter of the cells.

As soon as maceration is completed, water is added to the mixture—the amount added being determined by the activity of the catalase in the material under consideration. In the case of apple bark and leaves, Heinicke (1923, 1924) added sufficient water to give a dilution of 1 to 50. In working with seeds, Crocker and Harrington (1918) used from 0.5 to 0.025 g. of material for each determination but found 0.14 g. the most desirable. This preparation should not stand for longer than an hour or two, since in most cases the activity of catalase deteriorates very rapidly on aging. This enzyme, as a rule, also loses its activity very rapidly when powdered material is stored in the desiccator. On that account, it is best to grind the material and prepare the mixture just previous to making the catalase determination.

The methods that have been employed in making catalase determinations have been, for the most part, modifications of the method described by Appleman (1910). A description of the apparatus and procedure used by Davis (1926) will serve to illustrate, in general, the methods that are followed (Fig. 37). The apparatus consists of a burette A with a two-way stopcock G at the upper end and a connection with a leveling bulb B at the lower end. An ordinary pinchcock burette may be used with a T-tube inserted in the upper end and connected directly at the bottom with the leveling tube (Heinicke, 1924). A wide-mouthed bottle C of approximately 100 cc. capacity is fitted with a three-holed stopper H. Through one of these openings a small separatory funnel D with a stopcock t is inserted, the top of the funnel being connected with the bottle by the rubber tube E and a glass tube inserted through a hole in the stopper. The bottle is connected through its third hole with the opening R at the top of the burette by means of the rubber tube F.

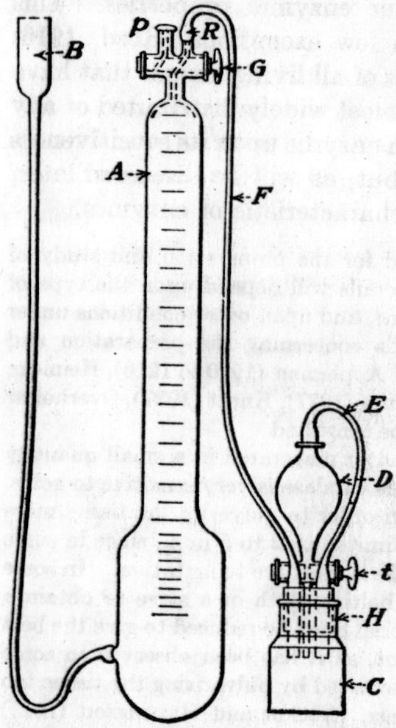

Fig. 37.—Apparatus for the determination of catalase. Description in the text. (*Redrawn from Davis, 1925.*)

After the apparatus is thus set up, a given portion of the mixture whose catalase activity is to be determined is transferred into the bottle C, and the stopper tightly inserted. The stopcock t being closed, 5 cc. of peroxide (dihydrogen) neutralized with 0.1 N sodium hydroxide is added to the separatory funnel D. The bottle is next placed in a water bath at a temperature of 20 to 25°C., and after the contents have reached this temperature the experiment is ready to proceed. The stopcock G is so adjusted that the burette A is connected with the outside air through P. The water level B of the leveling bulb is so placed that it corresponds to the water level in A, after which the exit P is closed and R is opened. The peroxide is then introduced into C through the stopcock t. The bottle C is then gently and uniformly shaken either by hand or by some mechanical apparatus. As the oxygen is liberated, it passes over through the tube F and causes the water level in A to fall. The water level of the leveling tube B is adjusted and the volume of oxygen liberated determined at any desired intervals.

2. *General Nature and Behavior.* Catalase is evidently different in its nature from the ordinary enzymes, since it is not unlimited in its power to decompose hydrogen peroxide, and the indications are that it is consumed in the reaction, and that a given amount is capable of decomposing a definite amount of hydrogen peroxide. According to Loew (1901), catalase exists in two forms, one insoluble in water and the other soluble, which he designated as α-catalase and β-catalase, respectively. Appleman (1910) found that the insoluble catalase of the potato could be separated from the soluble form by ordinary filter paper. Approximately 50 per cent of the total enzyme passed through the filter, indicating that the two types were present in approximately equal proportions. None of this catalase, however, would pass through a Chamberland-Pasteur

filter. It is believed by some (Becking and Hampton, 1920) that the two different catalases are different degrees of peptization of the same substance.

(*a*) *Temperature.*—Catalases from different sources show considerable variation in temperature relations. In the potato, Appleman (1910) found that the catalase was completely destroyed at 50°C. The velocity coefficient was 1.5 from 0 to 10°C., but at higher temperatures there was an apparent decrease in this coefficient, which was probably due to the actual destruction of the catalase as the temperature increased. Lantz (1927) found that there was a gradual accumulation of catalase in corn germinating at 10°C., so that ultimately the catalase content was nearly equal to that at 20 to 30°C. At 42°C. the catalase content was markedly reduced. The point of total destruction of catalase for most of the cases reported, however, ranges from 65 to 80°C.

Crocker and Harrington (1918) found that the air-dry seeds of Johnson grass were all killed and their catalase completely destroyed by heating at 100°C. for 5 hr. Heating at 81°C. for ½ hr. reduced the catalase activity by 30 per cent but improved the germination of the seeds. Heating at 81°C. for 17 hr. reduced the catalase activity approximately 90 per cent of its original value and killed all the seeds. The heating of the seeds of *Amaranthus retroflexus* at 81°C. from 4 to 48 hr. reduced the catalase activity but little, while the power of germination was destroyed. The catalase in the air-dry seeds of *Amaranthus* is thus comparatively heat stable, while the substances essential to vitality are comparatively thermolabile.

It was found by Green, M'Endarfer, Orth, and Burge (1929) that cold weather decreased and hot weather increased the catalase content in pine needles. Carrick (1929) found that catalase activity in the vascular tissue of McIntosh apples was increased after an exposure to a temperature of −7.5°C. for 3.6- and 9-hr. periods, but was decreased after a like exposure for 20 hr. It was noted by Leggatt (1929) in spring wheat that all the thermolabile catalase was not destroyed until heated at 100°C. for 90 min. Lantz (1930) reported that the catalase content of germinating corn diminished rapidly at 42°C. Gonzalez (1933) found that the rise in temperature due to smudge heating increased the activity of catalase in the leaves, buds, bark, and wood of mango trees. Arighi, Joslyn, and Marsh (1933) stated that the catalase content of peas and spinach varies with the temperature used in blanching. It is higher in those treated at 40°C. than in those treated at 20 or 50°C.

(*b*) *Water Content.*—The effect of drying the material upon catalase activity has been observed in numerous cases. Crocker and Harrington (1918) noted that drying the seeds of peach and Johnson grass after they had been in the germinator reduced the catalase activity. Lantz (1927), however, found that the drying of corn seedlings at room temperature *in vacuo* over sulphuric acid did not affect their catalase content. Heinicke (1924) reported that the maceration of dry tissue of apple bark resulted in a marked retardation of its catalase activity, but that maceration of wet tissue had little influence on its activity. The longer the tissue is ground, however, the greater is the reduction in its catalase activity. The addition of water in grinding may tend to neutralize the medium, or it may reduce the heat generated in grinding. Knott (1925) found that catalase preparations deteriorate rapidly at 10 to 20°C., and that placing the bottles on ice immediately after maceration and dilution appears to be the best method of keeping preparations of spinach and celery. Tomato catalase, however, lost its activity in 24 hr. even when on ice.

(*c*) *Acidity.*—Loew (1901) and Appleman (1910) pointed out that catalase is very sensitive to acids and that the maximal activity and minimal destruction occur in a neutral or slightly alkaline medium. The ground plant tissues either contain or develop sufficient acid to injure greatly the catalase activity. This injury to the enzyme is avoided by grinding the tissue with calcium carbonate. The effect of the hydrogen-ion concentration upon the activity of catalase apparently varies with different tissues. Thus Harvey (1920) found that the juice expressed from overgrowths on *Ricinus*, *Bryophyllum*, and beets had a hydrogen-ion concentration consistently lower than that obtained from the healthy tissues regardless of whether these overgrowths were caused by freezing or by *Bacillus tumefaciens*. The tumor juice in all cases showed a catalase activity much greater than that of the healthy tissue. In the case of the mosaic leaves of tobacco, however, the catalase activity was greatly increased over that of the normal leaves, while the pH values of the diseased and healthy leaves were approximately the same. Weiss and Harvey (1921) found that the hydrogen-ion concentration of potato tubers affected with the wart disease was consistently higher than that of healthy tubers from the same plant, the values being represented by pH 6.00 and pH 6.49, respectively. The catalase activity, however, was much greater in the wart tissue, the values being represented by 17.9 cc. of oxygen for diseased and 7.8 cc. for the healthy tissue. Overholser (1928) noted that when the pH value of the pulp of pears was reduced to 3.55 by the addition of 0.1 *N* hydrochloric acid the catalase was almost completely checked. As the pH value was raised by the addition of 0.1 *N* sodium hydroxide, the catalase activity increased to a maximum, with a pH of from 6.00 to 6.50.

(*d*) *Light and Darkness.*—It was observed by Euler, Myrbäck, and Myrbäck (1930) that the catalase content for various plants grown in the dark was lower from the fourth to the eighth day than that of those grown in the light. After the tenth day, however, the amount of enzyme in the two types of plants tended to become equal. Schoppel (1933) noted in the germination in the dark of the light-sensitive seeds of *Nicotiana tabacum*, that catalase activity slowly increased, reached a maximum in about 30 hr., and then decreased. In the light there was a marked increase in catalase activity after 37 hr.

3. *Relation to Physiological Activity.*—Since catalase occurs in practically all living cells, the question naturally arises as to what function it performs. Loew (1901) thought that hydrogen peroxide might be produced in the cell as a result of the respiratory process and that its accumulation would be harmful. He believed that its destruction is brought about by the catalase; this would not only prevent injury, but the oxygen thus liberated could be again used in respiration. The formation of hydrogen peroxide in respiration, however, has not been proved, and even if it were formed it is probable that it would not be injurious to the plant, as indicated by the work of Bach and Chodat (1903), who were able to cultivate plants in a medium containing 0.68 per cent hydrogen peroxide. Catalase, however, acts only on hydrogen peroxide and not on organic peroxides. These compounds, rather than hydrogen peroxide, are concerned in respiration, and Reed (1916) considered that the function of catalase in protecting the organism against hydrogen peroxide is very limited if it exists at all. The relation of catalase activity to respiration and to the general metabolic activity or vitality of plants has been extensively studied and will be considered in the discussion of those phenomena.

(*a*) *Respiration.*—The great majority of workers have found some connection between catalase activity and respiration, although the data show some contradictions, but the exact nature of this connection is not known.

(1) *Correlation.*—The catalase activity and the intensity of respiration have been found to parallel each other in many cases, and on that account it has been concluded that the one is the cause of the other.

Appleman (1910, 1916, 1918) found that the catalase activity in the potato tuber and in fresh sweet corn showed a striking correlation with respiratory activity. Thus respiration was very high when the corn was first pulled, but this rate fell off rapidly with storage. The catalase activity in a collateral set of ears showed a decline with storage which was almost directly proportional to the decline in respiratory intensity. Crocker and Harrington (1918) found that the catalase activity of the embryo of Stoner wheat was approximately twenty-eight times that of the endosperm. Since it has been found by Burlakow (1898) that the respiratory intensity of the wheat embryo is about twenty times that of the endosperm, it would appear that here also catalase activity parallels high respiratory intensity. In *Spirogyra*, Burge and Burge (1924) also found a parallelism between respiration and catalase activity. Rosenberg (1910) and Zaleski and Rosenberg (1911) likewise considered that catalase stands in a direct relationship with the respiratory process.

Ranjan and Mallik (1931) stated that catalase activity was correlated with the amount of monosaccharoses and that it was influenced more by the formation of hexoses than by the actual amount present. They considered that respiratory and catalase activity are affected indirectly through hexose formation by either hydrolysis or synthesis. Gustafson, Clark, Shaw, and Warweg (1932) noted in the John Baer tomato that high catalase activity accompanied a high rate of respiration and growth. Ransom (1935) found in both the seeds and fruit of *Polygonum scandens* that the catalase activity and the rate of respiration followed curves similar to those of afterripening and of germination.

(2) *Divergence.*—Sherman (1921) found in the germination of *Crataegus* seeds that the catalase activity increased continuously up to the twelfth day, while the respiratory activity increased only to the sixth day. In the germination of *Amaranthus* the fluctuations of catalase activity and respiratory activity were frequently in opposite directions.

It was found by Morinaga (1925) that the catalase of germinating rice increased under aerobic conditions with a reduced quantity of oxygen but did not increase in the course of anaerobic germination. He concluded that in the case of rice the ratio of increase of catalase activity is a function of the free oxygen in the medium. Magness and Ballard (1926) found that there was an increase in catalase activity in Bartlett pears as they were held in storage after picking, followed by a decrease as the fruit approached prime eating conditions. The decrease in catalase, however, came much earlier in the life of the fruit than did the decrease in carbon dioxide output.

Rhine (1924) noted in the germination of the seeds of wheat, feterita, clover, mustard, radish, and buckwheat that there was always a drop in catalase at the onset of germination, the low point being reached in from ½ hr. to several days, depending upon the type of seeds. This decrease in catalase activity was nevertheless followed by a sharp rise. The intensity of respiration, however, increased rapidly from the first hour of germination. The early curve of respiration thus diverged widely from the curve of catalase. Similar observations were made by Lantz (1927) in germinating corn. He found some correlation between catalase activity and respiration at 20°C., but at 10 and 30°C. this correlation was not apparent, and he concluded that there is, in this case at least, no evidence to assume that catalase is the enzyme chiefly concerned in physiological oxidation.

Bunzel and Kenyon (1933) could find no parallelism between oxidase activity and catalase activity in various parts of the potato plant, and there was some evidence that these stood in some sort of a reciprocal relationship. The oxidase activity was much greater in the tuber than in the foliage, while the catalase activity was higher in the latter and lower in the former. Oxidase activity was greatest in early stages of growth of the foliage, while catalase activity was the lowest at this stage. Marks (1935) found that the catalases from certain species of marine plants were inactivated either directly or indirectly by free oxygen. In barley, Pope (1932, 1933) found a negative correlation between respiratory rate and catalase activity. To explain his results on the basis that respiratory rate and total catalase production are correlated, it would be necessary to assume that in barley a high respiratory rate is associated with proportionately greater consumption of the catalase-activating agency. Thus the surplus of catalase would be actually smaller in amount than when the rate of respiration is low. Miller, Guthrie, and Denny (1934) found in treating potato tubers to overcome dormancy that the catalase and peroxidase activities were not directly connected with the variation in the carbon dioxide output. The enzymes were activated markedly but the changes tended to follow, rather than to precede or to coincide with, the changes in the rate of carbon dioxide production.

Since the catalase activity drops markedly at the beginning of germination and then increases, while respiration increases from the first, Rhine (1924) believed that the most plausible theory of catalase formation in the plant seems to be that it is formed as an enzyme according to the theory of need—the presence of the substance which it attacks acting as a stimulus for the production of the enzyme. The decrease of catalase with germination could thus be explained by supposing the catalase reserve to be used up in attacking the respiratory products faster than it is produced. The high stimulation of the suddenly increased quantity of by-products would thus lead in a short time to the increased production of catalase observed. If this is the case, the catalase curve would follow but never precede the respiration curve.

(b) *Vitality*.—It has been shown in some cases that catalase activity parallels the general metabolic activity of the organism, and numerous experiments have been conducted to determine if the catalase activity can be used as an indicator of the vitality and vigor of the plant or plant parts. The relation of catalase activity to the maturity, dormancy, and vitality of seeds has been especially studied. Crocker and Harrington (1918) found in the seeds of Johnson grass and Sudan grass that the fall in catalase began with harvest and continued for an indefinite period. It was found in these seeds that a considerable fall in catalase occurred with little or no fall in vitality. Thus in 9-year-old seeds of Sudan grass the percentage of germination was 98 per cent, although the catalase activity had fallen to less than one-half that of the 1-year-old seeds. In the case of the seeds of Johnson grass and Sudan grass, the catalase activity seemed to be an indicator of age rather than of vitality. In the seeds of *Amaranthus retroflexus*, however, there was no regular fall of catalase activity with the age or vitality of the seed. In maple seed, Jones (1920) noted that after 8 months of storage under laboratory conditions the catalase activity had declined to only one-half that of the fresh seeds. In the seeds of various varieties of lettuce 20 years old, Davis (1925) found that the relation of catalase to vitality was not such as to give reliable data on the viability of seeds, as the catalase content of dead seeds frequently was sufficiently high to indicate fair germination. He found, however, that, if seeds were soaked for a time in warm water at a temperature not sufficiently high to affect their viability, the catalase of dead seeds disorganized very rapidly, while that of the viable seeds was not affected. These observations are contrary to those reported by Nemec and Duchon (1921, 1922, 1923) in their work with seeds of oats and peas. They considered that the viability of seeds could be determined by means of the catalase content of these seeds, since they found that the ratio of catalase of poor seeds to that of viable seeds also expressed very closely the percentage germination of the seeds. The disappearance of catalase from seeds during and after death no doubt bears a close relation to such factors as temperature, moisture, and aeration to which the seeds have been subjected during that time (Davis, 1925).

The catalase activity as a rule increases markedly during germination, and in the seeds of Sudan grass and Johnson grass it doubled during that time (Crocker and Harrington, 1918). As is well known, the lower seeds of *Xanthium* under natural conditions germinate much more readily than the upper ones. It was found by Shull and Davis (1923) that, in the field during the germination season, the lower seeds increased in catalase activity while the upper lagged behind, showing practically no increase. They concluded that the catalase differences in *Xanthium* seeds are in harmony with the physiological differences that operate to bring about delayed germination of upper seeds with intact coats. Davis (1930) noted in the seeds of *Xanthium* and *Ambrosia* that during the period in which the embryo became dormant both the respiration and catalase activity were reduced. In the case of *Xanthium* during the period of afterripening or the removal of dormancy at low temperature, there was a rise in the catalase activity and also in the respiratory activity of the seed. Sherman (1921) noted in the seeds of *Acer saccharum* and *Juniperus virginiana* that catalase activity increased as dormancy ended and germination began.

Tyson (1930) found in sugar beets that catalase activity was correlated with the vigor, growth, and size of the plant. Neller (1931) reported that the catalase activity of the roots of the bindweed (*Convolvulus arvensis*) was greatly decreased after the aerial parts of the plant had been sprayed with chlorates. Landon (1934) believed that the measurement of the degree of variation in catalase activity is useful in studying the herbicidal properties of chemicals, provided that the toxic action peculiar to that specific chemical is considered. Leggatt (1933) found that the vitality of wheat seed may be estimated rather accurately from the determination of total and of thermostable catalase. A determination of this type may be completed within a few hours, while a germination test requires 12 days. It was reported by Baldwin (1935) that there was a close relationship between the viability of certain coniferous seeds and the "catalase quotient." He defined the catalase quotient as the ratio of the volume of oxygen evolved from seeds that had been in the germinator for a few days to the oxygen evolved from resting seeds.

The relation of catalase activity to the general vigor and response of the plant has been studied in the apple tree by Heinicke (1922, 1923, 1924). He found that the catalase activity in apple-leaf tissue was influenced by the factors that affected the nutritive or physiological conditions of it. Thus the wilting of leaves reduced the catalase activity. Leaves from trees growing in sod showed less catalase activity than did those from trees under cultivation, while nitrogenous fertilizers increased the catalase activity in leaves from trees in sod. Ringing reduced the activity of catalase. The catalase activity was less in those leaves grown in the shade as compared with those having good exposure to light. Heinicke found that the catalase activity of the phloem was greater than that of the xylem. The activity of young phloem and xylem was greater than the old phloem and xylem, respectively. Catalase in dormant vegetative buds was greater in amount than in dormant flower buds. Ezell and Crist (1927), however, in studying the effect of nutritive conditions upon the oxidase and catalase activity found in the case of radish, lettuce, and spinach plants that the correlation between the activity of catalase and growth or size of the plants was better than with oxidase and was significantly negative. Heinicke (1923) believed that the activity of catalase is a more sensitive measure of the metabolic status of the tissues than the usual chemical analysis, and that it may serve, along with other measures, as an indicator of the physiological responses of plants to various cultural conditions or treatments. If it could so be used, it would be a convenient indicator. The determinations are quickly and easily made, and numerous samples can be taken at frequent intervals from the same source without serious injury to the experimental subject. The effect of disturbances in physiological functions upon the action of catalase was observed by Moore and Willaman (1917) in tomato plants that had been fumigated with hydrocyanic acid. There was a reduction in the activity of catalase and oxidase, but within a few hours after the fumigation the oxidase activity had returned to normal, while the catalase and respiratory activities exceeded the normal.

(c) *Growth.*—It was reported by Lopriore (1928) that catalase is especially abundant in sexual organs, and that it increases as the time for fertilization approaches. Ripe pollen is richer in this enzyme than the unripe pollen. Camp (1929) noted in 12 species of plants that the tissues related to the male structures whether vegetative or floral showed a distinctly greater catalase activity than the tissues related to the female structures.

The relation of catalase to growth is uncertain in that some investigators have found a positive and others a negative correlation. Thus Harding (1930) found that catalase activity was correlated with the fertilization of Grimes Golden apple trees in that the fruit from those which had been treated with nitrate was consistently higher in catalase. The degree of catalase activity indicated early in the season whether or not tissue breakdown in the first was to occur, since immediately prior to this condition a high catalase activity was registered. Similar observations were made by Neller (1931) with regard to breakdown in Jonathan apples. Garner, Bacon, Bowling, and Brown (1934) noted in the tobacco plant that the catalase activity stood in direct relation to the vegetative activity in the leaf and reached its maximum in the later stages of leaf growth.

Haber (1928) found in the tomato plant that catalase activity in various plant parts was negatively correlated with growth and yield. Chance (1931) could find no parallelism between the catalase activity of inbred strains of corn seedlings and that of the corresponding crosses. Pope (1932, 1933) observed in barley that catalase activity was roughly proportional to the reciprocal growth rate. There were, however, three very definite elevations of catalase activity in the general curve. These were during early germination, during the development of crown roots, and during early jointing. Each of these peaks thus occurred at the time of the inception of a new and definite stage in the functional activity of the plant.

Euler, Gard, and Rislund (1931) found that low catalase and low chlorophyll amounts occur together. Since they considered that catalase is a porphyrin derivative it must have certain groups in common with chlorophyll. Newton and Brown (1931) reported that the catalase activity of the expressed juice of the leaves of winter wheat during the later summer and autumn was directly proportional to the winter hardiness of the varieties studied.

(*d*) *Afterripening.*—Denny, Miller, and Guthrie (1930) studied the effects of ethylene chlorohydrin, sodium thiocyanate, and thiourea upon the action of the enzymes of potatoes that had been treated with these chemicals to break the rest period. All these chemicals appeared to increase the activity of the catalase, peroxidase, and reducing agents in the juice. The increases in catalase and peroxidase began in 24 hr. after treatment with ethylene chlorohydrin, but the response to the other two chemicals was less marked and occurred more slowly. The effect was apparently on the potato tissues and not on the enzyme, since the addition of these chemicals to the expressed juice produced no effect. Flemion (1931) found that the stratified seeds of *Sorbus aucuparia* during afterripening increased in catalase content until at the termination of the process it was ten times that of the dry controls. The catalase content of the seedlings, however, was 20 times that of the seeds. No condition was found under which germination took place without an increase in catalase activity. Haut (1932) noted that as the afterripening of seeds of the apple, peach, and cherry progressed at low temperature, there was a distinct increase in catalase activity.

Bibliography

Allison, F. E., S. R. Hoover, and D. Burk, 1933: A respiration coenzyme, *Sci.*, **78**: 217–218.

Annett, H. E., 1922: The enzymes of the latex of the Indian poppy (*Papaver somniferum*), *Biochem. J.*, **16**: 765–770.

Appleman, C. O., 1910: Some observations on catalase, *Bot. Gaz.*, **50**: 182–192.

———, 1911: Physiological behavior of enzymes and carbohydrate transformation in after-ripening of the potato tuber, *Bot. Gaz.*, **52**: 306–315.

———, 1915: Relation of catalase and oxidases to respiration in plants, *Md. Sta. Bull.* 191.

———, 1916: Relation of oxidases and catalase to respiration in plants, *Am. J. Bot.*, **3**: 223–234.

———, 1918: Respiration and catalase activity in sweet corn, *Am. J. Bot.*, **5**: 207–209.

Archbold, H. K., and A. M. Barter, 1934: Chemical studies in the physiology of apples. XV, The relation of carbon dioxide output to the loss of sugar and acid in Bramley's seedling apples during storage, *Ann. Bot.*, **48**: 957–966.

Arighi, A. L., M. A. Joslyn, and G. L. Marsh, 1936: Enzyme activity in frozen vegetables, *Ind. and Eng. Chem.*, **28**: 595–598.

Aso, K., 1903: On the chemical nature of the oxydases, *Bull. Col. Agr. Tokyo*, **5**: 481–491; *Bot. Cent.*, **96**: 106 (1904).

———, 1905: On the nature of oxidases, *Beih. bot. Cent.*, I, **18**: 319–326.

Atkins, W. R. G., 1915: The oxidases and their inhibitors in plant tissues. II, The flowers and leaves of the Iris, *Notes Bot. School Trinity Col. Dublin*, **2**: 210–212.

———, 1915: Oxidases and their inhibitors in plant tissues. IV, The flowers of Iris, *Notes Bot. School Trinity Col. Dublin*, **2**: 236–246.

———, 1916: "Some Recent Researches in Plant Physiology," Chap. 13, The Oxidases in Relation to Pigmentation and the Anthocyanin Pigments, Whittaker and Co., London, pp. 250–292.

Aubert, E., 1892: Recherches sur la respiration et l'assimilation des plantes grasses, *Rev. Gén. Bot.*, **4**: 203–282, 321–331, 337–353, 421–441, 497–502, 558–568.

AUDUS, L. J., 1935: Mechanical stimulation and respiration rate in the cherry laurel, *New Phytol.*, **34**: 386–402.

BABCOCK, S. M., 1912: Metabolic water: Its production and role in vital phenomena, *Wis. Sta. Tech. Bull.* 22.

BACH, A., 1897: Du rôle des peroxydes dans les phénomènes d'oxydation lente, *C. R. Acad. Sci. Paris*, **124**: 951–954.

——— and R. CHODAT, 1903: Ueber den gegenwärtigen Stand der Lehre pflanzlicher Oxydationsfermente, *Biochem. Cent.*, **1**: 417–421.

——— and ———, 1903: Untersuchungen über die Rolle der Peroxyde in der Chemie der lebenden Zelle. IV, Ueber Peroxydase, *Ber. chem. Ges.*, **36**: 600–605.

——— and ———, 1903: Untersuchungen über die Rolle der Peroxyde in der Chemie der lebenden Zelle. V, Zerlegung der sogenannten Oxydasen in Oxygenasen und Peroxydasen, *Ber. chem. Ges.*, **36**: 606–609.

BAILEY, C. H., 1917: The catalase activity of American wheat flours, *J. Biol. Chem.*, **32**: 539–545.

———, 1917: The moisture content of heating wheat, *J. Am. Soc. Agron.*, **9**: 248–252.

——— and A. M. GURJAR, 1918: Respiration of stored wheat, *J. Agr. Res.*, **12**: 685–713.

———, 1919, 1920: The storage of wheat, *Operative Miller*, **24**: 352, 381–382; **25**: 5–6.

——— and A. M. GURJAR, 1920: Respiration of cereal plants and grains. II, Respiration of sprouted wheat. III, Respiration of rice paddy and milled rice. IV, The respiration of frosted wheat plants, *J. Biol. Chem.*, **44**: 5–7, 9–12, 13–15.

———, 1921: Respiration of shelled corn, *Minn. Sta. Tech. Bull.* 3.

BAILEY, E. M., 1912: Biochemical and bacteriological studies on the banana, *J. Am. Chem. Soc.*, **34**: 1706–1730.

BAILEY, I. W., 1910: Oxidizing enzymes and their relation to sap stain in lumber, *Bot. Gaz.*, **50**: 142–147.

BAKKE, A. L., and N. L. NOECKER, 1933: The relation of moisture to respiration and heating in stored oats, *Iowa Sta. Res. Bull.* 165.

BALCH, W. B., 1933: A large respirometer, *Sci.*, **77**: 116–117.

BALDWIN, H. I., 1935: Catalase activity as a measure of viability of tree seeds, *Am. J. Bot.*, **22**: 635–645.

BALLS, A. K., and W. S. HALE, 1934: On peroxidase: Role of peroxidase in the darkening of apples, *Am. Chem. Soc., Division Agr. and Food Chem. and Division Biol. Chem.*, 3 pp. Quoted by Denny, 1935.

——— and ———, 1935: Peroxidase in the darkening of apples, *Ind. and Eng. Chem.*, **27**: 335–337.

BARKER, J., 1933: Analytical studies in plant respiration. IV-V, The relation of the respiration of potatoes to the concentration of sugars and to the accumulation of a depressant at low temperatures. I, The effect of temperature history on the respiration/sugar relation. II, The form of the normal respiration/sugar relation and the mechanism of depression, *Proc. Roy. Soc.*, B, **112**: 316–358.

———, 1935: A note on the effect of handling on the respiration of potatoes, *New Phytol.*, **34**: 407–408.

———, 1936: Analytic studies in plant respiration. VI, The relation of the respiration of potatoes to the concentration of sugars and to the accumulation of a depressant at low temperatures. III, The relation of the respiration to the concentration of sucrose, *Proc. Roy. Soc.*, B, **119**: 453–473.

BARNES, C. R., 1905: The theory of respiration, *Bot. Gaz.*, **39**: 81–98.

BARTLETT, A. W., 1920: Note on an improved method for demonstrating the absorption of oxygen in respiration, *New Phytol.*, **19**: 151–152.

Beaumont, J. H., and J. J. Willaman, 1924: Preliminary report on the respiration of apple twigs during the winter, *Proc. Am. Soc. Hort. Sci.*, **21**: 99–104.

———, ———, and W. A. DeLong, 1927: Notes on apparatus for low temperature respiration studies, *Plant Physiol.*, **2**: 487–495.

Becker, H., 1912: Über die Keimung verschiedenartiger Früchte und Samen bei derselben Spezies, *Beih. bot. Cent.*, I, **29**: 21–143.

Becker, Z. E., 1936: A comparison between the action of carbonic acid and other acids upon the living cell, *Protoplasma*, **25**: 161–175.

Becking, L. G. M. B., and H. C. Hampton, 1920: Measurement of the catalytic power of catalase, *Am. J. Bot.*, **7**: 261–275.

Bělehrádek, J., and M. Bělehrádková, 1929: Influence of age on the temperature coefficient of the respiration rate in leaves of *Scolopendrium scolopendrium*, *New Phytol.*, **28**: 313–317.

Bennet-Clark, T. A., 1932: A method of automatically recording the oxygen intake of living tissues, *Sci. Proc. Roy. Dub. Soc.*, **20**: 281–291.

———, 1932: The respiratory quotients of succulent plants, *Sci. Proc. Roy. Dub. Soc.*, **20**, 293.

———, 1933: The role of organic acids in plant metabolism. II, *New Phytol.*, **32**: 128–159.

Bennett, J. P., and E. T. Bartholomew, 1924: The respiration of potato tubers in relation to the occurrence of black heart, *Calif. Sta. Tech. Paper* 14.

Benoy, M. P., 1929: The respiration factor in the deterioration of fresh vegetables at room temperature, *J. Agr. Res.*, **39**: 75–80.

——— and J. E. Webster, 1930: Note on the relation of rate of respiration to chemical composition in fresh vegetables, *Plant Physiol.*, **5**: 181–182.

Bergman, H. F., 1921: The effect of cloudiness on the oxygen content of water and its significance in cranberry culture, *Am. J. Bot.*, **8**: 50–58.

———, 1925: The respiratory activity of various parts of the cranberry plant in relation to flooding injury, *Am. J. Bot.*, **12**: 641–660.

———, 1929: Changes in the rate of respiration of the fruits of the cultivated blueberry during ripening, *Sci.*, **70**: 15.

———, 1930: Respiration of cranberry plants in relation to water injury, *Proc. Ann. Meeting Am. Cranberry Growers' Assoc.*, **60**: 21–25; *Exp. Sta. Rec.*, **69**: 343.

———, 1930: The oxygen content of water in winter flooded bogs under ice, *Wis. State Cranberry Growers' Assoc.*, **44**: 32–40.

Bertrand, G., 1894: Sur la latex de l'arbre à laque, *C. R. Acad. Sci. Paris*, **118**: 1215–1218.

———, 1895: Sur la laccase et le pouvoir oxydant de cette diastase, *C. R. Acad. Sci. Paris*, **120**: 266–269.

———, 1896: Sur les rapports qui existent entre la constitution chimique des composés organiques et leur oxydabilité sous l'influence de la laccase, *C. R. Acad. Sci. Paris*, **122**: 1132–1134.

———, 1896: Sur les rapports qui existent entre la constitution chimique des composés organiques et leur oxydabilité sous l'influence de la laccase, *Bull. Soc. Chim. Paris*, **15**: 793. Quoted by Appleman, 1916.

———, 1896: Sur une nouvelle oxydase, ou ferment soluble oxydant, d'origine végétale, *C. R. Acad. Sci. Paris*, **122**: 1215–1217.

Bezagu, M., 1919: Variations de la respiration des cellules de feuille avec l'âge, *C. R. Acad. Sci. Paris*, **169**: 701–702.

Bigelow, W. D., H. C. Gore, and B. J. Howard, 1905: Studies on apples. I, Storage, respiration and growth, *U. S. Bur. Chem. Bull.* 94.

BLACKMAN, F. F., and P. PARIJA, 1928: Analytic studies in plant respiration. I, Respiration of a population of senescent ripening apples, *Proc. Roy. Soc.*, B, **103**: 412–446.

———, 1928: Analytic studies in plant respiration. III, Formation of a catalytic system for the respiration of apples and its relation to oxygen, *Proc. Roy. Soc.*, B, **103**: 491–523.

———, 1930, 1931: Respiration and oxygen concentration, *Fifth Int. Bot. Congr. Cambridge* (1930); *Rept. Proc.* Pub. Cambridge (1931).

BOAS, F., 1936: Ionenwirkung, Katalasetätigkeit und Leistung der Zelle, *Angew. Bot.*, **18**: 13–16.

BOEHM, J. A., 1887: Ueber die Respiration der Kartoffel, *Bot. Ztg.*, **45**: 671–675, 680–691.

BOLAS, B. D., and R. MELVILLE, 1933: The influence of environment on the growth and metabolism of the tomato plant. I, Methods, technique and preliminary results, *Ann. Bot.*, **47**: 673–688.

BONAZZI, A., 1931: Studies in sugar cane physiology. II, Sugar cane gases, *Bot. Gaz.*, **92**: 442–445.

BONNIER, G., and L. MANGIN, 1884: Sur les variations de la respiration des graines germantes avec le développement, *Bull. Soc. Bot. France*, **31**: 306–309.

BRIGGS, G. E., and A. H. K. PETRIE, 1931: Respiration as a factor in the ionic equilibria between plant tissues and external solutions, *Proc. Roy. Soc.*, B, **108**: 317–326.

BÜCHNER, E., 1897: Alkoholische Gärung ohne Hefezellen, *Ber. chem. Ges.*, **30**: 117–124, 1110–1113.

BUNZEL, H. H., 1912: The measurement of the oxidase content of plant juices, *U. S. Bur. Plant Ind. Bull.* 238.

———, 1913: A biochemical study of the curly-top of sugar beets, *U. S. Bur. Plant Ind. Bull.* 277.

———, 1914: A simplified and inexpensive oxidase apparatus, *J. Biol. Chem.*, **17**: 409–411.

———, 1914: Oxidases in healthy and in curly-dwarf potatoes, *J. Agr. Res.*, **2**: 373–404.

———, 1915: On alfalfa laccase, *J. Biol. Chem.*, **20**: 697–706.

———, 1916: The mode of action of the oxidases, *J. Biol. Chem.*, **24**: 91–102.

———, 1916: The relative oxidase activity of different organs of the same plant, *J. Biol. Chem.*, **24**: 103–110.

———, 1917: The relationship existing between the oxidase activity of plant juices and their hydrogen ion concentrations with a note on the cause of oxidase activity in plant tissues, *J. Biol. Chem.*, **28**: 315–333.

——— and H. HASSELBRING, 1917: The supposed action of potassium permanganate with plant peroxidases, *Bot. Gaz.*, **63**: 225–228.

———, 1930: A simple apparatus for measuring catalase activity in plant and animal tissues, *Sci.*, **72**: 505–506.

——— and M. B. KENYON, 1933: On potato catalase, *Bull. Torrey Bot. Club*, **60**: 469–474.

BURGE, W. E., and E. L. BURGE, 1924: Effect of temperature and light on catalase content of *Spirogyra*, *Bot. Gaz.*, **77**: 220–224.

BURLAKOW, G., 1898: Ueber Athmung des Keimes des Weizens, *Triticum vulgare*, abst., *Bot. Cent.*, **74**: 323–324.

BURROUGHS, A. M., 1922: Changes in the respiration rate of ripening apples, *Proc. Am. Soc. Hort. Sci.*, **19**: 225–235.

BUSHNELL, J., 1925: The relation of temperature to growth and respiration in the potato plant, *Minn. Sta. Tech. Bull.* 34.

———, 1935: Sensitivity of the potato plant to soil aeration, *J. Am. Soc. Agron.*, **27**: 251–253.

Butler, O., 1913: A note on the significance of sugar in the tubers of *Solanum tuberosum*, *Bull. Torrey Bot. Club.*, **40**: 110–118.

Camp, W. H., 1929: Catalase activity and sex in plants, *Am. J. Bot.*, **16**: 221–224.

Cannon, W. A., 1932: On the variation of the oxygen content of cultural solutions, *Sci.*, **75**: 108–109.

——— and E. A. Purer, 1935: On the removal of oxygen from water by cut branches, *Sci.*, **81**: 100.

Carrick, D. B., 1928: The effect of freezing on the respiration of the apple, *N. Y. Sta.* (*Cornell*) *Mem.* 110.

———, 1929: The effect of freezing on the catalase activity of apple fruits, *N. Y. Sta.* (*Cornell*) *Mem.* 122.

———, 1930: Some cold-storage and freezing studies on the fruit of the vinifera grape, *N. Y. Sta.* (*Cornell*) *Mem.* 131.

Cerighelli, R., 1924: Sur le quotient respiratoire de la racine et ses variations au cours du développement de la plante, *C. R. Acad. Sci. Paris*, **178**: 645–647.

Chance, H. L., 1931: The relation between catalase activity and vigor in inbred strains and crosses of corn seedlings, *Am. J. Bot.*, **18**: 696–703.

Christensen, R. J., and R. Samisch, 1934: Effect of high frequency sound waves on oxidase activity, *Plant Physiol.*, **9**: 385–386.

Clark, E. D., 1910: "The Plant Oxidases," *thesis*, *Columbia Univ.*, New York.

———, 1911: The nature and function of the plant oxidases, *Torreya*, **11**: 23–31, 55–61, 84–92, 101–110.

Clausen, H., 1890: Beiträge zur Kenntniss der Athmung der Gewächse, *Landw. Jahrb.*, **19**: 893–930.

Clements, H. F., 1935: Morphology and physiology of the pome lenticels of *Pyrus malus*, *Bot. Gaz.*, **97**: 101–117.

Coleman, D. A., B. E. Rothgeb, and C. H. Fellows, 1928: Respiration of sorghum grains, *U. S. Dept. Agr. Tech. Bull.* 100.

Combes, R., 1913: Production expérimentale d'une anthocyane identique à celle qui se forme dans les feuilles rouges et automne, en partant d'un composé extrait des feuilles vertes, *C. R. Acad. Sci. Paris*, **157**: 1002–1005.

Coupin, H., 1925: Sur les peroxydases dans les graines sèches, *C. R. Acad. Sci. Paris*, **180**: 685–687.

Crocker, W., and W. E. Davis, 1914: Delayed germination in seeds of *Alisma plantago*, *Bot. Gaz.*, **57**: 285–322.

——— and G. T. Harrington, 1918: Catalase and oxidase content of seeds in relation to their dormancy, age, vitality and respiration, *J. Agr. Res.*, **15**: 137–174.

Cruess, W. V., and W. Y. Fong, 1929: The effect of pH value and hydrogen peroxide concentration of fruit oxidase activity, *Plant Physiol.*, **4**: 363–366.

———, R. N. Jeffrey, and H. M. Pancoast, 1932: Observations on the effect of pH value on oxidase activity, abst., *Am. J. Bot.*, **19**: 841.

Darsie, M. L., C. Elliott, and G. J. Peirce, 1914: A study of the germinating power of seeds, *Bot. Gaz.*, **58**: 101–137.

Dastur, R. H., and R. M. Desai, 1935: The carbon dioxide carbohydrate ratio in the aerobic and the anaerobic respiration of rice, *Ann. Bot.*, **49**: 53–69.

Davis, W. B., and C. G. Church, 1931: The effect of ethylene on the chemical composition and the respiration of the ripening Japanese persimmons, *J. Agr. Res.*, **42**: 165–183.

Davis, W. C., 1931: Phenolase activity in relation to seed viability, *Plant Physiol.*, **6**: 127–137.

Davis, W. E., and R. C. Rose, 1912: The effect of external conditions upon the after-ripening of the seeds of *Crataegus mollis*, *Bot. Gaz.*, **54**: 49–62.

———, 1925: The use of catalase as a means of determining the viability of seeds, *Proc. 18th Ann. Meeting Assoc. Off. Seed Analysts N.A.*, 6–12.

———, 1925: Unpub. data, *Kans. State Col.*

———, 1930: Primary dormancy, after-ripening and the development of secondary dormancy in embryos of *Ambrosia trifida*, *Am. J. Bot.*, **17**: 58–76.

———, 1930: The development of dormancy in seeds of cocklebur (*Xanthium*), *Am. J. Bot.*, **17**: 77–87.

De Boer, S. R. A., 1930: The effect of ionized air on the rate of respiration of fungi, *Ann. Bot.*, **44**: 989–1001.

Deleano, N. T., 1912: Studien über den Atmungsstoffwechsel abgeschnittener Laubblätter, *Jahrb. wiss. Bot.*, **51**: 541–592.

———, 1912: Untersuchungen über die in Weinblättern enthaltenen Kohlenhydrate und stickstoffhaltigen Körper, *Z. physiol. Chem.*, **80**: 79–94.

De Long, W. A., J. H. Beaumont, and J. J. Willaman, 1930: Respiration of apple twigs in relation to winter hardiness, *Plant Physiol.*, **5**: 509–534.

Denny, F. E., 1924: Effect of ethylene upon respiration of lemons, *Bot. Gaz.*, **77**: 322–330.

———, L. P. Miller, and J. D. Guthrie, 1930: Enzyme activities of juices from potatoes treated with chemicals that break the rest period, *Am. J. Bot.*, **17**: 483–509.

———, 1935: Thiourea prevents browning of plant tissues and juices, *Contr. Boyce Thompson Inst.*, **7**: 55–61.

De Saussure, T., 1804: "Recherches chimiques sur la végétation," Paris.

Dexter, S. T., 1934: Respiratory rate and enzyme activity as related to the hardened condition of plants, *Plant Physiol.*, **9**: 831–837.

Dowd, O. J., 1933: Preliminary studies on the internal atmosphere of apples, *Proc. Am. Soc. Hort. Sci.*, **30**: 162–163.

Doyle, J., and P. O'Connor, 1930: Seasonal changes in the catalase content of conifer leaves, *Ann. Bot.*, **44**: 907–917.

Drain, B. D., 1923: Oxidase activity in varieties of apples, *Proc. Am. Soc. Hort. Sci.*, **20**: 25–30.

———, 1926: Temperature and respiratory enzymes of apples, *Bot. Gaz.*, **82**: 183–195.

Duvel, J. W. T., 1904: The vitality and germination of seeds, *U. S. Bur. Plant Ind. Bull.* 58.

Eaves, C. A., 1935: Preliminary study of the effect of a series of temperature changes upon respiratory activity of apples during the post-climacteric in senescent decline, *Sci. Agr.*, **16**: 28–39.

Edwards, T. I., 1933: The germination and growth of *Peltandria virginica* in the absence of oxygen, *Bull. Torrey Bot. Club*, **60**: 573–582.

Elliot, K. A. C., 1932: Oxidations catalyzed by horse radish and milk-peroxidases, *Biochem. J.*, **26**: 1281–1290.

Emerson, P., 1930: A method for measuring time-rate carbon dioxide production, *J. Am. Soc. Agron.*, **22**: 819–820.

Emerson, R., 1927: The effect of certain respiratory inhibitors on the respiration of *Chlorella*, *J. Gen. Physiol.*, **10**: 469–477.

Ermakoff, A. I., and N. N. Iwanoff, 1931: Über die Atmung der Samen von Ölpflanzen, *Biochem. Z.*, **231**: 79–91.

Euler, H., and I. Bolin, 1909: Zur Kenntniss biologisch wichtiger Oxydation, *Z. physiol. Chem.*, **61**: 79–92; *Bot. Gaz.*, **49**: 395 (1910).

———, K. Myrbäck, and S. Myrbäck, 1930: Zum Bestimmen der Katalase in Pflanzenmaterial, *Z. physiol. Chem.*, **186**: 212–222.

———, U. Gard, and G. Rislund, 1931: Catalase and sugar determinations in chlorophyll-deficient plants, *Z. physiol. Chem.*, **203**: 165–177; *Chem. Abst.*, **26**: 1317.

Everest, A. E., 1914: The production of anthocyanins and anthocyanidins. I, II, *Proc. Roy. Soc.*, B, **88**: 326–332; **87**: 444–452.

Ewart, A. J., 1915: A comparative study of oxidation by catalysts of organic and inorganic origin, *Proc. Roy. Soc.*, B, **88**: 284–320.

Ezell, B. D., and J. W. Crist, 1927: Effect of certain nutrient conditions on activity of oxidase and catalase, *Mich. Sta. Tech. Bull.* 78.

Falk, K. G., G. McGuire, and E. Blout, 1919: Studies on enzyme action. XVII, The oxidase, peroxidase, catalase and amylase of fresh and dehydrated vegetables, *J. Biol. Chem.*, **38**: 229–244.

Fidler, J. C., 1933: Studies in zymasis. IV, The accumulation of zymastic products in apples during senescence, *Biochem. J.*, **27**: 1614–1621.

———, 1933: Studies in zymasis. V, Seasonal fluctuations in zymasis and in carbon dioxide/alcohol number ratios in apples in the absence of oxygen, *Biochem. J.*, **27**: 1622–1628.

———, 1934: Studies in zymasis. VII, The estimation of ethyl alcohol and acetaldehyde content of apples, *Biochem. J.*, **28**: 1107–1120.

Flemion, F., 1931: After-ripening germination and vitality of seeds of *Sorbus aucuparia*, *Contr. Boyce Thompson Inst.*, **3**: 413–439.

Fong, W. Y., and M. V. Cruess, 1929: The effect of pH value on the inactivation temperature of fruit oxidase, *Plant Physiol.*, **4**: 537–543.

——— and ———, 1929: Comparison of several indicators for fruit oxidase, *Am. J. Bot.*, **16**: 799–802.

———, 1931: The effect of pH value on the inactivation temperature of fruit oxidase, *Trans. Sci. Soc. China*, **7**: 88–94; *Chem. Abst.*, **26**: 496.

Fraymouth, J., 1928: The moisture relations of terrestrial algae. III, The respiration of certain lower plants, including terrestrial algae with special reference to the influence of drought, *Ann. Bot.*, **42**: 75–100.

French, C. S., H. I. Kohn, and P. S. Tang, 1934: Temperature characteristics for the metabolism of *Chlorella*. II, The rate of respiration of cultures of *Chlorella pyrenoidosa* as a function of time and of temperature, *J. Gen. Physiol.*, **18**: 193–207.

———, 1934: Temperature characteristics for the metabolism of *Chlorella*. III, The catalytic decomposition of hydrogen peroxide by *Chlorella pyrenoidosa*, *J. Gen. Physiol.*, **18**: 209–213.

Friedrich, R., 1908: Ueber die Stoffwechselvorgänge infolge der Verletzung von Pflanzen, *Cent. Bakt.*, II, **21**: 330–349.

Gallagher, P. H., 1923: Mechanism of oxidation in the plant. I, The oxygenase of Bach and Chodat: Function of lecithins in respiration, *Biochem. J.*, **17**: 515–529.

———, 1924: Mechanism of oxidation in the plant. II, An investigation of substances which are capable of behaving as peroxydases, *Biochem, J.*, **18**: 29–38.

———, 1924: Mechanism of oxidation in the plant. III, Peroxydase. Observations on the thermostability of the peroxydase of the mangold, *Biochem. J.*, **18**: 39–46.

Ganong, W. F., 1908: "A Laboratory Course in Plant Physiology," Henry Holt and Company, New York, 129 pp.

Garner, W. W., C. W. Bacon, J. D. Bowling, and D. E. Brown, 1934: The nitrogen nutrition of tobacco, *U. S. Dept. Agr. Tech. Bull.* 414.

GENEVOIS, L., 1927: Über Atmung und Gärung in grünen Pflanzen, *Biochem. Z.*, **186**: 461–473.

———, 1928: Über Atmung und Gärung in grünen Pflanzen. II, Der Stoffwechsel der Phanerogamen, *Biochem. Z.*, **191**: 147–158.

GERBER, C., 1900: Étude comparée de la respiration des graines oléagineuses pendant leur développement et pendant leur germination, *Actes 1st Cong. Internat. Bot. Paris*, **1900**: 59–101.

GERHARDT, F., and B. D. EZELL, 1934: Retention of carbon dioxide gas in the intercellular atmosphere of pears and apples, *Sci.*, **80**: 253–254.

GERHART, A. R., 1930: Respiration in strawberry fruits, *Bot. Gaz.*, **89**: 40–64.

GETCHELL, R. W., and J. H. WALTON, 1931: Some factors influencing the activity of peroxidase, *J. Biol. Chem.*, **91**: 419–433.

GILMAN, J. C., and D. H. BARRON, 1930: Effect of molds on temperature of stored grain, *Plant Physiol.*, **5**: 565–573.

GINSBURG, J. M., 1925: A modified respiration apparatus for plant and soil studies, *Soil Sci.*, **19**: 411–417.

GODLEWSKI, E., 1882: Beiträge zur Kenntniss der Pflanzenathmung, *Jahrb. wiss. Bot.*, **13**: 491–543.

GODWIN, H., 1935: The effect of handling on the respiration of cherry laurel leaves, *New Phytol.*, **34**: 403–406.

GONZALEZ, L. G., 1933: Influence of smudging on the respiration and catalase activity of the mango, *Mangifera indica*, *Philippine Agr.*, **21**: 533–540; *Exp. Sta. Rec.*, **70**: 482.

GORE, H. C., 1911: Studies on fruit respiration, *U. S. Bur. Chem. Bull.* 142.

———, 1914: Changes in composition of peel and pulp of ripening bananas, *J. Agr. Res.*, **3**: 187–205.

GORTNER, R. A., 1910: A contribution to the study of oxidases, *J. Chem. Soc. (London)*, **97**: 110–120.

GRĂCANIN, M., 1926: Ein Beitrag zur Kenntnis der Katalasewirkung bei autotrophen Pflanzen, *Biochem. Z.*, **168**: 429–442.

GREEN, F. C., M. E. M'ENDARFER, O. S. ORTH, and W. E. BURGE, 1929: Further study on the effect of summer and winter temperatures on the catalase of pine needles: a reply to criticism, *Proc. Roy. Irish Acad.*, **39B**: 156–159; *Biol. Abst.*, **5**: 2677.

GREEN, J. R., and A. H. JOHNSON, 1931: Effect of petroleum oils on the respiration of bean leaves, *Plant Physiol.*, **6**: 149–159.

———, 1936: Effect of petroleum oils on the respiration of bean plants, apple twigs and leaves, and barley seedlings, *Plant Physiol.*, **11**: 101–113.

GREGORY, F. G., and F. J. RICHARDS, 1929: Physiological studies in plant nutrition. I, The effect of manurial deficiency on the respiration, assimilation and assimilation rate in barley, *Ann. Bot.*, **43**: 119–161.

GRONER, M. G., 1936: Respiration of green and chlorophyll-deficient types in maize, *Am. J. Bot.*, **23**: 381–385.

GRÜSS, J., 1907: Enzymewirkungen aus Wundrand der Kartoffelknolle, *Z. Pflanzenkrank.*, **17**: 69–70.

GURJAR, A. M., 1917: The adaptation of Truog's method for the determination of carbon dioxid to plant respiration studies, *Plant World*, **20**: 288–293.

GUSTAFSON, F. G., 1929: Respiration of tomato fruits, *Plant Physiol.*, **4**: 349–357.

———, 1930: Intramolecular respiration of tomato fruits, *Am. J. Bot.*, **17**: 1011–1028.

———, 1932: Anaerobic respiration of cacti, *Am. J. Bot.*, **19**: 823–834.

———, I. CLARK, D. A. SHAW, and E. WARWEG, 1932: Catalase activity in tomato fruits at different stages of their development, *Plant Physiol.*, **7**: 155–164.

———, 1934: Production of alcohol and acetaldehyde by tomatoes, *Plant Physiol.*, **9**: 359–367.

———, 1934: Äthylalkohol und Acetaldehyd in gewissen Arten von Kakteen, *Biochem. Z.*, **272**: 172–179.

———, 1936: Influence of oxygen and carbon dioxide concentrations on the respiration of tomato fruits, *Am. J. Bot.*, **23**: 441–445.

GUTHRIE, J. D., 1930: An iodimetric method for determining oxidase activity, *J. Am. Chem. Soc.*, **52**: 3614–3618.

———, 1931: A method for the determination of peroxidase activity, *J. Am. Chem. Soc.*, **53**: 242–244

———, 1931: The inhibiting effect of oxidase on the reduction of sulphur by potato and gladiolus juice, *Contr. Boyce Thompson Inst.*, **3**: 125–130.

———, 1931: The effect of various chemical treatments of dormant potato tubers on the peroxidase, catalase, pH and reducing properties of the expressed juice, *Contr. Boyce Thompson Inst.*, **3**: 499–508.

———, F. E. DENNY, and L. P. MILLER, 1932: Effect of ethylene chlorhydrin treatments on the catalase, peroxidase, pH and sulphydryl content of Gladiolus corms, *Contr. Boyce Thompson Inst.*, **4**: 131–140.

HAAS, A. R. C., 1916: A simple and rapid method of studying respiration by the detection of exceedingly minute quantities of carbon dioxide, *Sci.*, **44**: 105–108.

———, 1917: Rapid respiration after death, *Proc. Nat. Acad. Sci.*, **3**: 688–691.

———, 1917: Anesthesia and respiration, *Sci.*, **46**: 462–464.

———, 1919: Respiration after death, *Bot. Gaz.*, **67**: 347–366.

———, 1919: Effect of anesthetics upon respiration, *Bot. Gaz.*, **67**: 377–405.

———, 1930: Oxidation-reduction indicators as a means of determining overheating in walnuts during dehydration, *Bot. Gaz.*, **89**: 200–204.

HABER, E. S., 1928: Catalase and oxidase of the tomato as influenced by the soil reaction, *Iowa State Col. J. Sci.*, **3**: 29–39; *Bot. Abst.*, **4**: 49.

HAFENRICHTER, A. L., 1928: Respiration of the soybean, *Bot. Gaz.*, **85**: 271–299.

HALLER, M. H., P. L. HARDING, J. M. LUTZ, and D. H. ROSE, 1931: The respiration of some fruits in relation to temperature, *Proc. Am. Soc. Hort. Sci.*, **28**: 583–589.

———, ———, and D. H. ROSE, 1932: The interrelation of firmness, dry weight and respiration in strawberries, *Proc. Am. Soc. Hort. Sci.*, **29**: 330–334.

HANES, C. S., and J. BARKER, 1931: The physiological action of cyanide. I, The effects of cyanide on the respiration and sugar content of the potato at 15°C., *Proc. Roy. Soc.*, B, **108**: 95–119.

HARDING, P. L., 1929: Respiration studies of Grimes apples under various controlled temperatures, *Proc. Am. Soc. Hort. Sci.*, **26**: 319–324.

———, T. J. MANEY, and H. H. PLAGGE, 1929: Apparatus for the determination of carbon dioxide in the respiration of apples, *Sci.*, **70**: 125–126.

———, 1930: Relation of catalase activity to temperature, respiration and nitrogen fertilization of Grime's Golden apples, *Proc. Am. Soc. Hort. Sci.*, **27**: 37–42.

———, 1935: Physiological behavior of Grimes Golden apples in storage, *Iowa Sta. Tech. Bull.* 182.

HARLEY, C. P., 1929: Relation of picking time to acetaldehyde content and core breakdown of Bartlett pears, *J. Agr. Res.*, **39**: 483–493.

HARRINGTON, G. T., 1923: Respiration of apple seeds, *J. Agr. Res.*, **23**: 117–130.

——— and W. CROCKER, 1923: A new and efficient respirometer for seeds and other small objects: Directions for its use, *J. Agr. Res.*, **23**: 101–115.

HARRIS, G. H., 1929: Studies on tree root activities. I, An apparatus for studying root respiration and factors which influenced it, *Sci. Agr.*, **9**: 553–565.

———, 1930: The influence of top on root as determined by root respiration of young fruit trees, *Proc. Am. Soc. Hort. Sci.*, **26**: 329–334.

———, 1930: Studies on tree root activities. II, Some factors which influence tree root respiration, *Sci. Agr.*, **10**: 564–585.

———, 1930: Studies on tree root activities. III, *Sci. Agr.*, **11**: 191–199.

HARVEY, R. B., 1919: Apparatus for the measurement of oxidase and catalase activity, *J. Gen. Physiol.*, **2**: 253–254.

———, 1920: Hydrogen ion changes in the mosaic disease of tobacco plants and their relation to catalase, *J. Biol. Chem.*, **42**: 397–400.

———, 1920: Relation of catalase, oxidase and H concentrations to the formation of overgrowths, *Am. J. Bot.*, **7**: 211–223.

———, 1924: Enzymes of thermal algae, *Sci.*, **60**: 481–482.

HASSELBRING, H., and L. ALSBERG, 1910: Studies upon oxidases, *Sci.*, **31**: 637.

——— and L. A. HAWKINS, 1915: Respiration experiments with sweet potatoes, *J. Agr. Res.*, **5**: 509–519.

HAUT, I. C., 1932: Catalase activity in relation to the after ripening of fruit tree seeds, *Proc. Am. Soc. Hort. Sci.*, **29**: 375–379.

HEINICKE, A. J., 1922: Catalase activity as an indicator of nutritive condition of fruit tree tissues, *Proc. Am. Soc. Hort. Sci.*, **19**: 209–214.

———, 1923: Factors influencing catalase activity in apple leaf tissue, *N. Y. Sta. (Cornell) Mem.* 62.

———, 1924: Catalase, activity in dormant apple twigs: its relation to the condition of the tissue, respiration and other factors, *N. Y. Sta. (Cornell) Mem.* 74.

———, 1933: Apparatus for making autographic records of catalase activity of plant tissues, and the procedure involved, *J. Agr. Res.*, **46**: 1137–1143.

HERZFELD, E., and R. KLINGER, 1919: Chemische Studien zur Physiologie und Pathologie. VI, Zur Biochemie der Oxydationen. (Zellatmung; Oxydationsfermente; zur Theorie der Narkose), *Biochem. Z.*, **93**: 324–352; *Bot. Abst.*, **8**: 95 (1921).

HILL, G. R., 1913: Respiration of fruits and growing plant tissues in certain gases, with reference to ventilation and fruit storage. *N. Y. Sta. (Cornell) Bull.* 330.

HOFFMAN, E. J., 1935: Oxidation and gas formation in the spontaneous heating of hay, *J. Agr. Res.*, **51**: 527–546.

HOFFMAN, M. B., 1935: The effect of lime sulphur spray on the respiration rate of apple leaves, *Proc. Am. Soc. Hort. Sci.*, **33**: 173–176.

HOPKINS, E. F., 1924: Relation of low temperatures to respiration and carbohydrate changes in potato tubers, *Bot. Gaz.*, **78**: 311–326.

———, 1927: Variations in sugar content in potato tubers caused by wounding and its possible relation to respiration, *Bot. Gaz.*, **84**: 75–88.

HOTTES, C. F., and A. L. HAFENRICHTER, 1928: A constant rate aspirator, *Sci.*, **67**: 320–321.

HOVER, J. M., and F. G. GUSTAFSON, 1926: Rate of respiration as related to age, *J. Gen. Physiol.*, **10**: 33–39.

HUTCHINSON, D. M., 1922: Comparative studies on respiration. XXIII, The effect of adrenalin on the production of CO_2 by animals and by plants, *Am. J. Physiol.*, **62**: 192–196.

ILJIN, W. S., 1923: Einfluss des Welkens auf die Atmung der Pflanzen, *Flora*, **116**: 379–403.

INAMDAR, R. S., and B. N. SINGH, 1927: Studies in the respiration of tropical plants. I, Seasonal variations in aerobic and anaerobic respiration in the leaves of *Artocarpus integrifolia*, *J. Indian Bot. Soc.*, **6**: 133–212; *Exp. Sta. Rec.*, **61**: 816–817.

INMAN, O. L., 1921: Comparative studies on respiration. XVI, Effects of hypotonic and hypertonic solutions upon respiration, *J. Gen. Physiol.*, **3**: 533–539.

———, 1921: Comparative studies on respiration. XVII, Decreased respiration and recovery, *J. Gen. Physiol.*, **3**: 663–667.

IRVING, A. A., 1911: The effect of chloroform upon respiration and assimilation, *Ann. Bot.*, **25**: 1077–1099.

IVANOW, S. L., 1913: Der Einfluss von Oelvorräten der Samen und der Temperatur auf den Respirationscoefficient CO_2/O_2, *Russ. J. Exp. Landw.*, **14**: 75–87.

JACQUOT, R., and A. MAYER, 1926: Hydration et respiration des graines, *Ann. Physiol. Physiochem. Biol.*, **2**: 408–425; *Biol. Abst.*, **1**: 611–612.

JEFFREY, R. N., and M. V. CRUESS, 1933: Gasometric method of estimating oxidase activity, *Plant Physiol.*, **8**: 327–331.

JOHANSSON, N., 1933: The relation between the respiration of the tree stem and its growth, *Biol. Abst.*, **8**: 50.

JOHNSTONE, G. R., 1925: Effect of wounding on respiration and the exchange of gases, *Bot. Gaz.*, **79**: 339–340.

———, 1925: Physiological studies of two varieties of *Ipomoea batatas*, *Bot. Gaz.*, **80**: 145–167.

JONES, H. A., 1920: Physiological study of maple seeds, *Bot. Gaz.*, **69**: 127–152.

JONES, J. W., 1933: Effect of reduced oxygen pressure on rice germination, *J. Am. Soc. Agron.*, **25**: 69–81.

JOSLYN, M. A., and G. L. MARSH, 1933: The role of peroxidase in the deterioration of frozen fruits and vegetables, *Sci.*, **78**: 174–175.

KARCHEVSKI, M., 1903: Influence of the fluctuation of temperature on the respiration of seeds and embryos of wheat, *Exp. Sta. Rec.*, **14**: 839.

KARLSON, A., 1925: Comparative studies on respiration. XXVIII, The effects of anesthetics on the production of carbon dioxide by wheat under aerobic and anaerobic conditions, *Am. J. Bot.*, **12**: 619–624.

KASTLE, J. H., and A. S. LOVENHART, 1901: On the nature of certain of the oxidizing ferments, *Am. Chem. J.*, **26**: 539–566.

———, 1910: The oxidases and other oxygen-catalysts concerned in biological oxidations, *Public Health and Marine Hospital Service Hygienic Lab. Bull.* 59.

——— and G. D. BUCKNER, 1917: Evidence of the action of oxidases within the growing plant, *J. Am. Chem. Soc.*, **39**: 478–482.

KEEBLE, F., and E. F. ARMSTRONG, 1912: The distribution of oxidases in plants and their role in the formation of pigments, *Proc. Roy. Soc.*, B, **85**: 214–218.

———, ———, and W. N. JONES, 1913: The formation of anthocyan pigments of plants. IV, The chromogens, *Proc. Roy. Soc.*, B, **86**: 308–317.

———, ———, and ———, 1914: The formation of anthocyan pigments of plants. VI, *Proc. Roy. Soc.*, B, **87**: 113–131.

KELLEY, V. W., 1930: Effect of certain hydrocarbon oils on respiration of foliage and dormant twigs of the apple, *Ill. Sta. Bull.* 348.

KERTESZ, Z. I., 1933: The oxidase system of a non-bearing yellow peach, *N. Y. Sta. (Geneva) Tech. Bull.* 219.

———, 1933: Some carbohydrate changes in shelled green peas, *N. Y. Sta. (Geneva) Bull.* 622.

———, 1933: Inactivating the respiration of cannery peas by heat, *The Canner*, Feb. 25, 1933.

———, 1934: The browning of yellow peaches. Enzyme studies reveal why yellow peach varieties show different extent of darkening, *Fruit Products J.*, **13**: 304–306.

———, 1934: Carbon dioxide content of the gas from pea pods, *Plant Physiol.*, **9**: 339–350.

Kidd, F., 1917: The controlling influence of carbon dioxide. III, The retarding effect of carbon dioxide on respiration, *Proc. Roy. Soc.*, B, **89**: 136–157.

———, C. West, and G. E. Briggs, 1921: A quantitative analysis of the growth of *Helianthus annuus*. I, The respiration of the plant and of its parts throughout the life cycle, *Proc. Roy. Soc.*, B, **92**: 368–384.

——— and ———, 1929: Physiology of fruit. I, Changes in the respiratory activity of apples during their senescence at different temperatures, *Proc. Roy. Soc.*, B, **106**: 93–109.

Kimbrough, W. D., 1924: The relation of respiration to storage and transportation of potatoes, *Proc. Ann. Meeting Potato Assoc. Am.*, **11**: 53–54.

———, 1925: A study of respiration in potatoes with special reference to storage and transportation, *Md. Sta. Bull.* 276.

Knight, R. C., and J. H. Priestley, 1914: The respiration of plants under various electrical conditions, *Ann. Bot.*, **28**: 135–163.

Knott, J. E., 1925: Some conditions influencing the determination of catalase activity in plant tissue, *Proc. Am. Soc. Hort. Sci.*, **22**: 398–407.

———, 1927: Catalase in relation to growth and to other changes in plant tissue, *N. Y. Sta. (Cornell) Mem.* 106.

Knudson, L., 1916: Influence of certain carbohydrates on green plants, *N. Y. Sta. (Cornell) Mem.* 9.

Kolkwitz, R., 1901: Ueber die Athmung ruhender Samen, *Ber. bot. Ges.*, **19**: 285–287.

Kostytschew, S., 1908: Über die Anteilnahme der Zymase am Atmungsprozesse der Samenpflanzen, *Biochem. Z.*, **15**: 164–195.

———, 1910: Über den Vorgang der Zuckeroxydation bei der Pflanzenatmung, *Z. physiol. Chem.*, **67**: 116–137.

———, 1927: "Plant Respiration." English translation by C. J. Lyons. P. Blakiston's Son & Company, Philadelphia.

Lall, G., 1934: Chemical studies in the physiology of apples. XIV, A method of estimating chemical change and rate of respiration in stored apples, *Ann. Bot.*, **48**: 273–292.

Landon, R. H., 1934: The effect of certain chemicals on the catalase activity in plants, *Am. J. Bot.*, **21**: 583–591.

Langdon, S. C., and W. R. Gailey, 1920: Carbon monoxide a respiration product of *Nereocystis leutkeana*, *Bot. Gaz.*, **70**: 230–239.

Lantz, C. W., 1927: Respiration in corn with special reference to catalase, *Am. J. Bot.*, **14**: 85–105.

———, 1930: Temperature and catalase activity in germinating corn, *Proc. Iowa Acad. Sci.*, **35**: 147–149.

Larmour, R. K., J. S. Clayton, and C. L. Wrenshall, 1935: A study of respiration and heating of damp wheat, *Can. J. Res.*, **12**: 627–645.

Leach, W., 1932: Further experimental methods in connexion with the use of the katharometer for the measurement of respiration, *Ann. Bot.*, **46**: 583–596.

——— and K. W. Kent, 1934: Researches on plant respiration. III, The relationship between the respiration in air and in nitrogen of certain seeds during germination. (a) Seeds in which fats constitute the chief food reserve, *Proc. Roy. Soc.*, B, **116**: 150–169.

———, 1936: Researches on plant respiration. IV, The relation between the respiration in air and in nitrogen of certain seeds during germination. (b) Seeds in which carbohydrates constitute the chief food reserve, *Proc. Roy. Soc.*, B, **119**: 507–521.

Leggatt, C. W., 1929: Catalase activity as a measure of seed viability, *Sci. Agr.*, **10**: 73–110.

———, 1933: A further note on catalase activity as a measure of seed vitality, *Can. J. Res.*, **9**: 571–573.

LeVan, W. C., 1930: The effect of metals on the respiration of *Lupinus albus*, *Am. J. Bot.*, **17**: 381–395.

Linossier, M. G., 1898: Contribution à l'étude des ferments oxydants. Sur la peroxydase du pus, *C. R. Soc. Biol.*, **50**: 373–375.

Lipman, C. B., 1936: Normal viability of seeds and bacterial spores after exposure to temperatures near the absolute zero, *Plant Physiol.*, **11**: 201–205.

Livingston, B. E., 1933: A computation concerning the importance of respiration water to young wheat seedlings, abst., *Am. J. Bot.*, **20**: 671.

———, 1934: Possible importance of respiration water to young wheat seedlings, *Bull. Torrey Bot. Club*, **61**: 217–223.

Loehwing, W. F., 1934: Physiological aspects of the effect of continuous soil aeration on plant growth, *Plant Physiol.*, **9**: 567–583.

Loew, O., 1901: Catalase, a new enzyme of general occurrence, with special reference to the tobacco plant, *U. S. Dept. Agr. Rept.* 68.

Long, E. R., 1917: Desiccation and respiration in succulent plants, *Carnegie Inst. Wash. Year Book*, **16**: 80–82.

———, 1918: Further results in desiccation and respiration of *Echinocactus*, *Bot. Gaz.*, **65**: 354–358.

Lopriore, G., 1928: Die Katalase-Reaktion und die Biologie des Pollens, *Ber. bot. Ges.*, **46**: 413–423.

Lund, E. J., 1919: A simple method for measuring carbon dioxide produced by small organisms, *Biol. Bull.*, **36**: 105–114.

Luthra, J. C., 1924: Some experiments on the effects of dry and moist air on the rate of respiration and breakdown of ripe pears, *New Phytol.*, **23**: 131–142.

——— and I. S. Chima, 1931: Some studies in respiration and other metabolic activities in berries of the grape vine, *Indian J. Agr. Sci.*, **1**: 695–714; *Exp. Sta. Rec.*, **67**: 260.

Lutman, B. F., 1926: Respiration of potato tubers after injury, *Bull. Torrey Bot. Club*, **53**: 429–457.

Lyon, C. J., 1921: Comparative studies on respiration. XVIII, Respiration and antagonism in *Elodea*, *Am. J. Bot.*, **8**: 458–464.

———, 1924: The effect of phosphates on respiration, *J. Gen. Physiol.*, **6**: 299–306.

———, 1927: Phosphate ion as a promoter catalyst of respiration, *J. Gen. Physiol.*, **10**: 599–615.

———, 1927: Arsenate as a catalyst of oxidation, *J. Gen. Physiol.*, **10**: 617–622.

———, 1927: The role of phosphate in plant respiration, *Am. J. Bot.*, **14**: 274–284.

Lyon, M. E., 1928: The occurrence and behavior of embryoless wheat seeds, *J. Agr. Res.*, **36**: 631–637.

MacDougal, D. T., E. R. Long, and J. G. Brown, 1915: End results of desiccation and respiration in succulent plants, *Physiol. Res.*, **1**: 289–327.

Mack, W. B., 1930: The relation of temperature and partial pressure of oxygen to respiration and growth in germinating wheat, *Plant Physiol.*, **5**: 1–69.

Magness, J. R., 1920: Composition of gases in intercellular spaces of apples and potatoes, *Bot. Gaz.*, **70**: 308–316.

——— and H. C. Diehl, 1924: Physiological studies on apples in storage, *J. Agr. Res.*, **27**: 1–38.

——— and W. S. Ballard, 1926: The respiration of Bartlett pears, *J. Agr. Res.*, **32**: 801–833.

MAIGE, A., and G. NICOLAS, 1910: Recherches sur l'influence des variations de la turgescence sur la respiration de la cellule, *Rev. Gén. Bot.*, **22**: 409–422.

——— and ———, 1910: Recherches sur l'influence des solutions sucrées de divers degrés de concentration sur la respiration, la turgescence et al croissance de la cellule, *Ann. Sci. Nat. Bot.*, 9e ser., **12**: 315–368.

MALPIGHI, M., 1679: "Anatomes plantarum pars Altera," Londini.

MANEY, T. J., P. L. HARDING, and H. H. PLAGGE, 1929: A new type of respiration chamber, *Sci.*, **70**: 44.

MANGENOT, G., 1928: Sur la localisation cytologique des peroxydases et des oxydases, *C. R. Acad. Sci. Paris*, **186**: 710–712.

MAQUENNE, L., and E. DEMOUSSY, 1913: Sur la valeur des coefficients chlorophylliens et leur rapports avec les quotients respiratoires réels, *C. R. Acad. Sci. Paris*, **156**: 506–512.

——— and ———, 1921: Sur la respiration des feuilles dans le vide ou des atmosphères pauvres en oxygène, *C. R. Acad. Sci. Paris*, **173**: 373–377.

MARKS, G. M., 1935: The inactivation of catalases from certain marine plants by oxygen, *Biochem. J.*, **29**: 509–512.

MARSH, G., 1935: Kinetics of an intracellular system for respiration and bioelectric potential at flux equilibrium, *Plant Physiol.*, **10**: 681–698.

MATTHAEI, G. L. C., 1905: Experimental researches on vegetable assimilation and respiration. III, The effect of temperature on carbon dioxide assimilation, *Phil. Trans. Roy. Soc.*, B, **197**: 47–86.

McGINNIS, F. W., and G. S. TAYLOR, 1923: The effect of respiration upon the protein percentage of wheat, oats and barley, *J. Agr. Res.*, **24**: 1041–1048.

McHARGUE, J. S., 1920: The significance of the peroxidase reaction with reference to the viability of seeds, *J. Am. Chem. Soc.*, **42**: 612–615.

McKENZIE, K. A., 1932: Respiration studies with lettuce, *Proc. Am. Soc. Hort. Sci.*, **28**: 244–248.

MELDRUM, N. U., 1933: "Cellular respiration," Methuen and Company, Ltd., London, 116 pp.

MEYER, A., 1918: Eiweissstoffwechsel und Vergilben der Laubblätter von *Tropaeolum majus*, *Flora*, **11–12**: 85–127.

MICHAELS, W. H., 1931: Respiration of the shoot as affected by temperature changes of the root, *Bot. Gaz.*, **91**: 167–182.

———, 1932: Relation of lenticels and surface area to respiration in the potato tuber, *Bot. Gaz.*, **94**: 416–418.

MIDDLETON, N. I., 1927: The effect of ionized air on the rate of respiration of barley seedlings, *Ann. Bot.*, **41**: 345–356.

MILLER, E. C., 1910: A physiological study of the germination of *Helianthus annuus*, *Ann. Bot.*, **24**: 693–726.

MILLER, E. R., 1929: The insoluble tyrosinase of the velvet bean seed coat, *Plant Physiol.*, **4**: 507–517.

MILLER, L. P., 1932: Effect of various chemicals on the respiratory rate and sugar content of potato tubers, abst., *Am. J. Bot.*, **19**: 840.

———, 1933: Effect of various chemicals on the sugar content, respiratory rate and dormancy of potato tubers, *Contr. Boyce Thompson Inst.*, **5**: 213–234.

———, 1934: Time relations in effect of ethylene chlorohydrin in increasing and of ethyl alcohol in decreasing the respiration of potato tubers, *Contr. Boyce Thompson Inst.*, **6**: 123–128.

———, 1934: The effect of the alkyl halides on the respiration of potato tubers, *Contr. Boyce Thompson Inst.*, **6**: 279–296.

———, J. D. Guthrie, and F. E. Denny, 1934: Induced changes in respiration rates and time relations in the changes in internal factors, *Contr. Boyce Thompson Inst.*, **8**: 41–62.

———, 1935: Further experiments on the effect of halogenated alipathic compounds on the respiration of potato tubers, *Contr. Boyce Thompson Inst.*, **7**: 1–18.

Mitchell, J. W., 1935: A method of measuring respiration and carbon fixation of plants under controlled environmental conditions, *Bot. Gaz.*, **97**: 376–387.

Moore, B., and E. Whitley, 1909: The properties and classification of the oxidizing enzymes and analogies between enzymic activity and the effects of immune bodies and complements, *Biochem. J.*, **4**: 136–168.

Moore, W., and J. J. Willaman, 1917: Studies in greenhouse fumigation with hydrocyanic acid: physiological effects on the plant, *J. Agr. Res.*, **11**: 319–338.

Morgulis, S., M. Beber, and I. Rabkin, 1926: Studies on the effect of temperature on the catalase reaction. I, Effect of different hydrogen peroxide concentrations. II, Loss of catalase activity. III, Temperature effects at different hydrogen ion concentrations. IV, A theory of catalase reaction, *J. Biol. Chem.*, **68**: 521–563.

Morinaga, T., 1925: Catalase activity and the aerobic and anaerobic germination of rice, *Bot. Gaz.*, **79**: 73–85.

Morse, F. W., 1908: The effect of temperature on the respiration of apples, *J. Am. Chem. Soc.*, **30**: 876–881.

Müller-Thurgau, H., 1882: Ueber Zuckeranhäufung in Pflanzentheilen in Folge niederer Temperatur, *Landw. Jahrb.*, **11**: 751–828.

Navez, A. E., 1929: Respiration and geotropism in *Vicia faba*. I, *J. Gen. Physiol.*, **12**: 641–667.

Neller, J. R., 1931: Effect of chlorates upon the catalase activity of the roots of bindweed, *J. Agr. Res.*, **43**: 183–189.

———, 1931: Relation of catalase activity to physiological breakdown in Jonathan apples, *Plant Physiol.*, **6**: 347–354.

Nemec, A., and F. Duchon, 1921: Peut-on déterminer la valeur des semences par voie biochimique? *C. R. Acad. Sci. Paris*, **173**: 933–935.

——— and ———, 1922: Sur une méthode indicatrice permettant d'évaluer la vitalité des semences par voie biochemique, *C. R. Acad. Sci. Paris*, **174**: 632–634.

——— and ———, 1923: Sur une nouvelle méthode biochimique pour le détermination de la force vitale des semences, *Ann. Sci. Agron., Franc., et Etrangère*, **40**: 121–150.

Newcombe, F. C., and E. A. Bowerman, 1918: Behavior of plants in unventilated chambers, *Am. J. Bot.*, **5**: 284–295.

Newton, R., and W. R. Brown, 1931: Catalase activity of wheat leaf juice in relation to frost resistance, *Can. J. Res.*, **5**: 333–336.

——— and J. A. Anderson, 1931: Respiration of winter wheat plants at low temperatures, *Can. J. Res.*, **5**: 337–354.

Nicolas, G., 1918: Contribution à l'étude des variations de la respiration des végétaux avec l'âge, *Rev. Gén. Bot.*, **30**: 214–225.

Olney, A. J., 1926: Temperature and respiration of ripening of bananas, *Bot. Gaz.*, **82**: 415–462.

Onslow, M. W., 1919: Oxidizing enzymes. I, The nature of the "peroxid" naturally associated with certain direct oxidizing systems in plants, *Biochem. J.*, **13**: 1–9.

———, 1920: Oxidising enzymes. II, The nature of the enzymes associated with certain direct oxidising systems in plants, *Biochem. J.*, **14**: 535–540.

———, 1920: Oxidising enzymes. III, The oxidising enzymes of some common fruits, *Biochem. J.*, **14**: 541–547.

———, 1921: Oxidising enzymes. IV, The distribution of oxidising enzymes among the higher plants, *Biochem. J.*, **15**: 107–113.

———, 1921: Oxidising enzymes. V, Further observations on the oxidising enzymes of fruits, *Biochem. J.*, **15**: 113–118.

———, 1923: Oxidising enzymes. VI, A note on tyrosinase, *Biochem. J.*, **17**: 216–220.

———, 1924: The oxidising enzymes. VII, The oxygenase of the higher plants, *Biochem. J.*, **18**: 549.

——— and M. E. Robinson, 1925: Oxidising enzymes. VIII, The oxidation of certain parahydroxy-compounds by plant enzymes and its connection with "tyrosinase," *Biochem. J.*, **19**: 420–423.

——— and ———, 1928: Oxidizing enzymes. X, The relationship of oxygenase to tyrosinase, *Biochem. J.*, **22**: 1327–1332.

Oparin, A., and N. Djatschkow, 1928: Über die Fermentbildung in reifenden Samen, *Biochem. Z.*, **196**: 289–293.

Oppenheimer, C., 1913: *Die Fermente*, **11**: 754.

Osterhout, W. J. V., and A. R. C. Haas, 1917: An adaptation of Winkler's method to biological work, *J. Biol. Chem.*, **32**: 141–146.

———, 1917: The rôle of the nucleus in oxidation, *Sci.*, **46**: 367–369.

———, 1918: A method of studying respiration, *J. Gen. Physiol.*, **1**: 17–22.

———, 1918: An indicator method of measuring the consumption of oxygen, *J. Gen. Physiol.*, **1**: 167–171.

———, 1918: Comparative studies on respiration. I, Introduction, *J. Gen. Physiol.*, **1**: 171–181.

Ota, J., 1925: Continuous respiration studies of dormant seeds of *Xanthium*, *Bot. Gaz.*, **80**: 288–299.

Overholser, E. L., and W. V. Cruess, 1923: A study of the darkening of apple tissue, *Calif. Sta. Tech. Paper* 7.

———, 1928: A study of the catalase of the fruits of pear varieties, *Am. J. Bot.*, **15**: 285–307.

——— and L. L. Claypool, 1931: The relation of fertilizers to respiration and certain physical properties of strawberries, *Proc. Am. Soc. Hort. Sci.*, **28**: 220–224.

———, M. B. Hardy, and H. D. Locklin, 1931: Respiration studies of strawberries, *Plant Physiol.*, **6**: 549–558.

Pal, N. L., 1934: Studies on the respiration of conjugating *Spirogyra* with special reference to fat metabolism, *New Phytol.*, **33**: 241–273.

——— and U. N. Chatterji, 1936: Effect of insulin on plant respiration, *Nature*, **137**: 535–536.

Palladin, W., 1888: Ueber Eiweisszersetzung in den Pflanzen bei Abwesenheit von freiem Sauerstoff, *Ber. bot. Ges.*, **6**: 205–212.

———, 1888: Ueber Zersetzungsproducte der Eiweissstoffe in den Pflanzen bei Abwesenheit von freiem Sauerstoff, *Ber. bot. Ges.*, **6**: 296–304.

———, 1893: Recherches sur la respiration des feuilles vertes et des feuilles étiolées, *Rev. Gén. Bot.*, **5**: 449–473.

———, 1894: Sur le rôle des hydrates de carbone dans la résistance à l'asphyxie chez les plantes supérieures, *Rev. Gén. Bot.*, **6**: 201–209.

———, 1896: Recherches sur la corrélation entre la respiration des plantes et les substances azotées actives, *Rev. Gén. Bot.*, **8**: 225–248.

———, 1899: Influence de la lumière sur la formation des matières protéiques actives et sur l'énergie de la respiration des parties vertes des végétaux, *Rev. Gén. Bot.*, **11**: 81–105.

———, 1899: L'influence des changements de température sur la respiration des plantes, *Rev. Gén. Bot.*, **11**: 241–247.

——— and A. Komleff, 1902: Influence de la concentration des solutions sur l'énergie respiratoire et sur la transformation des substances dans les plantes, *Rev. Gén. Bot.*, **14**: 497–516.

——— and S. Kostytschew, 1906: Anaerobe Athmung, Alkoholgärung und Acetonbildung bei den Samenpflanzen, *Z. physiol. Chem.*, **48**: 214–239.

———, 1909: The theory of respiration of plants, *Biochem. Z.*, **18**: 151–206; *Exp. Sta. Rec.*, **21**: 532.

——— and A. M. Sheloumova, 1918: Effect of loss of water on respiration of plants, *Bull. Acad. Sci. Russ.*, 6 ser., No. 8, 801–808 pp.; *Exp. Sta. Rec.*, **39**: 731.

———, 1922: "Plant Physiology." English translation by Livingston. P. Blakiston's Son & Company, Philadelphia.

Parija, P., 1928: Analytic studies in plant respiration. II, The respiration of apples in nitrogen and its relation to respiration in air, *Proc. Roy. Soc.*, B, **103**: 446–491.

——— and A. B. Saran, 1934: The effect of light on the respiration of starved leaves, *Ann. Bot.*, **48**: 347–354.

Peirce, G. J., 1908: A new respiration calorimeter, *Bot. Gaz.*, **46**: 193–202.

———, 1909: What is the use of respiration? *Plant World*, **12**: 193–198.

———, 1912: The liberation of heat in respiration, *Bot. Gaz.*, **53**: 89–112.

Pfeffer, W., 1881–1885: Ueber intramoleculare Athmung, *Untersuch. Bot. Inst. Tübingen*, **1**: 636–685.

Pope, M. N., 1932: Catalase activity in relation to the growth curve in barley, *J. Agr. Res.*, **44**: 343–355.

———, 1933: Catalase activity and respiration in the leaves of growing barley, *J. Agr. Res.*, **46**: 35–41.

Portsmouth, G. B., 1934: The effect of ionized air on the respiration of green plants, *Ann. Bot.*, **48**: 1013–1027.

Potter, M. C., 1917: Note on a method of demonstrating the heat of respiration, *Ann. Bot.*, **31**: 435–438.

Ranjan, S., and A. K. Mallik, 1931: A study of the catalase reaction, with special reference to respiration in plants, *New Phytol.*, **30**: 355–381.

Ransom, E. R., 1935: The inter-relations of catalase, respiration, after-ripening and germination in some dormant seeds of the *Polygonaceae*, *Am. J. Bot.*, **22**: 815–825.

Ray, G. B., 1923: Comparative studies on respiration. XXIV, The effects of chloroform on the respiration of dead and living tissue, *J. Gen. Physiol.*, **5**: 469–477.

———, 1923: Comparative studies on respiration. XXV, The action of chloroform on the oxidation of some organic acids, *J. Gen. Physiol.*, **5**: 611–623.

———, 1923: Comparative studies on respiration. XXVI, The production of CO_2 from organic acids in relation to their iodine absorption, *J. Gen. Physiol.*, **5**: 623–629.

———, 1923: Comparative studies on respiration. XXVII, The mechanism of oxidation in relation to chloroform anesthesia, *J. Gen. Physiol.*, **5**: 741–749.

Reed, G. B., 1914: The oxidases of acid tissue, *Bot. Gaz.*, **57**: 528–530.

———, 1915: Evidence for the general distribution of oxidases in plants, *Bot. Gaz.*, **59**: 407–409.

———, 1916: The significance of color changes in oxidase reagents, *Bot. Gaz.*, **61**: 430–432.

———, 1916: The measurement of oxidation potential and its significance in the study of oxidases, *Bot. Gaz.*, **61**: 523–527.

———, 1916: On the mechanism of oxidase action, *Bot. Gaz.*, **62**: 53–64.

———, 1916: The mode of action of plant peroxidases, *Bot. Gaz.*, **62**: 233–238.

———, 1916: The separation of oxidase reactions from the catalase reaction, *Bot. Gaz.*, **62**: 303–310.

———, 1916: The relation between oxidase and catalase in plant tissues, *Bot. Gaz.*, **62**: 409–412.

———, 1916: The relation of oxidase reactions to changes in hydrogen ion concentration, *J. Biol. Chem.*, **27**: 299–302.

REGEIMBAL, L. O., G. A. VACHA, and R. B. HARVEY, 1927: The effect of ethylene on the respiration of bananas during ripening, *Plant Physiol.*, **2**: 357–359.

RHINE, L. E., 1924: Divergence of catalase and respiration in germination, *Bot. Gaz.*, **78**: 46–67.

RICHARDS, H. M., 1896: The respiration of wounded plants, *Ann. Bot.*, **10**: 531–582.

———, 1897: The evolution of heat by wounded plants, *Ann. Bot.*, **11**: 29–63.

———, 1915: Acidity and gas interchange in cacti, *Carnegie Inst. Wash. Pub.* 209.

RICHTER, A. A., E. I. DVORETSKAIA, and A. I. GRECHUSHNIKOV, 1929: Perseverance of cultivated plants. I, Transpiration and katabolism of organisms infected with fungous diseases, *J. Exp. Landw. Südosten Eur. Russlands* (*Russian*), **7**: 99–114; *Biol. Abst.*, **6**: 966.

RIGG, G. B., and B. S. HENRY, 1935: On the origin of the gases in the float of bladder kelp, *Am. J. Bot.*, **22**: 362–365.

RISCHAWI, L., 1876: Einige Versuche über die Athmung der Pflanzen, *Landw. Versuchs-Stat.*, **19**: 321–340.

ROBBINS, W. J., 1929: A study of the metabolism of roots, *Mo. Sta. Bull.* 272.

ROBERTS, R. H., and J. E. KRAUS, 1934: Respiratory types and photoperiodism, *Sci.*, **80**: 122–123.

ROBINSON, M. E., 1924: A comparison of certain oxidizing enzymes of the higher and lower plants, *Biochem. J.*, **18**:,543–548.

ROHMANN, E., and W. SPITZER, 1895: Ueber Oxydationswirkungen thierischer Gewebe, *Ber. chem. Ges.*, **28**: 567–572.

ROSE, D. H., 1915: Oxidation in healthy and diseased apple bark, *Bot. Gaz.*, **60**: 55–65.

———, 1919: Blister canker of apple trees: a physiological and chemical study, *Bot. Gaz.*, **67**: 105–146.

———, H. R. KRAYBILL, and R. C. ROSE, 1920: Effect of salts upon oxidase activity of apple bark, *Bot. Gaz.*, **69**: 218–236.

ROSENBERG, A., 1910: Über die Rolle der Katalase in den Pflanzen, *Ber. bot. Ges.*, **28**: 280–288.

ROSENE, H. F., and E. J. LUND, 1935: Linkage between output of electric energy by polar tissue and cell oxidation, *Plant Physiol.*, **10**: 27–47.

———, 1935: Proof of the principle of summation of cells E. M. F., *Plant Physiol.*, **10**: 209–224.

SACHS, J., 1882: "Vorlesungen über Pflanzenphysiologie," Leipzig.

SAMISCH, R., 1935: The location of oxidase in the apricot, *Am. J. Bot.*, **22**: 291–293.

SAPOZHNIKOVA, K. V., 1928: Respiration of wheat seed in ionized air, *C. R. Acad. Sci., U.R.S.S.*, **A8**: 119–124; *Biol. Abst.*, **5**: 1618.

SCHÖENBEIN, C. F., 1863: Ueber die katalytische Wirksamkeit organischer Materien und deren Verbreitung in der Pflanzen- und Thierwelt, *J. Prakt. Chem.*, **89**: 323–344.

SCHRÖPPEL, F., 1933: Catalase, peroxidase and respiration in the germination of the light-sensitive seeds of *Nicotiana tabacum*, *Beih. bot. Cent.*, **51**: 377–407; *Chem. Abst.*, **28**: 503.

SCHULZ, K. G., 1926: Vergleichende Untersuchungen über die Atmungsvorgänge bei verschiedenen Kartoffelsorten, *Landw. Versuchs-Stat.*, **105**: 23–73.

SHERMAN, H., 1921: Respiration of dormant seeds, *Bot. Gaz.*, **72**: 1–30.

SHILL, A. C., 1931: The respiration of *Citrus* as affected by hydrocyanic acid gas fumigation, *Univ. Calif. Pub.*, **5**: 167–180.

SHULL, C. A., 1911: The oxygen minimum and the germination of *Xanthium* seeds, *Bot. Gaz.*, **52**: 453–477.

———, 1914: The role of oxygen in germination, *Bot. Gaz.*, **57**: 64–69.

——— and W. B. DAVIS, 1923: Delayed germination and catalase activity in *Xanthium*, *Bot. Gaz.*, **75**: 268–281.

SINGH, B. N., 1930: Studies in the respiration of tropical plants, *Proc. Indian Sci. Cong.* (*Calcutta*), **17**: 306–310; *Exp. Sta. Rec.*, **66**: 619–620.

——— and K. V. VARADPANDE, 1930: Studies in the respiration of tropical plants. Effect of injecting water, glucose, and phosphates on a failing system of respiration in the leaves of *Artocarpus integrifolia* and its significance on the mechanism of respiration, *Proc. Indian Sci. Cong.*, **17**: 311–312; *Exp. Sta. Rec.*, **66**: 722–723.

SMITH, A. M., 1915: The respiration of partly dried plant organs, abst. in *Rept. Brit. Assoc. Adv. Sci.*, **85**: 725; *Exp. Sta. Rec.*, **36**: 824.

SMITH, E. P., 1921: Comparative studies on respiration. XIX, A preliminary stage in the process of ether anesthesia, *J. Gen. Physiol.*, **4**: 157–163.

———, 1922: Comparative studies on respiration. XXII, The effect of lactic acid on the respiration of wheat, *Am. J. Bot.*, **9**: 307–311.

SMITH, O., 1929: Effects of various treatments on the carbon dioxide and oxygen in dormant potato tubers, *Hilgardia*, **4**: 273–306.

SMYTH, E. S., 1934: A contribution to the physiology and ecology of *Peltigera canina* and *P. polydactyla*, *Ann. Bot.*, **48**: 781–818.

SPOEHR, H. A., 1915: Variations in respiratory activity in relation to sunlight, *Bot. Gaz.*, **59**: 366–387.

——— and J. M. MCGEE, 1923: Studies in plant respiration and photosynthesis, *Carnegie Inst. Wash. Pub.* 325.

——— and ———, 1924: The effect of fluctuations in the CO_2 content on the atmosphere on the rate of respiration of leaves, *Am. J. Bot.*, **11**: 493–502.

———, 1924: The oxidation of carbohydrates with air, *J. Am. Chem. Soc.*, **46**: 1494–1503.

STEWARD, F. C., R. WRIGHT, and W. E. BERRY, 1932: The absorption and accumulation of solutes by living plant cells. III, The respiration of cut discs of potato tuber in air and immersed in water with observations upon surface: volume effects and salt accumulation, *Protoplasma*, **16**: 576–611.

STICH, C., 1891: Die Athmung der Pflanzen bei verminderter Sauerstoffspannung und bei Verletzungen, *Flora*, **74**: 1–57.

STILES, W., and W. LEACH, 1931: On the use of the katharometer for the measurement of respiration, *Ann. Bot.*, **45**: 461–488.

——— and ———, 1932: "Respiration in Plants," Dial Press (Lincoln MacVeagh), Inc., New York and Oxford.

——— and ———, 1932: Researches on plant respiration. I, The course of respiration on *Lathyrus odoratus* during germination of the seed and the early development of the seedling, *Proc. Roy. Soc.*, B, **111**: 338–355.

——— and ———, 1933: Researches on plant respiration. II, Variations in the respiratory quotient during germination of seeds with different food reserves, *Proc. Roy. Soc.*, B, **113**: 405–429.

———, 1935: Respiration, *Bot. Rev.*, **1**: 249–268.

STOCKLASA, J., and F. CZERNY, 1903: Isolirung des die anaerobe Athmung der Zelle der höher organisjrten Pflanzen und Thiere bewirkenden Enzyms, *Ber. chem. Ges.*, **36**: 622–634.

STOWARD, F., 1908: On endospermic respiration in certain seeds, *Ann. Bot.*, **22**: 415–448.

SWANSON, C. O., 1934: Some factors involved in damage to wheat quality, *Cereal Chem.*, **11**: 173–199.

TANG, P., 1931: Temperature characteristics for the oxygen consumption of germinating seeds of *Lupinus albus* and *Zea mays*, *J. Gen. Physiol.*, **14**: 631–641.

———, 1931: Temperature characteristics for the production of CO_2 by germinating seeds of *Lupinus albus* and *Zea mays*, *J. Gen. Physiol.*, **15**: 87–95.

———, 1932: On the respiratory quotient of *Lupinus albus* as a function of temperature, *J. Gen. Physiol.*, **15**: 561–569.

———, 1932: A respirometer vessel for the study of metabolism of seeds, *J. Gen. Physiol.*, **15**: 571–574.

———, 1932: The effects of CO and light on the oxygen consumption and on the production of CO_2 by germinating seeds of *Lupinus albus*, *J. Gen. Physiol.*, **15**: 655–663.

———, 1932: Temperature characteristics for the anaerobic production of CO_2 by germinating seeds of *Lupinus albus*, *J. Gen. Physiol.*, **16**: 65–73.

TASHIRO, S., 1913: A new method and apparatus for the estimation of exceedingly minute quantities of carbon dioxide, *Am. J. Physiol.*, **32**: 137–145.

TERROINE, E. F., R. BONNET, and P. H. JOËSSEL, 1923: Influence de la température sur le rendement énergétique dans la germination, *C. R. Acad. Sci. Paris*, **177**: 212–214.

THODAY, D., 1913: On the capillary eudiometric apparatus of Bonnier and Mangin for the analysis of air in investigating the gaseous exchanges of plants, *Ann. Bot.*, **27**: 565–573.

———, 1913: The effect of chloroform on the respiratory exchanges of leaves, *Ann. Bot.*, **27**: 697–717.

THOMAS, H. S., 1918: Comparative studies on respiration. IV, The effect of ether on the respiration of wheat, *J. Gen. Physiol.*, **1**: 203–209.

THOMAS, L. A., 1931: The respiration of carrot tissue in buffer solutions, *Australian J. Exp. Biol. and Med. Sci.*, **8**: 225–236.

THOMAS, M., 1925: The controlling influence of carbon dioxide. V, A quantitative study of the production of ethyl alcohol and acetaldehyde by cells of the higher plants in relation to concentration of oxygen and CO_2, *Biochem. J.*, **19**: 927–947.

———, 1929, 1931: The production of ethyl alcohol and acetaldehyde by apples in relation to the injuries occurring in storage. I, Injuries to apples occurring in the absence of oxygen, and in certain mixtures of carbon dioxide and oxygen. II, Injuries to apples and pears occurring in the presence of oxygen and in the absence of accumulation of CO_2 in the storage atmosphere, *Ann. App. Biol.*, **16**: 444–457 (1929); **18**: 60–74 (1931).

——— and J. C. FIDLER, 1933: Studies in zymasis. VI, Zymasis by apples in relation to oxygen concentration, *Biochem. J.*, **27**: 1629–1642.

THORNTON, N. C., 1931: The effect of carbon dioxide on fruits and vegetables in storage, *Contr. Boyce Thompson Inst.*, **3**: 219–244.

———, 1932: Carbon dioxide storage. III, The influence of carbon dioxide on the oxygen uptake by fruits and vegetables, abst., *Am. J. Bot.*, **19**: 843.

———, 1932: Carbon dioxide storage. IV, The influence of carbon dioxide on the acidity of fruits and vegetables, abst., *Am. J. Bot.*, **19**: 843.

———, 1933: Carbon dioxide storage. III, The influence of carbon dioxide on the oxygen uptake by fruits and vegetables, *Contr. Boyce Thompson Inst.* **5**: 371–402.

———, 1933: Carbon dioxide storage. IV, The influence of carbon dioxide on the acidity of plant tissue, *Contr. Boyce Thompson Inst.*, **5**: 403–418.

———, 1935: Carbon dioxide storage. VIII, Chemical changes in potato tubers resulting from exposure to carbon dioxide, *Contr. Boyce Thompson Inst.*, **7**: 113–118.

TRAUBE, M., 1877: Die chemische Theorie der Fermentwirkungen und des Chemismus der Respiration, *Ber. chem. Ges.*, **10**: 1984–1992.

TRUE, R. H., and W. W. STOCKBERGER, 1916: Physiological observations on alkaloids, latex and oxidases in *Papaver somniferum*, *Am. J. Bot.*, **3**: 1–11.

TYSON, J., 1930: Influence of soil conditions, fertilizer treatments and light intensity on growth, chemical composition and enzymic activities of sugar beets, *Mich. Sta. Tech. Bull.* 108.

WAKABAYASHI, S., 1925: The injurious effect of submergence on the cranberry plant, *N. J. Sta. Bull.* 420.

WARBURG, O., 1921: Physikalische Chemie der Zellatmung, *Biochem. Z.*, **119**: 134–166.

———, 1925: Iron, the oxygen-carrier of respiration ferment, *Sci.*, **61**: 575–582.

———, 1928: The chemical constitution of respiration ferment, *Sci.*, **68**: 437–443.

WEISS, F., and R. B. HARVEY, 1921: Catalase, hydrogen-ion concentration, and growth in the potato wart disease, *J. Agr. Res.*, **21**: 589–592.

WERNER, H. O., 1935: A review of the literature on the physiological aspects of the storage of potatoes, *Ann. Rept. Neb. Sta. Bd. Agr.*, **1935**: 680–703; 16th *Ann. Rept. Neb. Potato Improv. Assoc.*, **1934–1935**: 36–59.

WHELDALE, M., 1910: Plant oxidases and the chemical interrelationships of color varieties, *Prog. rei. Bot.*, **3**: 457.

WHIMSTER, K., 1927: The effect of ionized air on the assimilation and respiration of green leaves, *Ann. Bot.*, **41**: 357–374.

WHITE, J., 1909: The ferments and latent life of resting seeds, *Proc. Roy. Soc.* B, **81**: 417–442.

WHITEHEAD, T., 1934: The physiology of potato leaf roll. I, On the respiration of healthy and leaf-roll infected potatoes, *Ann. App. Biol.*, **31**: 48–77.

WILLAMAN, J. J., and J. H. BEAUMONT, 1928: The effect of accumulated carbon dioxide on plant respiration, *Plant Physiol.*, **3**: 45–61.

——— and W. R. BROWN, 1930: Carbon dioxide dissolved in plant sap and its effect on respiration measurements, *Plant Physiol.*, **5**: 535–542.

WILLSTÄTTER, R., and H. WEBER, 1926: Zur quantitativen Bestimmung der Peroxydase, *Liebigs Ann. Chem.*, **449**: 156–174.

WOODS, A. F., 1902: Observations on the mosaic disease of tobacco, *U. S. Bur. Plant Ind. Bull.* 18.

WRIGHT, R. C., 1932: Some physiological studies of potatoes in storage, *J. Agr. Res.*, **45**: 543–555.

WYND, F. L., 1932: An apparatus for the determination of carbon dioxide production in physiological plant studies, *Ann. Mo. Bot. Gard.*, **19**: 499–502.

———, 1935: An improved apparatus for the determination of carbon dioxide production in physiological plant studies, *Ann. Mo. Bot. Gard.*, **22**: 361–363.

——— and E. S. REYNOLDS, 1935: Studies in ultra-violet and respiratory phenomena. I. Review of work published before June 1935, *Ann. Mo. Bot. Gard.*, **22**: 771–835.

———, H. J. FULLER, and E. S. REYNOLDS, 1935: Studies in ultra-violet and respiratory phenomena. II, The effects of ultra-violet on respiration and respiratory enzymes of higher plants, *Ann. Mo. Bot. Gard.*, **22**: 837–852.

YEMM, E. W., 1935: The respiration of barley plants. I, Methods for the determination of carbohydrates in leaves, *Proc. Roy. Soc.*, B, **117**: 483–504.

———, 1935: The respiration of barley plants. II, Carbohydrate concentration and carbon dioxide production in starving leaves, *Proc. Roy. Soc.*, B, **117**: 504–525.

YOSHIDA, H., 1883: Chemistry of lacquer, *J. Chem. Soc. (London)*, **43**: 472–486.

YOSHISUKE, R., 1932: Davis' modified catalase method prophesying the germinating power of seeds, *Bull. Sci. Fakult. Terkeltura, Kjusu Imp. Univ.*, **5**: 103–114 (115–116 Eng. Summary); *Chem. Abst.*, **29**: 4797.

ZALESKI, W., and A. ROSENBERG, 1911: Zur Kenntnis der Rolle der Katalase in den Pflanzen, *Biochem. Z.*, **33**: 1–15; *Bot. Cent.*, **119**: 431 (1912).

———, 1915: Über die Alkoholoxydation durch die Samenpflanzen, *Biochem. Z.*, **69**: 289–293.

ZERBAN, F. W., 1918: The role of oxidases and of iron in the color changes of sugar cane juice, *J. Ind. Eng. Chem.*, **10**: 814–817.

ZIMMERMAN, P. W., 1930: Oxygen requirements for root growth of cuttings in water, *Am. J. Bot.*, **17**: 842–861.

CHAPTER XIV

THE PROCESS OF GROWTH IN PLANTS

I. NATURE OF GROWTH

Growth is one of the outstanding characteristics of living matter, but at the same time it is a process which is very difficult to define. In discussing the nature of growth, it should be remembered that it may be considered from the standpoint of the entire organism or from that of its individual parts. Thus growth may refer to the changes that occur in the protoplasm, in the cell wall, in the cell considered as a unit, in the organs or parts of organs, or in the entire plant considered as a whole. Any statements concerning growth, therefore, will be modified somewhat by the viewpoint from which they are considered.

Regardless of the standpoint from which growth is considered, the first step in the process is, with but few exceptions, the formation of new protoplasm. This step has been termed the "fundamental stage of growth." This formation of new living matter is always accompanied or immediately followed by a permanent increase in the amount of the nonliving materials that are always closely associated with the protoplasm. This increase in the amount of nonliving matter is plainly evident through the enlargement of the plant part and by an increase in its weight. These changes are due chiefly to an increased water content and to the deposition of materials in the cell walls. Since this stage in growth is readily observed, it has been termed the "evident stage of growth." This increase in the size and weight of the plant part under consideration is permanent as long as that part remains alive. In the light of these facts, growth has been defined by some authors as a permanent increase in weight, attended by a permanent change in form, these changes having been induced primarily by an increase in the quantity of protoplasm.

While there are instances where this definition of growth does not strictly describe the changes that occur, it is applicable in most cases. Two of the examples that have frequently been cited as cases where this definition does not apply are the sprouting potato tuber and germinating seeds. Thus the total dry weight of the tuber and sprout will be less than that of the original tuber, owing to loss in weight through respiration. The sprouts, however, have greatly increased in weight and changed markedly in form from the buds that produced them. The same relation exists in germinating seeds where the seedling and seed combined for several weeks contain less dry matter than the original seed. The plumule and radicle of the embryo, however, have increased their weight and have changed in form, although the seed as a whole has decreased in dry matter.

To some, however, growth implies more than simply an increase in weight and size. They believe that differentiation and development should also be included in the term. The development of a plant from the fertilized egg is not simply an increase in the weight and size of the fertilized egg; otherwise a mature plant would be only an enlarged egg and not a highly specialized individual. It is evident that if the differentiation of a plant or plant part is not considered as a phase of growth, it must at least be regarded as intimately associated with it.

The growth of an organism consists primarily of the formation of new cells, of their enlargement and specialization. In order, therefore, to understand growth in plants, we must consider in detail the manner in which a plant cell grows.

II. STAGES IN THE GROWTH OF A PLANT CELL

In the formation and development of a plant cell, three more or less distinct stages or types of activity are to be observed. The transition from one stage to another is not sudden and complete, but the different stages can readily be distinguished and can be seen to be differently affected by the immediate environment of the cells (Priestley, 1929). The different stages of growth which may be distinguished are (*a*) the embryonic or formative stage, (*b*) the stage of elongation, and (*c*) the stage of differentiation.

A. Embryonic or Formative

This stage includes that period in which the cell is formed from a preexisting cell. In the vascular plants, the formation of new cells is localized in more or less definite regions known as "meristems." These regions are fairly definitely confined to the tips of branches, near to the tips of roots, to the cambium, to the bases of the internodes, especially in the monocotyledons, and to the bases of the growing sheaths and leaves of monocotyledons.

The cells of the meristem which thus give rise to new cells are small, with dense protoplasmic contents and prominent nuclei. The nuclei are of the normal size of other plant cells, but in such small cells they bulk large in proportion to the rest of the protoplasm. The protoplasm is especially free from vacuoles and characteristic inclusions, no granular food reserves being evident, while the plastids are very minute or invisible. With the exception of the cambium, the cells may be 12 to 14 sided but frequently are cubical and are closely packed together with no intercellular spaces. The cell walls are the thinnest of any in the plant body, and, according to Priestley (1929), they frequently take stains, which indicate that they are impregnated with protoplasm. He considered that the walls of these cells are the channels of transport for the food supplies of this tissue.

In the meristematic cells, protein synthesis is dominant, while carbohydrate formation is at a minimum. In this formative stage, therefore, the protoplasm is primarily engaged in protoplasmic synthesis. According to Priestley (1929), this is maintained only when the liquid surrounding the cell has a hydrogen-ion concentration near the isoelectric point of the main constituent protein of the cell. When the mass of a meristematic cell has increased to a certain point, cell division follows, and the narrow range of cell size in these regions indicates that the division follows the attainment of a certain upward limit of the cell mass. MacDougal (1916 to 1921) came to the conclusion from his extensive experiments with growth that the first step in the process is an increase in the volume of the protoplasm by hydration. This increase in volume is usually but not always accompanied by the incorporation of materials other than water in the colloids of the protoplasm with the entailed additional capacity for absorbing water. The actual manner in which this incorporation occurs, however, is not known. MacDougal believed that the protoplasm of plants consists of a comparatively inert base of a polysaccharose character in colloidal combination with proteins, amino acids, lipins, and salts. The enlargement of protoplasmic mass due to hydration is therefore determined by the character of this colloidal mixture as well as by the quantity and type of salts and by the acidity or alkalinity of the medium. These factors might all be altered by the action of the metabolic products of the meristematic cells.

McDougal (1925) called this procedure the accretion stage of the cell. During this stage, proteins and carbohydrates are formed by synthesis, dehydration, and condensation.

Hottes (1929) found in the roots of *Vicia faba*, the tips of which were mechanically inhibited from further elongation, that nuclear and cell division ceased when the cell reached a minimum size. He considered that turgor is perhaps the principal inhibitor of nuclear division. Turner (1929) and Farr (1931, 1933) noted that the hair or fiber of cotton is formed by the outward extension of a single epidermal cell of the seed coat. These epidermal cells, which develop into fibers, are formed during a period of at least 10 days following the beginning of the opening of the flower.

Priestley (1930) and Priestley, Scott, and Malins (1933) noted that the cambium changes in the spring from a "granular gel" condition to a swollen "semifluid sol" condition. Some time later, cell division begins. It commences in the buds, proceeds downward in the cambium of the old wood, and finally reaches the younger roots, in which a slow cambial activity may persist for most of the season. Thus the initiation of cell division may be weeks later in the roots than in the trunk. The dormant cambial layer is usually about one cell in thickness, and, with the exception of their granular nature at this stage, there is little to distinguish these cells from the adjoining ones; thus separation of the bark and wood is difficult. The first sign of growth activity is a swelling of the cambial cells. The radial walls rapidly become thinner as they extend, and the contents of the cell appear to change from a solid to a liquid consistency. After this stage is attained, the bark readily slips over the wood, and separation occurs in the plane of the cambium. The new cells are produced almost entirely on the inner side of the cambium. The student is referred to Priestley (1930) for a thorough and critical discussion of the concept of sliding growth in the cambium.

According to Wright (1933) cambial growth in the trunk of *Pinus sylvestris* begins in May, before the opening of the buds, while the most vigorous growth occurs from May to early July. After this, slow but continuous growth occurs until October.

The retardation of growth of an organ may be directly connected with the lessening of the water capacity of the colloidal masses. This theory of the role of colloidal hydration in growth has been substantiated by the work of MacDougal (1920, 1921) and MacDougal and Spoehr (1917) with artificially prepared mixtures consisting largely of amorphous, condensed carbohydrates such as agar to which had been added a small proportion of albumen or amino acid. These mixtures were found to respond to the action of acids, alkalies, and salts in a manner similar to that of the plant.

Instances of growth are known in which water only has been added to the colloids of the protoplasm, but in these cases in all probability the solid particles are variously rearranged. MacDougal (1919) found no increase in dry matter in the leaves of *Crassulaceae*, in the joints of cacti, and in fruits, although growth as measured by the change in form was taking place. The proportion of water and solid matter thus underwent but little change, their incorporation being at a rate that kept them near the initial proportion.

B. Elongating

1. Nature.—After a new cell has been formed, the protoplasm tends toward carbohydrate synthesis rather than protein synthesis. The increased carbohydrate synthesis is indicated by the increase in the thickness of the cell walls and by the inclusion of starch grains in the protoplasm (Priestley, 1928). The amount of protoplasm increases little if at all during this stage of cell development.

The protoplasm of the new cell now exhibits syneresis similar to that shown by many colloids in which cavities or canals are formed throughout the colloidal mass. These syneretic cavities enlarge by the absorption of water to form the "vacuoles," by which name they are generally called. These vacuoles continue to expand by the absorption of water and eventually coalesce to form the vacuole of the cell. The expansion of these vacuoles and the consequent increase in volume of the cell take place primarily by the force of osmotic action. The vacuole continues to hold some of the colloidal material and in addition the various solutes that enter the cell or are formed therein (MacDougal, 1919).

MacDougal (1925) stated that the greatest distension of a cell does not occur in the isoelectric zone of its proteins but at the hydrogen-ion concentration at which the protoplasmic layer is the least permeable. Freeland (1933) noted in *Bryophyllum* that the hydrogen-ion concentration and total acidity of the margins of proliferating leaves increased for 4 days after the inception of this process and then decreased. The osmotic value in the margins of the proliferating leaves was in general higher than in the margins of the paired, inactive leaves. During proliferation, starch and reducing sugars accumulated in the growing foliar bud, and the amounts of total sugars, sucrose, amino nitrogen, amide nitrogen, and nitrate nitrogen increased in the leaf. In addition, there was an increase in the amounts of catalase, diastase, and oxidase in the foliar bud. Loomis (1932, 1934) reported that the growth of maize depends upon a liberal supply of water at the growing point. Such a supply is reduced, and hence growth is checked by the following factors, listed in order of effectiveness: direct sunlight, deficient soil moisture, and low relative humidity. Growth decreases rapidly as the temperature approaches 10°C. This may be due in part to a slowing of the chemical processes of cell division and to a decreased rate of translocation of food material.

During the elongating period, the newly formed cell increases in size, especially in length (Fig. 3). When this elongation takes place, the protoplasm, which has increased but little in amount over that of the newly formed cell, spreads out in a thin film along the cell wall. The volume of a plant cell may thus increase several hundred times and yet contain very little more protoplasm than before expansion began. In this respect, the plant cell differs from the animal cell, for in the latter no such increase in volume takes place without a corresponding increase in the amount of protoplasm. By this expansion of cells due to the absorption of water very rapid growth occurs without the expenditure of the energy necessary to manufacture relatively large quantities of protoplasm. This ability to grow rapidly is a distinct advantage to an organism where there is a time limit to its period of growth. The total period during which the elongation of the plant cell occurs is termed the "grand period of growth."

2. Rate.—Except in rare cases, growth in plants proceeds at such a rate that it cannot be detected by the naked eye. The most rapid growth in length is that of the filaments of stamens in the wheat flower. According to Andrews (1932) this growth may be 2 to 3.5 mm. per minute. If such a growth rate were maintained for 24 hr., the stamen would attain a length of 6 to 8 ft. Under favorable conditions the pollen tube of *Zea mays* may grow 6 in. in 24 hr. When an appreciable period of time is considered, the bamboo grows the most rapidly of any plant, and a total increase in the length of its stem of 2 ft. in 24 hr. is not infrequent. Under the conditions of summer, the squash vine grows approximately 6 in. during a 24-hr. day. Florell and Faulkner (1934) studied the daily increase in height of the peduncle of wheat at heading. The maximum daily elongation during this period ranged from 4.3 to 9.1 cm. Lichens, however, may grow as little as 2 to 5 mm. per year under ordinary, favorable conditions.

Based on the data obtained from various species of plants growing under many conditions, it is estimated that their average growth in length during the growing season does not exceed 0.3 in. for a 24-hr. period. Booth (1929) noted in oats that the growth in length of the grain was most rapid during the first 6 days after fertilization. Following this the rate of elongation gradually diminished until the maximum length was attained on the fifteenth day. The grain increased rapidly in width and thickness during the first 10 days and reached its limit in these dimensions before growth in length ceased. Dastur and Kapadia (1931) stated that the maximum axial growth of the tendril of *Momordica charantia* is 8.0 to 9.4 cm. in 24 hr., while that of *Lagenaria vulgaris* is 17.5 cm. during a like period.

C. Differentiating

About the time a cell reaches its maximal enlargement, it begins to undergo changes that will transform it into a specialized cell. It may become a sclerenchyma cell, a sieve tube, a tracheid, or any other of the numerous types of cells that compose a plant, depending upon its location and the resultant of internal and external factors. When a cell has once become specialized, it seldom if ever reverts to another form but remains the same type of cell as long as it exists. After a cell is differentiated, it functions solely in aiding other cells to grow and develop.

It was stated by Loomis (1932) that growth is an increase in size due to cell division and cell enlargement, while "differentiation" is the sum of the morphological changes that start during cell enlargement and end with the death of the cell.

Inulin occurs in *Helianthus tuberosus,* but not in *H. annuus.* According to Thoday (1933), grafts, made between the two species by Daniel (1921) and Colin (1922), showed that the transport of this carbohydrate occurred freely from one to the other, but each part retained its normal characteristic. In a double graft in which a piece of the sunflower was interpolated between the basal and apical parts of an artichoke, each of the three portions functioned in their usual manner. There must, therefore, be a specific difference in the metabolism of the parenchymatous cells of the two stems. Metabolic differences are shown in the various parts of the same plant. Thus in rhubarb, the rhizome contains very little acid and the nitrogen is present largely in the form of amino acids or amides and a small amount of ammonia. In the petioles, however, two-thirds of the total nitrogen is in the form of ammonia, and acids are abundant.

McNair (1934) believed that as a general rule the more highly the plant is organized, the more complex are likely to be its chemical products. Thus he believed that the volatile oils of the families highest in evolutionary development have constituents with a large number of double bonds, more aromatic compounds, and more compounds with small amounts of substances of low molecular weight.

III. GROWTH OF PLANT ORGANS

In the great majority of plants the regions of cell expansion are strictly localized. Thus the following are the regions in the higher plants where increase in the size of the cells occurs: (*a*) Near the root tip. The elongating region is restricted to a portion of the root not over 10 mm. in extent. The region of maximal elongation generally occurs only about 3 mm. back of the root tip. (*b*) At the tips of stems. The stem, as a rule, has a longer elongating region than the root. (*c*) At the cambium. (*d*) At the base of the internodes in the monocotyledons. (*e*) At the base of the sheaths and leaves of monocotyledons. In the leaves of dicotyledons, growth is rather evenly distributed throughout the entire surface but with slightly more activity at the basal region than at the apical one. In general, growth is more localized in a mature plant than in a younger one and more so in a perennial than in an annual plant.

Since each cell of a plant passes through the three more or less distinct stages in its growth and development, each plant part or organ in its development passes through the same stages. Thus each organ or plant part has its formative stage, its elongative stage or grand period of growth, and its stage of differentiation or maturation. The enlargement of any plant part begins at a slow rate that gradually increases to a maximum, after which the rate progressively decreases until enlargement ceases. This behavior of growth is illustrated by the rate of elongation of the growing zone of a root of *Vicia faba*, as quoted by Raber (1928).

Day	Growth, millimeter	Day	Growth, millimeter
1	2	6	14
2	4	7	7
3	18	8	4
4	18	9	0
5	17		

From these data it is seen that elongation during the first 2 days is relatively slow but reaches a maximum during the next 3 days, after which it rapidly declines to zero.

The higher plants, as a rule, continue to develop new organs as long as they live, forming stems, leaves, roots, flowers, and seeds in a somewhat indefinite repetition. Even in an annual plant no definite number of organs is developed. This is in marked contrast to most animals, which develop a single set of organs that serve them throughout their life. Although growth in an annual plant is limited to one season, this process in perennial plants continues as long as the plant lives. Thus a tree that is several thousand years old will continue to produce organs and increase in height and thickness of stem.

IV. NATURE OF GROWTH CURVES

The rate of growth of plants or plant parts is commonly measured by their increase in extent or by their increase in weight. As has been previously mentioned, the rate of growth of a plant or any portion of it is relatively slow at first, increases rapidly to a maximum, and finally decreases until it ceases. This behavior is especially noticeable when the elongation of plant parts is considered. When the rate of growth is plotted against time, a typical S curve is obtained.

Martini, Harlan, and Pope (1930) did not believe that the S curve of growth is wholly applicable to the development of barley kernels. They considered that factors other than growth must be considered.

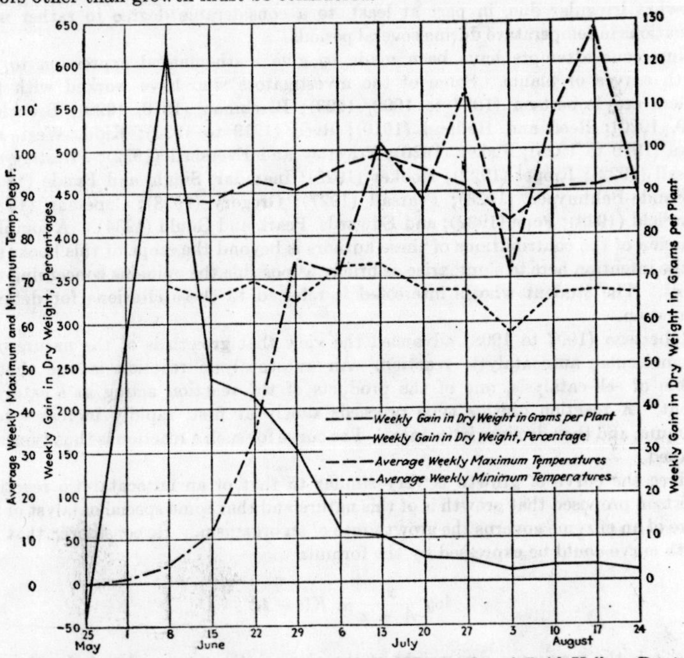

FIG. 38.—Graphs showing the weekly gain in dry weight of Reid Yellow Dent corn grown at Manhattan, Kans., in 1925, together with the weekly averages of the maximum and minimum temperatures.

There are various ways in which the growth increment may be expressed, but there is no consensus of opinion as to which is the best. Hackenberg (1908) and others used the term "substance quotient" in expressing the increment of growth. The substance quotient is the ratio of the dry weight at any given period to the dry weight of the preceding period. The rate may also be expressed as the percentage increase of dry weight or of extent for a given period over that of the preceding one. The increase in leaf surface or in leaf weight is also considered by some to be a good indicator of growth increment.

Pearl, Winsor, and Miner (1928) stated that the growth of an etiolated seedling from a sterilized seed in the dark on a nonfood, sterile medium is an expression of the "inherent vitality" of that individual. Inherent vitality is defined as the total potential capacity of an organism to perform vital actions in the complete absence of exogenous derivation of food or energy. The "total vitality" is the capacity of an organism to perform vital activities in the presence of its endogenous sources and of sufficient exogenous supplies.

In the table on page 1027 are given data obtained by Miller (1925) in the growth of the corn plant at Manhattan, Kans. The weekly increments of growth are therein expressed by the increase in dry weight, the percentage increase in dry weight during each period over that of the succeeding period, the increase in the leaf surface, and the increase in the height of the plant for each period. The rate of transpiration per plant and the weather conditions in regard to temperature are also included. A portion of this data is plotted in Fig. 38. The increments in growth in this case are somewhat irregular due, in part at least, to a considerable degree to rather wide fluctuations in temperature during several periods.

Numerous attempts have been made to give mathematical expression to the growth curves of plants. Some of the investigators who have worked with this problem are Robertson (1907 to 1908, 1923); Blackman (1919, 1920); Brenchley (1919, 1920); Reed and Holland (1919); Reed (1919 to 1923); Kidd, West, and Briggs (1919 to 1928); Fisher (1920); Priestley and Evershed (1922); Priestley and Pearsall (1922); Rippel (1922); Hooker (1925); Inamdar, Singh, and Pande (1925); van Sande-Bakhuyzen (1926); Pearsall (1927); Gregory (1928); Copeman (1928); Porterfield (1928); Pope (1932); and Edwards, Pearl, and Gould (1934). A complete discussion of the contributions of these authors is beyond the scope of this book, and it is the intention here to summarize as briefly as possible the opinions concerning this subject. The student who is interested is referred to these citations for detailed information.

Robertson (1907 to 1908) advanced the view that growth is of the nature of a monomolecular autocatalytic reaction. An autocatalytic reaction is one that is capable of self-catalysis, one of the products of the reaction acting as a catalytic reagent. A reaction of this type proceeds slowly at first, rapidly increases to a maximum, and then declines in its rate. The curve for such a reaction is thus typically S shaped.

Since the curve of growth is very similar to that of an autocatalytic reaction, Robertson proposed that growth is of this nature and that some special catalyst of the nature of an enzyme governs the growth rate of an organism. He considered that the growth curve could be expressed by the formula

$$\log \frac{x}{A - x} = K(t - t_1)$$

where A is the maximum dry weight of the plant, x the dry weight of the plant at any time, t the time at which the weight of the plant is one-half the final dry weight, and K a constant. Robertson considered that the greatest increase in volume or weight in any unit of time in any growth cycle takes place when the total growth due to the cycle is one-half accomplished. Reed and Holland (1919) and Reed (1919 to 1923) from their observations on the elongation of the sunflower, apricot, and other plants, and Murneek (1925) from observations on the tomato and cotton plants concluded that the growth rate approximated the course of an autocatalytic reaction. A close similarity between the observed and calculated growth values led them to the conclusion that the growth rate is governed by constant internal factors rather than by external factors. They obtained no strong correlation of growth either with temperature or with transpiration summation. In the case of apricot, Reed (1919)

WEEKLY GROWTH CHANGES OF REID YELLOW DENT CORN AT MANHATTAN, KANS. DURING GROWING SEASON OF 1925
Averages of Five Plants

Date	Average dry weight per plant, grams	Weekly gain in dry weight per plant		Total leaf surface per plant, grams	Weekly increase in leaf surface per plant		Rate of growth in height, inches	Transpiration per plant, cc.	Average weekly maximum temperature, °F.	Average weekly minimum temperature, °F.
		Grams	Per cent		Grams	Per cent				
May 11	Seed planted. Average dry weight of seed 0.3 g.									
11–25	0.16	0.14[1]	47.0[1]	70.5	....	.....	3.5	59		
25–June 1	0.65	0.49	306.2	348.1	277.6	393.7	2.5	102		
June 1–8	4.66	4.01	617.0	1,554.8	1,206.7	346.9	5.0	1,079	88.8	68.7
8–15	15.74	11.08	237.7	4,260.7	2,705.9	174.0	6.0	2,874	88.4	65.1
15–22	49.30	33.56	213.2	9,891.5	5,620.8	131.9	14.0	4,110	92.7	69.6
22–29	123.70	74.40	151.0	17,069.2	7,277.7	73.5	17.0	8,290	90.4	64.8
29–July 6	195.00	71.30	57.6	21,585.8	4,516.6	26.4	17.0	22,610	95.8	72.9
July 6–13	297.00	102.02	52.3	26,699.2	5,113.4	23.6	18.0	27,220	99.4	73.6
13–20	385.50	88.48	29.8	28,629.4	1,930.2	7.2	33.0[2]	30,420	95.3	67.9
20–27	499.70	114.20	29.6	Full leaf development				22,190	90.1	67.6
27–Aug. 3	570.80	71.10	14.2					15,530	83.1	58.7
Aug. 3–10	679.00	108.20	18.9					21,610	90.7	66.3
10–17	807.50	128.50	18.9					15,890	91.0	67.4
17–24	895.90	88.40	10.9					17,130	93.9	66.3

[1] Loss.
[2] Tasseling.

considered that an organism grows at a definite rate which is at any moment proportional to the amount of growth yet to be made.

Blackman (1919, 1920) suggested that the growth of an annual plant can be treated as a process following the compound-interest law expressed by the formula

$$W_1 = W_0 e^{rt}$$

or, written in another manner,

$$\log_e \frac{W_1}{W_0} = rt$$

In this formula W_1 is the dry weight of the plant at the end of the time t, W_0 the initial seedling or seed rate, e the base of the natural logarithms, and r the rate of interest or increase. Blackman termed this rate of increase in dry matter the "efficiency index of production," since it represents the efficiency of the plant as a producer of new material and gives a measurement of the plant's economy. It may also be termed the "economy constant" of the plant and is comparable to the velocity constant of a chemical reaction. The rate of increase of dry material increases during daylight according to external conditions and during the night becomes a negative quantity. On that account, Blackman (1920) stated that the efficiency index when applied to plants growing under natural conditions is only an average measure of the activity of the plant in the production of new material. He also stated that caution must be exercised in making deductions as to a plant's general economy from the efficiency index calculated over periods. The efficiency index nevertheless gives a measure of the plant's average efficiency during any particular period.

Kidd, West, and Briggs (1920) and Briggs, Kidd, and West (1920) calculated the growth rate by means of the formula

$$\frac{R}{100} = \frac{W_2 - W_1}{W_1}$$

where W_1 and W_2 are the dry weights, respectively, of the plants at the beginning and end of the period under consideration. This formula gives the rate of increase in dry weight for the period as a percentage of the dry weight at the beginning of the period. It was adversely criticized by Fisher (1920), who believed that the growth rate should be calculated as a percentage of the mean value for the period.

In a consideration of growth formulas, it should be noted that the rate of change at any time is conditioned by the state of the system at that time; that the rate of change of the system is conditioned by its previous history as well as by the state of the general environment (Gregory, 1928).

V. PERIODIC ELONGATION

According to Friesner (1920), it was first observed by Sachs (1874) that plants exhibit a rhythm of elongation. He found that plants exposed to the alternation of light and darkness showed a maximal elongation shortly after sunrise and the minimum shortly after sunset. Prantl (1873) found a similar behavior in the increase in width of leaves and under normal conditions the maximum was reached in the morning from six to nine and the minimum in the evening from six to nine. He observed that by shifting the hours of illumination and darkness he could alter the time of maximal and minimal expansion, since for each change in the time of illumination and darkness there was a corresponding change in the times of maximal and minimal expansion. These investigators could find no evidence of periodicity when the plants were grown in the dark from the beginning, and Baranetzky (1878) observed that plants which exhibit this regular periodicity lose this when placed in darkness. The time for its complete loss varied from 2 to 3 days and in one case to 14 days. He also observed that plants grown from the beginning in darkness did not exhibit this periodicity except in the case of the shoots of *Brassica rapa,* where it continued to exist.

MacMillan (1891), however, noted that the increase in the diameter of the potato tuber developing in continuous darkness is not regular but rhythmic. The maxima of growth may occur from one to four times during a 24-hr. period. These are not of long duration and are followed by periods of slower growth or entire cessation of growth. The maxima of some days were greater absolute maxima than those of other days, which indicate a grand period of growth for the tuber. MacMillan believed that there is some connection between the periodic growth of the tuber and the periodic growth of the aerial stem, but that there is also an independent periodicity in the growth of the tuber, which is obscured and modified by the secondary induced periodicity, which is related to the aerial stem and its mode of growth.

It was observed by Lewis (1901) that the roots of *Allium cepa* showed two waves in their rate of cell division, the maxima coming at midnight and noon and the minima at 4 A.M. and 4 P.M. His work was substantiated by Kellicott (1904), who also observed that the periods of rapid cell division coincided with a low rate of elongation and that during rapid elongation the rate of cell division was the lowest.

Friesner (1920) in the study of the rate of elongation of the radicles of *Cucurbita pepo, Lupinus albus, Pisum sativum, Vicia faba,* and *Allium cepa* showed that elongation in all these plants proceeded in a wavelike fashion, two to four waves being exhibited in a 24-hr. period. He also found that the curve of cell division in these plants exhibited a number of oscillations and that the maxima of elongation were near the minima of cell division, and *vice versa.* He made the important discovery that the exact time of the occurrence of the maximal and minimal cell division is dependent upon the time of the initiation of activity and not upon the time of day.

The question now arises as to what are the causes of the rhythm found in both the elongation and the cell division of plants. It apparently is not due to external influences of changes in illumination and temperature, since this occurs in darkness under conditions of constant temperature. It is apparently not due to heredity, since it is generally lost when the plants are placed in continuous darkness and uniform temperatures. Friesner (1920) considered that when a large number of cells are undergoing mitosis, the total energy available within the tip is directed more to mitosis than toward elongation, so that the one process will be near to its maximum when the other is near to its minimum. These processes alternate with each other in the individual cell and in the root as a whole. The fact that these rhythms have a definite interval and that the maxima and minima in the different curves depend for the time of their appearance upon the time when activity begins indicates that this alternation between mitosis and elongation is entirely an internal cause and is not related to external conditions. Friesner suggested that a distinction should be made between the terms "periodicity" and "rhythm." He considered that the term "periodicity" as used by the earlier writers means a regular oscillation which is caused by the alternation of day and night or by other external changes and which is lost when the environmental conditions are rendered constant. The term "rhythm" he would restrict to mean any oscillation in activity which is definite and regular and not related to any external influence.

VI. GROWTH-PROMOTING AND GROWTH-INHIBITING SUBSTANCES

It was assumed by Sachs that specific organ-forming substances are needed for growth and that the exhaustion of these substances in the growing regions is the cause of the inhibition of the growth of dormant buds. Pfeffer (1900) considered that regulatory influences exerted by enzymes and other chemical substances play a role in growth. Fitting (1905–1906) studied the conduction of the photostimulus in the coleoptile of oats, and Boysen-Jensen (1910) concluded that this stimulus is conducted by the downward movement of a growth-promoting substance on the shaded side of this organ. Armstrong and Armstrong (1911) stressed the function of substances resembling hormones in the metabolism of plants, and Gericke (1924) concluded from nutritional studies of wheat that plants have a physiology which is subject to growth-inhibiting and growth-stimulating agencies.

It is now generally considered that the growth of plants and plant parts is promoted or inhibited by certain substances other than the inorganic and organic compounds that have been shown to be necessary for plant life. Most of these substances are apparently manufactured by the plant, although some seem to be obtained directly from the soil. These substances have been variously termed "vitamins," "hormones," "growth-promoting substances," or "growth-inhibiting substances."

A. Growth-promoting Substances

The fact that small amounts of organic extracts appear to stimulate plant growth to a marked degree has been taken to mean that substances other than those used directly in food building are necessary for the proper functioning of the plant. Thus the application of animal manure to the soil greatly increases the development of the plant. It has generally been assumed that this stimulation was due to the compounds of nitrogen, phosphorus, and potassium that it contained. Breazeale (1927), however, found that this stimulating property rests largely in the water-soluble organic matter and not in the inorganic compounds that it contains. He believed that the beneficial effects of manure and crop rotation are due in large part to the fact that their vegetative matter decomposing develops or sets free certain organic compounds that are essential to the growth of plants.

The cause of the dominance of the apical bud on stems and tubers and of the polarity of stems has always been a question of great interest, and much of the literature on growth-promoting and growth-inhibiting substances has been devoted to theories that attempt to explain such behavior. In extensive studies concerning the development of roots, buds, and shoots in *Bryophyllum calycinum*, Loeb (1915 to 1924) proposed several theories to account for the growth of certain plant organs and for the inhibition of others. One of these theories assumes that the growth of the bud depends upon the flow of certain substances from the leaf to the bud. The bud that receives these substances first will grow out first and thereby prevent the flow to the other buds whose growth is thereby inhibited. The apparent inhibition of growth in certain regions is thus due simply to the fact that under certain conditions the substances required for growth flow to some other place and are retained there. The removal of this inhibition consists in creating conditions that force the substances to flow to other places at which growth will then occur.

Appleman (1918) obtained evidence from sprouting potato tubers which indicates the presence of growth-promoting substances therein. Thus the vigor of potato sprouts bears a direct relation to the size of the seed piece or, in other words, to the amount of tissue surrounding the eye. When a certain minimum is reached, the vigor of the sprout decreases as the size of the seed piece is reduced, the plant resulting therefrom remaining weak during its vegetative period and yielding a small crop of tubers. The weak sprouts are not due to the lack of ordinary food materials, since a sufficient quantity of these for future growth remains in the tuber piece. It has also been shown that the ordinary inorganic nutrients are not the materials that are lacking under these conditions. Appleman concluded that the potato tuber contains a limited amount of a special growth-promoting substance, and that if the amount of tissue surrounding the growing bud is too small, there is not enough of this substance available for normal growth.

1. Metaxenia.—Any effect produced on the endosperm of an angiosperm by pollen from a plant having a different kind of endosperm is called "xenia." Swingle (1928) suggested the term "metaxenia" for the direct effect of the pollen on the parts of the seed and fruit lying outside the embryo and endosperm. This effect of the pollen has been especially noted and studied in the date by Swingle (1928, 1931) and Nixon (1928, 1934, 1935). They reported that the pollen affects the shape of the seed, and the size, shape, character and time of ripening of the fruit. The time of ripening can be varied as much as 10 days, depending upon the source of the pollen that is used for fertilization. Swingle believed that the simplest, and the most probable, explanation of this behavior is that the embryo or endosperm, or both, secrete hormones, or soluble substances analogous to them. These diffuse into the tissues that constitute the fruit and there exert a specific effect varying according to the particular male parent used to fecundate the embryo and endosperm.

Nebel (1930, 1934) reported that apples grown on the same tree but originating from different kinds of pollen were unlike. These differences though small were indicated in the length, color, weight of fruit, and length of seed. He noted also that the chemical composition of the fruit may be influenced by metaxenia, as indicated by differences in pH value, and in the amount of sugar and titratable acid. Hibbard (1933) and Degman and Auchter (1934) with the apple, and Tufts and Hansen 1933) with the pear, however, could detect no effects due to metaxenia.

Harrison (1931) stated that he had observed metaxenia in cotton as indicated by the length of time required for the fruit to mature, by the length of lint, and by the quality of the fiber on the seed.

2. Vitamins.—The term "vitamin" is used by animal physiologists to denote certain substances that are necessary in small amounts for the proper functioning of the animal organism. Williams (1928) suggested that the word "nutrilite" be used to designate all those vitaminlike substances which function in minute amounts in the general nutrition of organisms.

The vitamins present in animal tissues and in products of animal activity have for the most part not originated there but have been obtained from a vegetative source. Clark (1929) found that diatoms and fresh-water algae can form vitamins from purely inorganic constituents. The synthesis of vitamins by bacteria is in doubt. Yeasts contain a large amount of the water-soluble vitamin B, but it is not known whether it is necessary for the metabolism and reproduction of this plant. In India it has been observed that rice from different localities varied in nutritive value, and this has been correlated with soil differences. In millet the vitamin B content of the grain and in wheat the vitamin A content of the grain were highest when grown on manured land. There are apparently substances in manure that enable the plant to manufacture vitamins A and B (Rowlands and Wilkinson, 1930; Richardson and Mayfield, 1932). Mineral fertilizers, however, influence the production of vitamins little or not at all.

Haber, Nelson and Swanson (1931) found that the content of vitamin A in the leaves of *Coleus* was much higher when the plants were grown in fertile soil than when grown in sand. Clark and Thomas (1934) reported that if conditions are favorable to the growth of green plants, the formation of vitamin A is not affected by the absence of microorganisms. Within certain limits the variation in the amount of light has little or no effect on the production of this vitamin. Gunderson and Skinner (1934) stated that the unicellular alga, *Chlorococcum sp.*, synthesized vitamin A, or its provitamin, in large quantities, and vitamins B and G in lesser amounts but not vitamin C when grown in complete darkness on a medium of inorganic salts and dextrose. Vitamins A, B and G thus must have been elaborated by the cells without light, bacterial symbiosis, or complex nutrients. Richardson and Mayfield (1932), and Virtanen (1936) stated that the quantity, potency, or available amount of a certain vitamin in a plant of a given species is not uniform. This production appears to vary with the variety of plant, its maturity, and the climatic and soil conditions of its habitat.

It was observed by MacLeod, Armstrong, Heap, and Tolbert (1935) in five varieties of sweet potatoes that the content of vitamin A increased as much as 3 and 4 per cent after a storage period of 2 months. This increase in vitamin A after harvest indicates either (1) that the carotene in the sweet potato is not fully developed when the roots are harvested and thus has not yet developed into the precursor of vitamin A, or (2) that the carotene responsible for the formation of vitamin A is present in a form less available to the animal body when the roots are first harvested than after a period of storage. Dove and Murphy (1936) stated that there is a wide difference in the vitamin C content of apples, and that the ascorbic acid content of the leaves has a direct relation to the vitamin C content of the fruit.

Although it is evident that plants manufacture the vitamins that are so essential to the growth and development of animals, the evidence as to their role in the plant body is very limited. The fact that they are frequently present in the largest amounts in those parts of the plant where the metabolic processes are the most active has been considered to indicate that they play a role in the nutrition of the plant (Tschirch, 1921). Havas (1935) added 3- to 5-cc. portions of ascorbic acid in varying concentrations (1 to 10,000, 5 to 10,000, 2.5 to 1,000, and 5 to 1,000 parts of water) to seeds germinating under sterile conditions. The two lowest concentrations proved to be the stimulatory ones. Their addition caused no stimulation of germination but produced an acceleration of growth indicated by an increase of 25 to 30 per cent in the length and the weight of shoots, and an increase of 50 per cent in the weight of the roots. The seedlings of oats were much less sensitive to this treatment than those of wheat. Hausen (1935) added ascorbic acid to nutrient solutions and obtained 35 to 75 per cent increases in the dry weight of the plants. The treated plants also showed a much higher content of vitamin C than the controls. Virtanen (1936) found that the addition of crystalline vitamin C to the medium in sterile cultures of peas caused an increase of 40 to 100 per cent in the dry weight. It is thus considered that vitamins are essential to the best growth and development of green plants.

3. Growth Hormones.—According to Snow (1932) the best known of the "growth substances" that normally occur in plants is at present the substance formed by the tip of the coleoptile in seedlings of members of the grass family. This growth substance, "auxin," "growth regulator," "growth hormone," or "phytohormone" accelerates the elongation of the coleoptile.

a. Historical.—The first indication that the tip of the coleoptile normally produces such a substance was obtained by Paal (1914, 1918) while investigating the transmission of a phototropic stimulus across a discontinuity. After excising the tips of the coleoptiles of *Coix lacryma*, he replaced them eccentrically on the stump and found that the coleoptiles curved strongly away from the side covered by this tip. He assumed this indicated that the tip produced a growth-accelerating substance which passed across the moist discontinuity and down that side of the coleoptile. The result was that the treated side grew faster than the other and caused the coleoptile to curve. Söding (1926, 1929) substantiated the presence and behavior of a similar substance in the coleoptiles of the seedlings of the oat. Went (1928), however, fully established the presence of this growth substance in the tips of coleoptiles of various seedlings and devised means for its quantitative determination. From 1928 to the present, an enormous amount of work has been done to determine the occurrence, nature, transport and the stimulatory action of growth substances. The facts obtained must of necessity be discussed here only briefly, and the student is referred for a detailed review of the literature to Snow (1932); Went (1935); Stiles (1935); Thimann (1935); Boysen-Jensen, Avery, and Burkholder (1936); and Went and Thimann (1937).

b. Occurrence and Preparation.—Growth substances have been obtained from many plant and animal sources. Thus they occur in the coleoptiles of the seedlings of various grains, in seedlings of various other species of plants, in buds, root tips, leaves, green algae, mammalian urine, and saliva. It was found by Nielsen (1930, 1931), Boysen-Jensen (1931, 1932), and Nielsen and Hartelius (1932) that relatively large quantities of growth substances which accelerate the growth of coleoptiles but retard or completely arrest the growth of roots are produced by various fungi and bacteria. Two genera, *Rhizopus* and *Aspergillus*, produce relatively large amounts of these growth substances to which Nielsen gave the general name, "rhizopin." Seubert (1925) found that growth substances are associated with malt extract, pepsin, and diastase. Thimann and Dolk (1933) found that aeration greatly increased the production of growth substance by *Rhizopus*.

Some cytological work has been done to determine which cells of a certain region produce the growth substances. Tetley and Priestley (1927) could find no definite region in the coleoptile of barley for the production of growth substances, since all the cells appeared vacuolated and inactive. They concluded that the response of the coleoptile to light is not the result of any chemical substance but rather of differential permeability of tissue on the two sides of an unequally light-stimulated organ. Perry (1932) found that the structure of the epidermal layer of the coleoptile suggests a secretory function. Over the tip of the coleoptile the cells making up the dermatogen are numerous with large, deep-staining nuclei, dense cytoplasm, small vacuoles, and many granules. Continuous with these cells and extending down the sides of the coleoptile are the epidermal cells, which become longer as their distance from the tip increases. He suggested that the growth substance is produced in the tip cells, and that it is transported to the region of bending by the rapid streaming of cytoplasm in the relatively long epidermal cells. They could find no evidence of the secretion or translocation of growth substance in unequally illuminated coleoptiles. Avery and Burkholder (1936) studied cytologically the development of the coleoptile of *Avena* and concluded that cell division is not involved in coleoptile growth at the time that it is used in tests for growth substance.

The growth substances are obtained by the diffusion method and by the extraction method. In the diffusion method, the growth substance diffuses into water or into an aqueous gel. The latter type of medium was first used by Went (1928) and is now the most universally used. The parts that contain the growth substances are decapitated, and the cut end is placed on 3 per cent agar, into which the substances diffuse. After this diffusion has progressed a sufficient time, the agar is cut into cubical blocks of a standard size and used for experimentation.

The extraction method may be illustrated by describing the procedure used by Thimann (1934). The fresh material is killed by immersion in chloroform, after which about one-fifth its volume of *N* hydrochloric acid is added, and the mixture thoroughly ground. The chloroform layer is removed, fresh chloroform is added, and the acidified tissue is again ground. The chloroform is evaporated and the lipoidal material is extracted in a small volume of water. This extracted material, which contains the growth substance, may then be mixed with agar and its action determined by the usual methods.

According to Went and Thimann (1937) the extraction method of Thimann (1934) demonstrated that growth substance is distributed throughout the seedling plant, although the greatest amount occurs at the tips. It also showed that the growth substances are present in the plant in the free-moving form and in the bound form. The chloroform method determines the total amount of growth substance present at a given time, while the diffusion method is successful only if there is continual production of the material.

c. Determination of Action.—Two general methods have been used for determining the action of growth substances. The one that has been the most universally used was first reported by Went (1928) and may be called the "coleoptile method." This method involves a determination of the degree of curvature of a decapitated coleoptile on which has been placed the growth substance being tested. A block of agar containing the growth substance is applied unilaterally to the decapitated coleoptile sometime during a 40-min. period following its decapitation. The time allowed for action varies from 110 to 120 min., after which the degree of curvature of the coleoptile is determined. It was established by Went (1928) that there is a direct proportionality between the growth substance and the curvature produced. Thimann and Bonner (1932) reported that the curvature is proportional to the concentration of the growth substance and not to the total amount present.

Various units have been proposed for determining the degree of curvature due to the growth substance. Kögl and Haagen-Smit (1931) introduced the term "Avena unit" or "A. E." (Avena Einheit). This unit is defined as the amount of growth substance that, when contained in a block of 3 per cent agar measuring 2 by 2 by 0.5 mm., imparts a curvature of 10 deg. in 2 hr. at 22 to 23°C. to one decapitated coleoptile of *Avena* upon which the agar block is placed eccentrically. They found that from the tips of the coleoptiles of barley, there diffused outward a solution containing 300 A. E. per milligram. Urine contained 400 A. E. per milligram. From urine these workers isolated a pure crystalline substance that had an efficiency of 30,000,000 A. E. per milligram. One milligram of this substance thus is sufficient to impart a curvature of 10 deg. to 30,000,000 *Avena* coleoptiles.

Bonner (1932) described a unit of growth substance as that amount of "rhizopin" which when dissolved in 1 cc. of water and mixed with 1 cc. of 3 per cent agar in the form of agar blocks of standard technique and applied to the side of a decapitated *Avena* coleoptile will produce in 110 min. a deflection of 1 deg.

In about 3 hours after the coleoptile is decapitated, the uppermost part of the stump begins to act physiologically like a tip, in that it begins to produce growth substance (Söding, 1929). This behavior is known as the "physiological regeneration of the tip. To avoid this complication in experimental work the coleoptiles are decapitated a second time after 2 hr. Physiological regeneration of the stump is much retarded by replacing the original tip.

The "pea method" for the quantitative determination of growth substances was devised by Went (1934). The seedlings of *Pisum sativum* were grown in sand cultures in the dark until the epicotyls were 5 to 20 cm. in length. Pieces 2 to 20 cm. in length were cut from the stem about 5 mm. below the growing point and the top of each piece of stem was split longitudinally for 1 to 3 cm. If such a piece of stem is immersed in an aqueous solution, the two halves bend outward. If a growth substance is present in the solution, the free ends of the two halves begin to curve inward after approximately 1 hr. at 25°C. The inward curvature is greater the more concentrated the growth substance, and equilibrium is generally attained in about 6 hr. By using a range of concentrations of the substance to be tested and finding the lowest concentration in which a recognizable reaction occurs, the test can be made quantitative.

d. Mode of Action.—Proof has been furnished by Heyn (1931, 1934), Heyn and van Overbeek (1931), Söding (1931), Thimann and Bonner (1933), Bonner (1933, 1934), and Went (1935) that the application of growth substances to coleoptiles increases the plasticity of their cell walls. Opinion differs as to whether the initiation of growth is caused by an increase in turgor or by an increase in elasticity or plasticity of the walls; and also as to whether or not the laying-down of the new particles is itself a necessary accompaniment of growth. Growth substances seemingly induce only the elongation of the cells that have been formed already, for in the zone of elongation of the coleoptile the cells have ceased to divide. Althoslechner (1934) believed that certain "hormones" are concerned in initiating cell division. Elasticity, however, may or may not increase. The changes in elasticity are correlated with growth but not with growth substances. Increased elasticity is a result of growth, while increased plasticity is the cause of growth. It was observed by duBuy (1936) that with increasing age, the response of *Avena* coleoptiles to growth substances decreased. Plants are only sensitive to these materials at a certain stage of their development.

The amount of growth substance that enters a given region is too small to bear any simple stoichiometrical relation to the substances that are formed in the cell wall during the resultant growth. The growth substance is thus a true hormone in that it is active in minute amounts and in an indirect manner. It has been found that if growth substances are applied to plasmolyzed cells no increase in plasticity occurs, and hence no growth occurs. Since the increase in plasticity occurs only when the protoplasm is in contact with the cell wall, it is inferred that growth substance acts on the cell wall through the protoplasm. This belief is further substantiated by the fact that growth substances may increase the rate of respiration as much as 27 per cent. It was observed by Dolk and Thimann (1932) that if growth substance is supplied to the *Avena* coleoptile in a solution buffered to an alkaline pH, its effect in causing cell elongation is abnormally small. It is common experience that solutions of growth substance must be acid to give the maximal effect of cell elongation. Bonner (1934) found that acid solutions increase the growth rate by increasing the plasticity of the cell wall. There is no increase in cell acidity, however, under the influence of growth substance. It is considered that the increase of growth rate in acid buffers is due to the conversion of growth substances, already in the plant, from an inactive to an active form.

e. Chemistry.—The chemistry of the growth substances has been studied by many workers, including Kögl and his helpers (1931, 1933), Dolk and Thimann (1932), Went (1934), and Thimann and Went (1934). It was suggested by Kögl that the term "auxin" be used for any phytohormone that causes cell enlargement. He distinguished three auxins—auxin *a*, auxin *b* and heteroauxin. Auxin *a* is a monocyclic, trihydroxycarboxylic acid with one double bond and with the formula $C_{18}H_{32}O_5$. It is stable in acid but not in alkali, has a molecular weight of 328, and is heat- and light-stable. Auxin *b* is a monocyclic, hydroxyketocarboxylic acid with one double bond and with the formula $C_{18}H_{30}O_4$. It is stable in light and heat, has a molecular weight of 310, and is destroyed by both acids and alkalies. Heteroauxin is β-indole acetic acid, or 3-indole acetic acid, and has the formula $C_{10}H_9O_2N$. It is destroyed by acids but is stable in alkali and has a molecular weight of 175. This is the auxin that is present in yeast and other fungi.

Hitchcock (1935) and Zimmerman and Wilcoxon (1935) showed that there are numerous organic compounds that may be classed as growth substances. These compounds were dissolved in solutions in which cut stems were placed, were injected into plant parts, or were absorbed by lanolin which was rubbed on the stem or leaf. The principal plant responses initiated by these substances were the local initiation of adventitious roots or stems, proliferations, swelling and bending of stems, acceleration of growth and epinasty of leaves. These compounds moved both upward and downward in the stem. Some of the substances used were indole-3-*n*-propionic acid, phenylacrylic acid, phenolpropionic acid, carbon monoxide, ethylene, propylene, acetylene, α-naphthaleneacetic acid, β-naphthaleneacetic acid, fluoreneacetic acid, anthraceneacetic acid, and α-naphthylacetonitrile. They considered that there are probably many compounds other than those now known which would be equally effective on growth. If plants manufacture their own growth substances, it is not probable that any one plant can produce all those compounds that are now known to be effective. Likewise it is not logical to assume that all plants naturally make and use one and the same growth substance. The growth of a given plant may not be regulated always by the same substance but this may vary with the environmental conditions of the plant.

Wallace (1927) found that 1/100 p.p.m. of ethylene in air induced intumescences in the twigs of the Transparent apple. Williams, Lyman, Goodyear, Truesdail, and Holaday (1933) reported that a single acid substance apparently occurs in all tissues, and that it is capable of stimulating the growth of yeast. They called this substance "pantothenic acid," a term meaning "from everywhere." Tincker (1935) failed to obtain accelerated growth or flowering by presenting small quantities of ketohydroxyoestrin and theelol to flowering plants in different ways. Havas and Caldwell (1935) reported that glandular extracts had little or no effect on the growth and development of plants. Levine (1934) reported that different portions of plants which had been painted or injected with the carcinogenic agents of animals showed a proliferation of tissue around the treated parts giving rise to the swellings of the stem accompanied with small irregular masses of new tissues. Davis (1934) found that plants of *Vicia faba*, *Allium*, *Narcissus*, and *Pisum sativum* which were injected with a solution of thyroid material in water showed a stimulation in the time of flowering and in the height and number of flowering stalks.

The classification of ethylene as a growth substance has been questioned by Thimann (1934) and Michener (1935). Since, in observed cases, ethylene affects the production of growth substances but not its transport or utilization, they believed that ethylene influences only the enzyme concerned in the production of growth substance.

f. Transport and Polarity.—According to Snow (1932) the transport of the growth substance in coleoptiles is polar since it travels only in a morphologically downward direction. If the lower end of a portion of the coleoptile is placed on an agar block and a cube of agar containing the growth substance is placed on the upper end, the growth substance diffuses downward into the agar block below. If the cylinder is inverted, however, no translocation of growth substance occurs. If the cube of agar containing the growth substance is placed at the bottom of the inverted coleoptile it will diffuse upward into the pure agar on the other end. The growth substance thus must be transported in a manner different from that by which the common organic substances are moved for they travel in either direction according to the circumstances. It has been demonstrated that the growth substance does not travel in the two small vascular bundles to the coleoptile. Avery (1935) noted in leaves of *Nicotiana* that the transport of auxin is polar, in that it moves from the leaf tip to base through the midrib, to which it is conveyed by the lateral veins. Snow (1933) found that the cambial stimulus from the leaves can pass across a protoplasmic discontinuity, which indicates that the stimulus is caused by a hormone.

van Weij (1932) distinguished between the "velocity of transport" and the "intensity of transport" of growth substances. The velocity of transport is measured by the time that elapses before the first appreciable trace of growth substance reaches the basal end after traveling through a coleoptile of known length. The intensity of transport is measured by the amount of growth substance that subsequently reaches the lower end per unit of time. The intensity of transport through paths of different lengths is approximately the same. This behavior differs from simple diffusion, since in diffusion the intensity varies inversely with the length of the path. The intensity of transport varies with the temperature, reaching a maximum at about 30°C. and falling off above or below this point. The velocity of transport, however, is nearly independent of temperature. The concentration of growth substance in the blocks of agar on the lower ends of the coleoptiles continues to rise until it is much higher than in the upper blocks in which the growth substance is contained at the start. The increase in amount does not appreciably retard the diffusion into the lower part. Koch (1934) stated that the transverse transport of growth substance in the *Avena* coleoptile was slower than the longitudinal transport.

g. Tropisms.—Numerous investigators have believed that the geotropic reactions of stems and roots are controlled in part at least by growth substances. Thus it was shown by Hawker (1932, 1933) that if the root tips of *Vicia faba* were placed on blocks of gelatin, and these blocks subsequently placed eccentrically on the cut ends of decapitated roots, the latter curved to the sides covered by the blocks. This showed that these root tips excreted a growth-retarding substance, since blocks of pure gelatin similarily placed caused no curvatures. These results supported the theory of Cholodny (1924, 1926, 1927, 1928) and Went (1928) that the opposite curvatures of root and coleoptile in response to gravity are caused by the growth substance formed by the tips of these organs which ordinarily travel in straight lines to the elongating zones and arrive there in equal concentrations on all sides. Growth is thus equal on all sides of the stem and root and they extend in a straight line. As a result, however, of the stimulus of gravity, these substances are somewhat diverted so that they travel obliquely and accumulate in greater concentration on the lower side of the elongating zone. This procedure accelerates growth on the lower side of the stem as compared to the upper and the stem grows away from the stimulus of gravity. In the root, growth is retarded on the lower side and this organ responds in a positive manner to the stimulus. This general theory was substantiated by Keeble, Nelson, and Snow (1929); Navez and Robinson (1932); and Boysen-Jensen (1933), who found a greater quantity of growth substance on the lower than on the upper side of a root that had been geotropically induced for 2 to 4 hr.

It has generally been assumed that a growth substance must be responsible for the response of plant organs to light. Sierp and Seybold (1926) found that in the coleoptile of the oats the uppermost portion of the tip back to a distance of ¼ mm. is sensitive to light, and that this sensitivity decreases rapidly towards the base. Navez (1933) found that the apical portions of seedlings of *Lupinus albus*, grown under illumination, released into agar blocks twice as much growth substance as those grown in the dark. According to Boysen-Jensen (1936), it has been proved that the concentration of growth hormone is decreased on the illuminated side and increased on the shaded side of a bending organ. From studies of the *Avena* coleoptile it is assumed that unilateral illumination does not affect the formation of growth substance but brings about its displacement toward the shaded side during the course of its downward movement. The subsequent rate of growth on each side is proportional, within limits, to the concentration of the growth substance present. The cause of light exercising an influence upon the distribution of the growth substance is not understood. Keeble and Nelson (1935) believed that negative and positive traumatic curvatures may be interpreted in terms of growth substances and their concentration on opposite sides of a wounded root.

h. Root Formation.—According to Snow (1932) it was first reported by Cholodny (1924, 1926, 1928) that the coleoptile tips of maize retard the growth of decapitated maize roots when placed upon them. These observations were confirmed by Keeble, Nelson, and Snow (1931). The same workers in 1930 observed that if the decapitated roots of the seedlings of *Zea mays* and *Pisum sativum* were washed from 15 to 20 min. they grew much more rapidly than those not so treated. They considered that the wound substances which are produced at the cut surface and which retard growth are largely removed by washing. Hammett (1929) showed that the portion of the root in which mitotic activity occurred contained an acid-stable, alkali-labile substance that stimulated root growth.

Went (1929) reported that he extracted a heat-stable substance from germinating seeds and barley leaves which promoted the formation of new roots when it was applied to cuttings. In 1934 he gave the name "rhizocalin" to this substance. In 1934 he described a method for the quantitative determination of this substance. Rhizocalin mixed with lanolin is an excellent material for the induction of root formation. Navez (1933) found that the decapitation of a root of the *Lupinus* seedling stopped its elongation for a varying period of time. After this cessation there was a period of renewed elongation at a rate lower than the original one. The tipping of the decapitated root with a tip of the coleoptile of *Zea mays* or *Avena* induced a decrease in the rate of elongation of the root. The diffusate of root tips has a similar action on the rate of root elongation. The evidence indicates that the growth substances of the coleoptile tip or root tip inhibit the rate of elongation of the decapitated root. Boysen-Jensen (1933) obtained growth substance from the root tips of *Zea curagua* and *Vicia faba*. It appeared to be concentrated in the first 6 mm. of the root tip. It was noted by Kögl, Haagen-Smit, and Erxleben (1934) that the immersion of roots of the oats seedlings into a dilute solution of auxin *a* greatly inhibited their longitudinal growth. It was stated by Cholodny (1934) that the decapitated *Avena* coleoptiles gave significant curvatures in 2 hr. after the application of the root tips. The application of the root tips of *Lupinus*, *Cucurbita*, and *Helianthus*, however, gave no response. Loo and Loo (1935) found that the extracts of willow roots and of mulberry leaves stimulated the elongation of the primary root of *Zea mays*.

Zimmerman, Crocker, and Hitchcock (1932, 1933) exposed a large number of species of plants to carbon monoxide. They found that under the conditions of the

experiment, this gas induced the initiation of roots in numerous young stems. This gas also stimulated the growth of preexisting root primordia in the older part of stems. The root hairs also were more abundant on the roots exposed to this gas than on the controls. Butyric acid, acetylene, propylene and ethylene gave similar results in inducing the formation of root primordia. Zimmerman and Hitchcock (1933) called these chemicals "root-forming substances," since they are specific for the formation of adventitious roots. When only the root tip was treated, new roots formed directly back of this tip. If the root was treated along the region of elongation, root formation was limited to this region. If a zone 8 in. back of the tip was treated, elongation was retarded and new roots were formed all the way down to the tip region. Thus the indications are that the root-forming substance has a direct action on the tissues with which it comes in contact. Hitchcock and Zimmerman (1936) noted that when the cuttings or shoots of *Ilex, Taxus, Hibiscus, Acer,* and *Chrysanthemum* were treated with preparations of indoleacetic, indolepropionic, indolebutyric, or naphthaleneacetic acids, earlier rooting was induced, more roots were produced, and the roots emerged from a greater area of the stem than was observed in the controls.

Cooper (1935) mixed 1 part of pure, synthetic β-indylacetic acid with 2,000 parts lanolin and applied 10 mg. of this paste unilaterally on small areas of the cuttings of lemon, lantana, fig and *Tradescantia.* In all cases the cuttings showed a much greater root development than the untreated ones.

Zimmerman and Hitchcock (1935) applied lanolin preparation of α-naphthaleneacetic acid, indolebutyric acid, indoleacetic acid, indolepropionic acid, Δ-3-indolevaleric acid, and phenylacetic acid to growing roots. When these substances were thus applied along the region of elongation, new branch roots appeared through the epidermis in 3 to 5 days. They also caused swelling and retardation in the elongation of these roots. Application of the growth substance back of the region of elongation was relatively ineffective. During the same year these workers reported that the application of heteroauxin, indolepropionic, indolebutyric, naphthaleneacetic, phenylacetic, and phenylpropionic acids to the soil induced responses on the plants (tomato and tobacco) that were similar to those produced by the application of these substances in lanolin or oil to the aerial parts of the plants. Davies, Atkins, and Hudson (1936) reported that when willow twigs were set in Pfeffer's solution, containing various growth substances in concentrations of 1 part per 100,000, the order in which both roots and shoots appeared was: ascorbic acid, β-indylacetic acid, β-indylpropionic acid, and control. Zimmerman, Hitchcock, and Wilcoxon (1936) added the nine esters—methyl α-naphthalene acetate, ethyl α-naphthalene acetate, methyl phenyl acetate, ethyl phenyl acetate, *n*-butyl acetate, isobutyl acetate, methyl β-indole acetate, methyl β-indole propionate, and methyl β-indole butyrate—to the list of growth substances. These esters when absorbed by the roots from water solution and moved upward cause formative responses in the aerial parts.

i. General.—It was noted by Wallace (1931) that the concentration of ether necessary to prevent the evening closure of the leaves of *Mimosa pudica, Oxalis stricta,* and *Marsilia macropus* was different from that necessary to prevent their opening in the morning. The optimum temperature for the sensitivity of *Mimosa* was 40°C.; the minimum, 14°C.; and the maximum, 60°C. Sensitivity appeared independent of relative humidity. The period of greatest sensitivity was about 5 A.M., while the period of least sensitivity was from 1 to 7 P.M. The leaves were not sensitive after an exposure to darkness of 4 to 5 days. von Euler and Philipson (1932) reported that the coleoptile of *Avena* yielded an extract containing a substance which stimulated by 20-fold the growth of the yeast, *Saccharomyces cerevisiae.*

Snow (1935) found that the application of 1 to 2 p.p.m. of pure synthetic heteroauxin and of pure auxin *a* in gelatin to the decapitated stems or hypocotyls of young sunflowers greatly activated cambial growth. He concluded that cambial growth is activated by the same growth substance that is formed by young leaves and promotes extension in the stems. Leonian (1936) noted that a growth substance extracted from garden peas greatly stimulated the growth of certain of the unicellular green algae.

Levine (1936) reported that scars and swellings are produced when certain injured parts of the stems of tobacco, tomato, sunflower, *Ricinus*, and *Datura* are painted with numerous substances including glycine, glutamic acid, and cystine. LaRue (1936) reported that certain growth substances in agar or in lanolin inhibited the development of the abscission layer in the petioles of *Coleus* when applied to these parts.

Some investigators, however, do not consider that it is necessary to assume the existence of special growth-controlling substances to explain growth behavior. Gustafson (1927) believed that growth, especially in fruits, is influenced mainly by nutrition. As the cells begin rapidly to divide, rapid growth ensues, provided that food is supplied; and when the supply of food becomes limited, cell formation ceases and the fruit begins to mature. He considered that the growth of shoots and roots can be explained in the same manner. Thus, in the tomato when the nutrients were not diverted to fruit formation, the rate of growth of the shoot continued at approximately a uniform rate in contrast with an increasingly lower rate when fruiting was taking place. Pearsall (1923) stated that, although there appears to be some evidence in plants of the effect of specific substances, it is not known whether we are dealing with specific substances that are products of cell metabolism or simply with combinations of direct effects.

B. Growth-inhibiting Substances

The absorption by the plant of compounds from the soil which inhibit plant growth has been discussed in detail in Chap. III. The discussion here will deal especially with the inhibitors that retard the development of stems and roots.

1. Buds and Leaves.—Loeb (1915 to 1924) proposed a theory to account for the dominance of the apical stem upon the basis of the formation of growth-inhibiting substances by the plant. He assumed that the growing apex of a stem might form a definite substance which is antagonistic to bud growth and which, migrating toward the base, would impede or inhibit the growth of the lower buds. The apical buds are the first to be freed from this substance and begin to grow out, producing at the same time fresh supplies of the inhibitor, the concentration of which increases gradually toward the base of the stem. The subapical buds are therefore retarded in growth increasingly in this direction, until finally the point is reached where the concentration is great enough to inhibit lateral shoot growth completely.

Appleman (1918) obtained evidence that the terminal buds of the potato tuber inhibit the development of the lateral ones. Thus when all the eyes of a tuber are subjected to the same external conditions, the buds on the terminal or seed end will grow out first, the number of such buds developing depending upon the variety, size of the tuber, and the vigor of the terminal sprouts. The number of eyes on the tuber, however, bears no direct relation to the number of sprouts that will grow out from the terminal end. If the terminal buds grow normally, they inhibit the growth of the more basal ones. If, however, the terminal sprouts are destroyed or otherwise retarded, sprouts will appear on the stem half of the tuber, which otherwise would have remained dormant. If the tubers are cut transversely into halves, the suppressing influence of the terminal eyes is thus removed and the eyes on the stem end show an equal if not greater capacity for the production of vigorous sprouts than the terminal bud (Fig. 39).

FIG. 39.—A potato tuber bearing an axillary sprout *B* and a terminal sprout *A*. The axillary branch developed because the tuber was notched and thus the growth-inhibiting substance secreted by the terminal branch prevented from reaching the axillary region and inhibiting growth there.

Reed and Halma (1919) believed that the manner of growth of the young shoots of pear and Chinese lemon observed by them afforded evidence for the hypothesis that a growth-inhibiting substance is produced in the apical portion of the shoot, and that it travels downward toward the base, causing a condition of dormancy in the subapical buds. It was observed by Priestley and Evershed (1922) and Priestley and Pearsall (1922) that the growth curve of the roots formed from cuttings of tomato and *Tradescantia* showed a progressive series of S curves. In all cases the first curve ended just before the secondary root formation and the second reached its low level at the appearance of the tertiary roots. They attributed these phases of retardation to the operation of inhibiting factors such as accumulated end products that were left over from the previous growth.

Snow (1925, 1929, 1931, 1932) concluded that the axillary buds in *Phaseolus* are inhibited in development by a substance that originates at the growing tip. The terminal bud of *Vicia faba*, the youngest leaf, the next leaf, and the third leaf possessed respectively 120, 22, 15, and 3 units of growth substance. In *Pisum sativum* the inhibiting effect exerted by the shoot upon the axillary buds comes from three or four of the developing leaves. A leaf of this plant begins to inhibit when it has reached a length of 2.0 to 2.5 mm.; the inhibitory action begins to fall when a length of 20 mm. is attained and ceases entirely when it has reached a length of 40 mm.

It was observed by Keeble, Nelson, and Snow (1931) that the growth of the decapitated roots of maize was retarded by the tips of coleoptiles or the tips of roots. When these tips were placed eccentrically upon decapitated roots, they curved toward the side covered by the tip. This indicated that the growth substance excreted by these tips inhibits the elongation of the cells of the root. Thimann and Skoog (1933, 1934) found that the terminal bud of *Vicia faba* produced a growth substance similar to that obtained from *Rhizopus suinus*, but none was observed in the undeveloped lateral buds. When the terminal bud was decapitated, the lateral buds developed rapidly and each produced approximately one-half as much growth substance as the terminal bud of the intact plant. This behavior may account for the well-known fact that the axillary buds of the stem grow rapidly when the growing tip of the main stem is removed. Tukey and Brase (1933) noted that in "double-worked" fruit trees, the intermediate stem piece influenced the entire plant by producing a dwarfing effect. Van Overbeek (1935) considered that the difference in growth of normal corn and *nana*, a dwarf type, is due to the fact that dwarf corn produces a smaller amount of growth substance than the normal plant.

2. Fruits, Seeds, and Flowers.—The effect of flowering and fruiting upon the vegetative growth of plants has been studied by Murneek (1925, 1926, 1932, 1933) who also summarized the observations of Egorov (1915), Mason (1922) and Taranovsky (1923). It appears that in all types of plants an antagonism exists between vegetative growth and reproduction. In plants with determinate growth, as in the grains and grasses, the vegetative structures undergo rapid senile changes and die promptly upon fertilization. Those plants with an indeterminate type of growth, such as tomato, cotton, pumpkin, some legumes, and most perennial woody plants, do not show this clear-cut antagonism between vegetation and reproduction. Vegetative growth is curtailed or ceases when a maximum set of fruit or seed is being formed, but there is a resumption of vegetative development after the maturing of the fruit. Many apple trees that bear heavily during a given season frequently do not yield fruit the following year and may require several years for complete recovery.

Murneek suggested that some definite and powerful mechanism must be at the disposal of the reproductive organs to accomplish the transfer of the food supply to the newly formed embryo. There is nothing known regarding the nature of this mechanism beyond the conjecture that it may be either of a general physiological or of a specific enzymatic or hormonic character.

According to Murneek the cultivated apple exhibits four waves of abscission of nonsetting flowers and immature fruits, which occur at intervals of approximately 2 weeks and appear to be governed primarily by internal, hereditary factors. The causes for these abscissions may be (*a*) structurally and functionally abnormal flowers, (*b*) unpollinated or unfertilized flowers, (*c*) fertilized flowers in which the embryos abort early, and (*d*) the abscission of some fruits owing to the competition for a local food supply.

Mason (1922) observed in cotton that there was a marked retardation in the growth of both the central stem and lateral branches during flowering and fruiting. This retardation was proportional to the number of flowers that were formed. Kearney and Harrison (1932) found that when two species of cotton were pollinated with a mixture of pollen from both species, selective fertilization in favor of like pollen occurred. These results were explained on the assumption that the presence of like pollen induces a reaction in the stigmatic tissues which renders them less suitable for the development of unlike pollen. The development of the pollen tube appears necessary for the reaction.

McCollum (1934) noted in the cucumber that growing fruits exerted an inhibiting influence upon plant development until the seed coats of the developing seeds began to harden and mature. Parthenocarpic fruits did not produce a striking inhibitive effect on the plant. It is assumed that this inhibitory effect is due to the production of a growth-regulating substance by the developing, fertilized ovaries. Austin (1935) stated that the removal of the flowers from the soybean does not affect the growth of this plant the same as it does many others, but that the exfoliated plants cease growing at the same time as the controls. The changes that accompany the development of fruits in the normal soybean plant are: a diminution of water, a marked decrease in the potassium content in all parts, and a low amount of phosphorus in the stem tips.

3. Volatile Substances.—It was observed by Elmer (1932, 1936) that a volatile substance emanating from ripe apples inhibits the normal sprout development of germinating potatoes. The apical growth of the sprouts is practically stopped, and small stem tubers instead of normal sprouts may develop. Immature apples, however, did not have this effect. Pears and the fruit of the hawthorn also produced the same effect as the apples. Kidd and West (1932) reported that ripe apples, pears, peaches, tomatoes, and bananas produce a vapor that has the effect of stimulating at once the beginning of ripening. Smith and Gane (1933) and Kidd and West (1934) emphasized the need of having ripe fruit to obtain these effects. Nelson and Harvey (1935) believed that a gaseous, unsaturated hydrocarbon, or similar compound, is present in self-blanching celery during the natural blanching of the leaves.

Denny and Miller (1935), Denny (1935, 1936) and Gane (1935) obtained epinastic response of leaves from the emanations of fruits of the apple, pear, tomato, banana, cantaloupe, squash, and eggplant; of petals of geranium, verbena, hollyhock, and petunia; of anthers of the lily; of pistils of the unopened flowers of rose, squash, and hollyhock; of leaves of the rose, spinach, onion, Virginia creeper, potatoes, and tomato; of roots of the radish, turnip, and beet and of the stem of the tomato. Epinasty was noticed after 4 to 6 hours of exposure to these emanations. Some of the plants that are sensitive to the volatile products of ripe apples are *Mimosa pudica, Ricinus communis, Helianthus, Salvia,* and *Solanum lycopersicum.* Some of the seedling plants that are affected by these emanations are pea, radish, turnip, white mustard, cress, peanut, flax, and runner bean.

Elmer (1932, 1936), Botjes (1933), and Denny and Miller (1935) believed that this gaseous emanation is ethylene. Gane (1934) reported that ethylene was identified among the gaseous products produced by ripe apples.

VII. DORMANT OR RESTING PERIODS

The rest or dormant period in plants is a period when the plant or plant parts to all appearances do not grow. Woody plants especially exhibit this characteristic, and it is very common in seeds, bulbs, tubers, buds, and spores. Plants and plant parts vary greatly in the extent of this dormant period. In some instances it may last for only a few days, while in some seeds it may last to 100 years or more. Woody plants go into the dormant condition in the temperate zone at the close of the summer, while crocus, hyacinth, spring beauty, and tulip die to the ground in the spring, and their bulbs become dormant by early summer and almost never begin to grow again before fall, no matter what sort of weather prevails. The majority of seeds will not germinate at once when placed under the conditions that ordinarily produce germination. Howard (1915) found that, if 2 weeks is sufficient time to allow seeds for making immediate germination, fully 75 per cent of the species of Missouri exhibit dormancy. Of the 76 species of seed examined by him, 23.6 per cent grew in less than 3 weeks, 39.4 per cent in the fall or following spring, while 37 per cent did not germinate at all during the time of his experiments.

Duvel (1905) and Goss (1924) found that of the 107 species of seeds buried by them, 50 species grew after a period of 20 years, the germination ranging from 1 to almost 100 per cent. Ohga (1923) found that seeds of *Nelumbo nucifera* taken from strata of soil in a railroad cut and estimated to be from 120 to 400 years old still retained their vitality and readily germinated after their coats were filed to allow the intake of water. The rest or dormant period in the seeds of wild plants is more general and more persistent than in the cultivated ones.

Eastham (1914) found no loss in the vitality of oats after storage for 10 years. Sifton (1920) reported that wheat seed retained its vitality for 5 years, but that 75 per cent of its vitality was lost between the eleventh and fifteenth years. The longevity of oats was found to be greater, and 14 per cent of the kernels were alive after 19 years. The vitality of timothy seed began to decrease at once, and after 12 years it had decreased 88.5 per cent. Darlington (1931) tested the viability of seeds that had been kept for 50 years in sterile, dry soil in sealed pint bottles. He obtained 52, 38, 62, and 8 per cent germination, respectively, for *Rumex crispus, Oenothera biennis, Verbascum blattaria*, and *Brassica nigra*. Robertson and Lute (1933) in Colorado tested the viability of seeds that had been stored under ordinary conditions for 10 years, and found that wheat, barley, and oats had decreased in vitality during that time 7, 14, and 13 per cent, respectively. The vitality of Rosen rye and of soybeans had decreased only 10 per cent in 5 years, but after that time it declined rapidly. Black Amber sorgo decreased only 2 per cent in viability in 6 years. Yellow Dent corn showed a high rate of germination for the first four years then decreased in vitality 13 per cent in the fifth year, and 20 per cent after 8 years. Stevens (1935) found that the viability of good seed of sweet clover and alfalfa decreased steadily from an original 95 per cent to 60 per cent in 20 years. Red clover dropped to approximately 10 per cent. Soybeans declined steadily, yet after 9 years they had an average vitality of 30 per cent. Blake (1935) decided after testing the germination of the seeds of 42 species of grasses from the tall-grass prairie that there were wide fluctuations in their germinating power. As a general rule, it was low at harvest, highest in spring and autumn, and lowest in winter and midsummer. Most of these seeds lost their vitality in 5 to 6 years.

A. Causes of Dormancy in Plants and Plant Parts

According to Crocker (1916), dormancy is brought about by factors inhibitory to general processes preceding or accompanying growth. The rest period is considered by some to be a direct response to the changing external conditions, while others consider it to be the result of fixed heredity. It seems reasonable to consider it a result of a combination of these two factors. Thus, for example, the structure of a seed which possesses impermeable seed coats or an immature embryo which causes it to be dormant is apparently due primarily to hereditary factors. The environmental conditions under which the seed develops, however, will serve to alter to a considerable degree the nature of the seed structures. The conditions under which such a seed is placed after its separation from the plant will also determine to a marked degree the extent of its resting period.

The work of Howard (1910) indicates that in many cases dormancy is induced primarily by changes in the environment. Thus, of 283 species of woody plants brought into the greenhouse in November and January more than one-half grew easily inside 2 weeks without any treatment. In some species the winter rest is a habit induced by unfavorable outward circumstances, and if these conditions are removed some species can be induced to grow readily. The peach tree apparently belongs to this type of plants. It has a comparatively short dormant period and frequently becomes active during the warmer days of winter, with the result that the fruit buds are killed.

The causes of dormancy in seeds have been studied extensively. According to Crocker (1916), their dormancy is due to one or a combination of the following causes:

1. The possession of thick or hard coats that prevent the intake of water and probably also of oxygen. Some of the seeds that possess such coats are those of the legumes except peanut, several species of *Nelumbo*, velvet leaf, Kentucky coffee bean, asparagus, morning glory, and okra.

Some other seeds in which dormancy is either entirely or partly due to the impermeability of the seed coats to water are vetch (Jones, 1928), Kentucky coffee bean (Raleigh, 1930), *Tilia* (Spaeth, 1932, 1934; Barton, 1934), and cotoneaster (Giersbach, 1934). Johnson (1935) stated that delayed germination in the seeds of *Avena fatua* is determined by a condition of the seed coat that develops after fertilization. There is apparently a correlation between the germinability of these seeds and their position in the panicle.

The possession of coats that are hard or impermeable to water is one of the most common causes of dormancy in seeds. In some species all the seeds are hard, but more frequently only a portion of the seeds are hard, and these are of varying degrees of hardness. This in nature insures the distribution of germination of a single crop over a number of years. Most of the long-lived seeds have hard, impermeable coats.

In some cases, the seed coats are readily permeable to water but are so firm that they prevent the expansion of the embryo. The seeds of *Alisma plantago* and *Amaranthus retroflexus* are apparently inhibited in their germination on this account. Crocker and Davis (1914) reported that when the seeds of *Alisma plantago* lie in water, the embryo itself does not much more than one-fourth consummate its possible imbibitional and osmotic swelling. The embryo thus lives in water for years restrained in its swelling by the seed coat against which it must exert a pressure of approximately 100 atmospheres. When the carpel wall is removed, the seed swells rapidly and increases 40 per cent of its air-dry weight in the course of 2 hr. The elongation of the embryo when the seed coat is removed was 30 per cent of its length in 5 hr.

2. The possession of coats that interfere with the absorption of oxygen. The classical example of this type is the seed coats of the cocklebur (*Xanthium*). This character appears in both the upper and lower seeds of the bur but is more marked in the former. It is considered that the germination of many other seeds is greatly influenced by the rate of the diffusion of oxygen through the seed coats. Spaeth (1932) stated that the dormancy of the seeds of *Tilia* is due in part to the impermeability of the nucellar membrane to oxygen.

3. The possession of immature embryos. When some seeds are separated from the parent plant, the young plant in the seed has not yet reached its full embryonic development. This development is reached after the seed has fallen off, and several weeks or months must elapse before it becomes sufficiently developed to germinate. The seeds of

Ginkgo biloba, Gnetum gnemon, Erythronium denscanis, Ranunculus ficaria, Tilia, and *Cotoneaster* are some of those whose dormancy is due to an undeveloped embryo.

4. The possession of an embryo which, although to all appearances mature, must undergo certain changes before it will germinate. Most of the seeds of the *Rosaceae*, including peach, apple, cherry, plum, and hawthorn, are of this type. Before these seeds will germinate they must "afterripen," as will subsequently be discussed.

B. Overcoming Dormancy

1. Stems.—The principal methods that have been used in breaking the dormant period of these plant parts are (*a*) raising the temperature, (*b*) lowering the temperature, (*c*) treatment with gases and vapors, (*d*) immersion in chemical solutions, and (*e*) desiccation. Howard (1915) found that the most effective treatments for overcoming the dormancy of 65 species of herbaceous perennials with which he worked were freezing, desiccation, and ether vapors. With woody stems, freezing for 8 days followed by treatment with ether caused the quickest growth. Coville (1920) showed that the blueberry and other woody plants which have become dormant will not resume normal growth in the spring unless they have been subjected to chilling. Dormant plants that have not been chilled do not resume growth until long after those which have been exposed to winter weather. It was found by Denny and Stanton (1928) that the vapors of ethylene dichloride and ethylene chlorohydrin were very effective in breaking the rest period of the lilac, crab apple, flowering almond, *Azalea*, and others. A treatment of 24 to 48 hr. was usually sufficient. The gain in time varied from 2 weeks in the flowering almond to 2 months in the crab apple. Other chemicals which gave favorable results were propylene chlorohydrin, furfurole, and acetylene tetrachloride.

The effect of various treatments is apparently limited to that portion of the plant which has been treated. Thus, when a branch of blueberry was placed through a hole in the window and exposed to winter weather, it leafed out in the spring, while the portion that remained inside and was unchilled did not do so (Coville, 1920). By applying the various chemical treatments to individual buds of the lilac, Denny and Stanton (1928) showed that dormancy in this plant is not systemic but is localized in the buds. Thus of the pair of buds at the tip of a twig, one could be made to start growth by treating without breaking the dormancy of the other member of the pair.

Gardner (1929) noted that dormant pear trees which had been stored in the greenhouse at ordinary temperatures remained in that condition, while those which had been stored at 2°C. or outdoors during the winter grew after such treatment. Kramer (1934) reported that the treatment of the dormant seedlings of yellow poplar, red oak, and white oak with ethylene chlorohydrin broke the dormancy in 85 to 100 per cent of the cases. Mack and Livingston (1933) noted that the effects of ethylene on plants are influenced by oxygen and temperature relations, light, water, and the nature and physiological state of the tissues.

2. Bulbs and Tubers.—Denny (1930) and Denny and Miller (1934) stated that the treatment of the corms of gladiolus with 0.12 to 0.20 per cent ethylene chlorohydrin per liter in a closed container for 3 to 5 days hastened sprouting. The advance in time of germination ranged from 60 to 180 days, varying with the variety and stage of dormancy.

The treatment of potato tubers for the overcoming of dormancy has received considerable attention. According to Appleman (1918), treatment of the buds with ether or chloroform forces them to open 3 to 6 weeks ahead of normal. Treatments that furnish a better oxygen supply to the tissues also shorten the dormant period. Thus wounding of the tuber and the removal of the skin shorten the dormant period. If the tubers are wrapped in cotton saturated with hydrogen peroxide, the rest period is shortened. The hydrogen peroxide is absorbed by the tuber, and oxygen is released through the catalase that it contains.

Denny (1926, 1928) tested the value of a large number of chemicals for the hastening of germination in potatoes. The vapors of ethylene chlorohydrin and solutions of sodium and potassium thiocyanate gave excellent results. The treated tubers showed 100 per cent germination, and in 2 months the plants were 2 ft. high and tubers were forming, while the check tubers had not yet appeared above the ground. Favorable results were also obtained from trichloroethylene, dichloroethylene, carbon bisulphide, ethylene dichloride, and ethyl bromide. According to Vacha and Harvey (1927), ethylene also breaks dormancy in the potato.

Rosa (1928) reported that the more mature the potato tubers are at harvest, the shorter is their dormant period. Storage at 28 to 30°C. has a marked accelerating effect upon subsequent sprouting, as compared to storage at lower temperatures. The dormancy of fully matured potatoes was overcome in 6 to 15 days, depending on the variety, by a treatment with ethylene in concentrations of 1/400 to 1/2,200 of air. Wright and Peacock (1934) found that the rest period of immature potatoes was longer than that of mature ones. Miller (1933) showed that ammonium dithiocarbonate, thiosemicarbazide, hydrogen sulphide, and ethyl mercaptan are effective in breaking the dormancy of freshly harvested potatoes. Denny and Miller (1935) found that a concentration of ethylene chlorohydrin can be obtained which, when applied continuously for 7 days, will break the dormancy of potato tubers so that, when they are planted 2 weeks after treatment, satisfactory germination will be secured.

3. Seeds.—The treatments that have been used to hasten the germination of seeds are scarification, treatment with chemicals, alternation of temperatures, freezing and thawing, storage, and placing under conditions favorable to afterripening. Dormancy that is due to hard or thickened coats is overcome by scratching or filing the coats or by dissolving them with strong sulphuric acid so that water may be absorbed. The dormancy of hard-coated seeds is also overcome by placing them in the soil to be subjected to freezing and thawing during the winter season. Rose (1915) found that the hard-coated seeds of legumes, lettuce, mustard, okra, and snapdragon could be made to germinate more quickly by being blown against needle points. It was found by Jones (1928) that the seed coats of *Nelumbo lutea* do not absorb water after soaking in water for 18 months at room temperatures, but that when the seed coats were broken without injury to the embryo, the seeds quickly germinated. These seeds may be prepared for germination by treatment for 5 hr. with concentrated sulphuric acid, followed by washing in tap water and drying on a screen to eliminate immediate germination. The threshing machine acts as an effective abrasive agent on hard seeds and hastens their germination. Thus certain legume seeds when hulled by hand have as high as 90 per cent of hard seeds that will not immediately germinate, while there are only 20 per cent of such hard seeds when hulled by machine. The longevity of some seeds, however, is lessened by being scarified (Wolfe and Kipps, 1926).

Gracanin (1928) found that the speed of germination of the seeds of beet, rye, sunflower, *Festuca*, and others was accelerated by treatment with *o*-phosphoric acid. Deuber (1931) and Bramble (1932) showed that the dormant seeds of sugar maple, Norway maple, and certain varieties of oats could be stimulated into germination with solutions of thiourea and ethylene chlorohydrin and by the vapors of ethylene chlorohydrin. Baldwin (1932) improved the germination of the seeds of *Picea rubra* by immersion in absolute alcohol. Flemion (1934) stated that the germination of freshly harvested seeds of peach, apple, and hawthorn could be obtained by placing the embryos under germinating conditions after the removal of the outer and inner seed coats. Brown (1933) reported that the delinting of cotton seeds with sulphuric acid gave an increased rate of germination and a 20 per cent increase in yield over the controls.

a. Temperature.—It was found by Harrington (1923) that seeds of redtop, parsnip, celery, orchard grass, bluegrass, Bermuda grass, and Johnson grass germinate much better with favorable alternations of temperature than at constant temperatures, the exact alternation depending on the kind of seed. In alternating the temperatures, the upper temperature should be maintained only a small part of the day, never more than 8 and usually not more than 6 hr., and the change to the lower temperature should be fairly rapid. The best germination of the seeds of Johnson grass was obtained after being subjected to a temperature of 30°C. for 18 to 22 hr. and at 45°C. for 2 to 6 hr. each day. The best results were obtained for parsnip, celery, redtop, and orchard-grass seeds when placed at 20°C. for 16 to 18 hr. and at 30°C. for 6 to 8 hr. Morinagà (1926) reported that alternating temperatures were necessary to germinate the intact seeds of Bermuda grass and *Typha latifolia*, but that breaking the seed coats or treating with sulphuric acid allowed germination at constant temperatures.

Fivaz (1931) found that the seeds of *Ribes rotundifolia* and *Ribes cynosbati* germinated more promptly and more abundantly when subjected to daily alternations of temperature from 10 to 25°C. than when the temperature remained constant. Crocker and Barton (1931) reported that apple seeds germinate when the temperature fluctuates between 5 and 10°C. Flemion (1931) found that the dormant seeds of *Sorbus aucuparia* are rendered active by alternating the temperatures daily or weekly from 1 to 5°C.

Busse and Burnham (1930) found that the cotyledons and hypocotyls of the seedlings of cotton and flax showed abnormalities when the seeds were treated with liquid air. Busse (1930) stated that dormancy of seeds due to impermeable seed coats may frequently be broken by cooling these seeds to −190°C. with liquid air. Lipman and Lewis (1934) reported that seeds of various kinds which had been dried over calcium chloride and subjected to temperatures of −189 to −193°C. for 60 days, germinated when placed under suitable conditions. Tang (1931) found that with wheat seeds, higher temperatures in general gave higher percentages of germination until an optimum was reached, after which still higher temperatures gave lower percentages of germination.

According to Kondo and Okamura (1931), it is difficult to maintain the vitality of rice seeds containing 14 per cent moisture. By lowering the moisture content of rice over calcium chloride, however, vitality may be preserved readily at temperatures of 30 to 40°C. Robbins and Petsch (1932) noted that it required a temperature of 80°C. for 2 hr. to kill 75 per cent of the embryos when the moisture content of the grain was 5.2 per cent. However, a temperature of only 57°C. was necessary to kill the same percentage of embryos when the seed contained 30.5 per cent of moisture. According to Ludwig (1932) the minimum temperature for the germination of cotton seed is approximately 12°C. Barton (1932) found that the sealed storage of *Delphinium* seeds at any temperature is more effective than open storage. Under open storage conditions, temperatures of 8 and −15°C. are superior to room temperature for preserving the vitality of these seeds. Livingston and Haais (1933) obtained complete germination of "French" rice seed within 14 days at all low temperatures excepting 8 and 12°C., and excepting 41 to 55°C. for high temperatures. Edwards

(1934) stated that the optimum temperature for the germination of "Black Eyebrow" soybeans was between 33 and 36.5°C. The literature on the temperature relations of germination was reviewed critically by Edwards (1932).

b. Oxygen.—The supply of oxygen is an important factor in germination, and it is generally considered that the increased germination due to various treatments is in many cases caused by the favorable effect the treatment has upon the diffusion of oxygen through the membranes of the seed. The exact changes in the oxygen relations are, however, not understood. Morinaga (1926) noted that the seeds of cattail (*Typha latifolia*), which germinated poorly or not at all in air, germinated promptly when the oxygen supply was decreased by diluting the air with 48 to 80 per cent hydrogen or nitrogen. When 20 per cent oxygen was added to the air, germination was only 1.3 per cent. Some oxygen, however, is necessary for germination, since when all of it is removed the seeds will not germinate. Morinaga also found that Bermuda grass germinates better with a reduced oxygen supply, but the effects are not so marked as in the case of the seeds of *Typha latifolia*. The breaking of the seed coats of *Typha latifolia* removes the necessity of reducing the oxygen supply, and they germinate equally well with from 1 to 90 per cent of full atmospheric oxygen. Thornton (1935) found that as the temperature was increased from 21 to 31°C. the upper and lower seeds of the cocklebur required a lower partial pressure of oxygen for germination.

c. Light.—Light has an effect upon the germination of certain seeds. Thus Gardner (1921) reported that the seeds of *Rumex crispus*, *Datura stramonium*, and *Phoradendron flavescens* are light sensitive. The germination of the first and last were promoted by light, while that of *Datura stramonium* was hindered by light. It was noted by Morinaga (1926) that light, nitrate, and nitrite with alternating temperatures increased the germination of the seeds of Bermuda grass. The highest germination of the seeds of bluegrass was obtained with these three factors in combination. Light was found to be effective in the germination of celery seeds only when unfavorably high temperatures were used.

Andersen (1931) placed seeds of Canada bluegrass in a 0.02 *N* potassium nitrate solution in light at 30°C. for 6 hr. and then in darkness for 18 hr. for 28 days, and obtained a germination rate of 60 to 70 per cent. Seeds that were placed in the dark in a dilute solution of nitric acid gave a 60 to 70 per cent germination, while those in water in the dark gave a germination of only 20 to 30 per cent. Hutchings (1932) found that freshly collected seeds of the monkey flower (*Mimulus ringens*) are light sensitive. At low light intensities, the germination of this seed is roughly proportional to the intensity of light. Thus with intensities of 6.0, 3.5, and 1.5 foot-candles, there were germinations of 75, 34.5, and 11.5 per cent, respectively.

Axentjev (1930) found that the light-inhibiting effect on germination was entirely dependent on the presence and integrity of the seed coats in *Amaranthus retroflexus*, *Phacelia tanacetifolia*, *Androsace maxima*, and *Bromus squarrosus*, but not entirely due to the coats in *Cucumis melo*, *C. sativus*, *Cucurbita pepo* and *Nigella arvensis*. In light-stimulated seeds the light effect was entirely due to the integrity of the seed coats in *Rumex crispus* and *Epilobium hirsutum*, but the seed-coat condition played little or no part in the light stimulation of *Oenothera biennis* or *Silene densiflora*. The coats evidently act by restricting the oxygen available to the embryos. Apparently, in some cases, light interferes with oxidation processes within the seed, while in others it favors these processes.

Because lettuce seeds are frequently light sensitive, the germination of these seeds has attracted much attention. According to Shuck (1934, 1935), American lettuce seed is in a physiologically unstable condition that makes it particularly sensitive during germination to the effects of light, moisture, and temperature. In the laboratory, germination is promoted by the exposure of the seeds to light, by the use of a very moist substratum, and by starting the germination at a low temperature. The light requirement may be satisfied by continuous exposure to light under germinating conditions or by exposing the moist seeds to light before placing them in a dark chamber. He considered that light causes certain little-understood, photochemical changes within the seed. Shuck believed that there is a growth-inhibiting substance in lettuce seed, since after a sufficient number of seeds had been germinated in the dark on a cotton medium for 10 days, the germination of similar seeds was completely inhibited on this medium; but when this cotton was washed in water and again used as a substratum, 98 per cent of a similar lot of seeds germinated. Thompson (1935) found indications that immaturity at the time of harvest is an important factor contributing to the dormancy and light-sensitivity in lettuce seed. Much of this difficulty may be eliminated by using only fully matured seeds. Perhaps Borthwick and Robbins (1928) worked with fully matured seeds for they stated that the only requirements for the germination of lettuce seeds are an adequate supply of moisture, a temperature below 25°C., and good aeration. Thornton (1936) obtained good germination of lettuce seed at 35°C. in the presence of carbon dioxide in either light or darkness.

Flint (1934) found that the red, orange, and yellow rays of light were the most effective in promoting the germination of lettuce, while the green, blue, and violet rays inhibited germination. He found further that non-light-sensitive seeds could be made sensitive by subjecting them for a period, while moist, to blue light and then drying. Seeds thus treated would not germinate in the dark but would do so under red light. Flint and McAlister (1935) discovered a band in the region of 7600 Å which inhibits the germination of light-sensitive lettuce seed far more effectively than do similar inhibitory influences previously noted in the regions 4200 to 5200 Å. The relative effectiveness of radiation in the violet, blue, and green regions of the spectrum (4200 to 5200 Å) in inhibiting germination in light-sensitive lettuce seed is found to be the same as its relative effectiveness in causing phototropic response in the etiolated coleoptiles of oats.

C. Afterripening

The term "afterripening" refers to the series of chemical or physical changes, occurring within the plant or plant parts, which bring to a close the dormant period and make growth again possible (Jones, 1920). It is evident that these changes will vary for different plants and for different treatments, so that no prediction can be made with certainty concerning the changes that will occur.

1. Changes Accompanying.—Howard (1915) believed that the specific effect of all rest-période-breaking agents on dormant woody tissue is due to the stimulation of enzymes. He thought that the diastatic, proteolytic, fat-splitting, and oxidizing enzymes are stimulated into greater activity by the treatment with the various reagents. He considered that the rest period sets in on account of the inhibition of enzyme activity due to the overaccumulation of the products of their action.

In the potato tuber, Appleman (1914, 1916) could observe no changes in the amount of diastase and invertase under normal growing conditions until sprouting had occurred. The protein, lipoid and organic extractives, and the inorganic phosphorus also remained constant until that time. The abrupt changes in regard to these substances was concurrent with sprouting. The juices from the tubers at the end of the rest period caused a greater acceleration of the oxidation of pyrogallol than the juice from new immature tubers of the same varieties. It was observed by Appleman and Miller (1926) that the ripening and maturing process in immature potatoes may continue during storage, so that by the end of the rest period, immature potatoes large enough for seed have practically the same percentage composition and respiratory response as potatoes allowed to mature on the vine, provided that they are stored under the same conditions.

The stimulating effect produced by chilling in overcoming dormancy in woody plants is believed by Coville (1920) to be intimately associated with the transformation of starch into sugar, since tests for starch and sugar showed a decrease of starch and an increase of sugar as winter progressed. The theory offered in explanation of this is that the starch grains are at first separated by living cell membranes from the diastatic enzymes, but that, when the plant is chilled, the vital activity of these membranes is weakened, the enzymes are allowed to pass through, and the starch is changed to sugar. This change is regarded as the mechanism for creating high osmotic pressures to start the plants into growth.

The changes accompanying afterripening have been especially studied in seeds. The period of afterripening may be greatly shortened by proper treatment; thus the seeds of the hawthorn (*Crataegus mollis*) have a latent period of 1 year or more. Davis and Rose (1912) showed that if these seeds are removed from the carpels and kept very moist at a temperature of 5 to 6°C., the latent period may be shortened to 2.5 to 3 months. If the testas are removed and the embryo treated as mentioned, the dormant period may be reduced to 30 days. Eckerson (1913) studied microchemically the changes accompanying the afterripening of these seeds and observed that the first initial change in the embryo was an increase in acidity, which was correlated with an increased water-holding capacity and an increase in the activity of catalase and peroxidase. Near the end of the afterripening period there was a sudden increase in acidity and in the water content, and oxidase made its first appearance. The enzymes, acidity, and water absorption all increased in amount until the hypocotyl was 3 to 5 cm. in length. At that time the fats decreased and sugar appeared. Eckerson found that afterripening in this case could be greatly shortened by treating the embryos with dilute hydrochloric, acetic, or butyric acid. The water-holding capacity, acidity, and amount of peroxidase increased much more rapidly and oxidase much earlier in the embryos thus treated than in the untreated ones. This increase in acidity, water-holding capacity, and catalase and oxidase activity during afterripening was also observed by Rose (1919) in the seeds of *Tilia* and by Jones (1920) in the seeds of the sugar maple.

Pack (1921) in the study of the embryo of *Juniperus* has reported the most extensive data in regard to the changes accompanying afterripening. The following changes were observed: (1) rapid or complete imbibition followed by a steady slow decrease in water content until germination, (2) increased hydrogen-ion concentration in the embryo and an increment of titratable acid, (3) a steady and enormous increase in the degree of dispersion of stored fats, (4) a decrease in the amount of stored fat and protein with an increase of sugar, (5) the translocation of food in the form of fat or fatty acids from the endosperm to the embryo, (6) a sevenfold increase in the amino acid content, (7) an increase of the soluble proteins with a marked hydrolysis of the stored proteins, (8) a slight increase in the respiratory rate, (9) an increase in the respiratory quotient, (10) a decrease in intramolecular respiration, (11) a twofold increase in catalase activity, (12) a slight growth of the embryo, and (13) a rise in the vigor of the seeds as indicated by their resistance to fungous attacks. The changes that accompany afterripening in the seeds of *Juniperus* are thus represented by the accumulation of cell-building materials and enzymes.

The changes accompanying afterripening have been especially studied in recent years. A brief summary of the various changes that have been observed is given in the table on page 1056.

2. Ripening of Fruits and Vegetables.—In the ripening process certain chemical changes occur. Many of these changes are the same regardless of the manner in which the ripening occurs. According to the conditions under which it occurs, ripening of fruits and vegetables may be designated as natural or artificial.

a. Natural Ripening.—It was found by Rosa (1928) in the ripening of cantaloupes that there was a progressive increase in the percentage of dry matter, in total sugars, in soluble solids, and in the specific gravity of the juice. The sucrose increased more rapidly than the reducing sugars decreased, indicating that sugars continued to move into the fruit. The pectic material remained the same, while the protopectin was high in amount in the unripe melons but changed into pectin and pectic acid as the fruit ripened. Kertesz (1930) noted that, as soon as peas were shelled, their content of sucrose decreased, while the percentage of alcohol-insoluble residue increased.

Allen (1932) found that apples, peaches, pears, and apricots require sunlight for the development of red color. The softening of the flesh is one of the most important changes that occur during ripening. All the fruits analyzed gained significantly in sugar as they colored on the tree.

CHANGES OCCURRING DURING AFTERRIPENING

Investigator	Plant part	Treatment used	Results
Gardner (1929)	Bartlett pear trees	Cold	Increase in hexose sugars, sucrose, and organic acids and depression of freezing point of cell sap and in starch.
Denny (1930)	Potato tubers	Ethylene chlorohydrin Sodium thiocyanate	Increase in sucrose, decrease in starch. No change in reducing sugars.
	Lilac plants	Thiourea Ethylene chlorohydrin	Increase in catalase, invertase, moisture in buds, soluble nitrogen, and respiratory rate. Decrease in total sugars.
Denny and Miller (1932)	Potato tubers	Organic sulphur compounds	Increase in catalase and peroxidase, pH from 0.1 to 0.4 over controls, sucrose 2 to 4 times, respiration 4 to 5 times.
Hasselbring (1932)	Narcissus bulbs	Immersion in water at 43.5°C.	More available carbohydrate and acceleration of metabolic activity.
Denny (1933)	Gladiolus corms	Ethylene chlorohydrin	Increase in sucrose and soluble nitrogen. Decrease in reducing sugars and in insoluble nitrogen.
Evans (1933)	Seeds of *Magnolia grandiflora*	Chemical changes occurring between time of the rupturing of coat and protrusion of hypocotyl.	Embryo increased in water content by 50 to 60 per cent, epidermal cells increased in acidity. Starch appears in the cotyledons. Absorbs food from endosperm. Oils decrease in endosperm. Starch disappears in the endosperm.
Flemion (1933, 1934)	Seeds of:		
	Rhodotypos kerrioides	At 50°C.	Increase in catalase, peroxidase, lipase, titratable acid, sucrose, and water. Decrease in ether-soluble substances.
	Symphoricarpos racemosus		Increase several fold in catalase and peroxidase.
Guthrie (1933, 1934)	Potato tubers	Ethylene chlorohydrin	Slight increase in conductivity of tissue. Increased leaching of electrolytes. The expressed juice increased in pH from 6.0 to 7.0, and decreased in citric, sulphuric, and nitric acids. Isolated glutathione from juice.
Thornton (1933)	Potato tubers	40 to 50 per cent CO_2 with 20 per cent O_2 at 25°C.	Increased respiration, reducing action of expressed juice. Decreased glutathione during treatment but increased over controls when planted.

After picking they showed a decrease in starch and a gain in sugar. Caldwell (1934) stated that the young fruits of oranges, grapefruit, apples, cherries, strawberries, and others had a hydrogen-ion concentration approximately the same as the other tissues of the plant. As these fruits mature, however, the acid content increases as much as 80-fold in citrus fruits. This great increase in the hydrogen-ion concentration markedly increases the imbibitional capacity of the protoplasmic colloids and of certain cell-wall constituents.

b. Artificial Ripening.—The term "artificial ripening" has been used to designate the hastening of natural ripening by the application of some artificial stimulus. The artificial ripening and coloring of fruits have apparently been practiced for a long time, but during the past 20 years renewed interest has been taken in this problem and methods have been devised for artificial ripening on a large scale of citrus fruits, bananas, dates, persimmons, pineapples, cantaloupes, tomatoes, celery, and others. The first method consisted in placing kerosene stoves in the car or bin, and it was thought for some time that heat was the effective agent in the process. Later, however, it was found that the gases generated by the stoves were effective when distributed by pipes among the fruits and vegetables to be treated. Denny (1924) determined that the change in color in lemons was due to some unsaturated hydrocarbon compound. He tried ethylene and found that the coloring was very effectively and quickly done. Ethylene is now extensively employed in the ripening process. According to Chace and Denny (1924) and Harvey (1928), it has distinct advantages over the older methods of ripening. The time is shorter; there is less chance for spoiling, since a lower temperature is maintained; and a better color and flavor result. The best temperature for treatment is 65 to 70°F., and the best concentration 1 cu. ft. of gas to 1,000 cu. ft. of air. The time of treatment varies, but 48 hr. is generally sufficient to ripen the greenest fruits.

Harvey considered that the effect of the ethylene is to activate the enzymes which digest starch to sugar and oxidize acids and tannins. He found that the treated stalks of celery increased their sugar content 20 to 30 per cent over the checks. Increases in the sugar content were also observed in treated tomatoes and bananas. Regeimbal and Harvey (1927) noted that treated pineapples were sweeter and the activity of the proteolytic enzymes more marked than in the controls. It was demonstrated by Babb (1928) that the blanching of celery with ethylene does not affect its vitamin-B content. Other investigators, including Chace and Church (1927) and Hibbard (1930), however, could not observe an increased carbohydrate content in the fruits and vegetables treated with ethylene. Harvey (1928) suggested that the strength of the gas used in the treatment might have its influence upon the changes in the sugar content.

The effect of ethylene in artificial ripening has also been studied by Harvey (1925) on celery, Rosa (1925) on tomatoes, and Harvey (1926) on bananas. Wolfe (1931) stated that bananas in an atmosphere containing $\frac{1}{100}$ to 1/10,000 parts ethylene turn yellow, increase slightly in sugar, and decrease in starch. According to Davis and Church (1931), ethylene seems to act on Japanese persimmons by its effect on the general metabolism. The treatment stimulated softening, color development, and respiratory activity. Hartshorn (1931) reported that "the carbide treatment" hastened the ripening processes of green bananas, as shown by the rate of softening, respiration, starch hydrolysis, flavor, and color changes. Apparently the effect is mainly a shortening of the period of low activity that normally occurs at the beginning of the ripening processes. Dustman (1934) found that stored apples were not affected chemically when treated with ethylene, ethylene chlorohydrin and ultraviolet irradiation. Treatment with ethylene, however, hastened the color changes from green to yellow, and softened the tissues.

D. Regeneration

According to White (1931), the successful culturing of excised root tips was reported independently by Kotte (1922) and Robbins (1922). Chambers (1923) found that individual cells would separate and migrate from root fragments, provided the latter were of a definite size and were cut from a definite portion of the root. Robbins and Maneval (1923) found that the excised root tips of beans, morning glory, alfalfa, mustard, wheat, sunflower, and flax were capable of making considerable growth in sterilized Pfeffer's solution containing 2 per cent of dextrose, and that they were capable of growing in sterile Pfeffer's solution alone for only 12 days. White (1932) studied the effect of hydrogen-ion concentration, temperature, aeration, and illumination upon the growth of excised wheat roots. In 1933 he reported the best compounds of iron and sulphur to use in the culture of root and stem tips. He grew over 400 excised stem tips of *Stellaria media* in hanging-drop cultures for 3 weeks without changing the medium. During this time there occurred division and differentiation of the meristematic tissue into leaves, stems, and floral organs. In 1934, White grew excised root tips (10 mm. long) of the tomato for over a year, making 52 transfers in the following medium:

Salt[1]	Millimols	Salt	Millimols
$Ca(NO_3)_2$	0.60	KCl	0.87
$MgSO_4$	0.30	KH_2PO_4	0.09
KNO_3	0.80	Fe_2SO_4	0.006

[1] Plus 2 per cent of sucrose and the filtered extract of 0.01 per cent by weight of dried, brewer's yeast.

This solution was placed in 50-cc. flasks and immediately autoclaved at 15 lb. pressure for 20 min. One isolation of a root tip in vitro produced during the year 35,000 growing points and more than 400,000 mm. or approximately 1,300 ft. of tissue. Robbins, Bartley, and White (1936) found that if the root tips of corn and pea, 1 to 2 mm. in length, were divided longitudinally into equal parts, both pieces developed into complete roots in the nutrient medium used. Rea and Karper (1932) showed that the sorghums can readily be propagated from cuttings.

LaRue (1935) reported that the seeds of corn, wheat, and barley which were split lengthwise produced roots and shoots. The embryos of wheat, Sudan grass, and sorghum that were deprived of endosperm were capable of growing into complete, although weak plants. Wheat seedlings from which the shoots were removed produced new shoots, Sudan grass seedlings produced roots but no shoots, while barley and oat seedlings died. Later (1936) he found that pieces 0.5 mm. in length cut from the immature embryos of dandelion, wild lettuce, oxeye daisy, and tomato could be grown into complete plants in culture solutions. The cotyledons of the tomato excised from immature embryos grew into complete plants, but, in the other species, the cotyledons failed to regenerate. He used White's solution, omitting the yeast extract, but added synthetic heteroauxin. No tissues regenerated without either yeast extract or heteroauxin in the medium. All the plants grown in culture were found to be capable of growth in the soil.

VIII. FACTORS AFFECTING GROWTH OF PLANTS

According to Reed (1919), growth may be considered a function of two variables. The first is the genetic constitution of the individual, and the second is the resultant of all those factors which make up what is commonly called the "environment" of the organism. The factors of the first group are essentially internal, while those of the second are essentially external.

A. INTERNAL FACTORS

In the fertilized egg there are certain genetic tendencies which when expressed later during the life of the growing plant tend to produce under any set of external conditions a certain definite form of growth (Hanna, 1925). Thus a plant of a dwarf variety, although it may be made to vary slightly in its size and composition by its environment, cannot be made to attain any considerable size by any known cultural methods. The effect of inherent factors on growth is also illustrated by the similar growth rates of green and albino corn seedlings, which were observed by Kempton (1924). That sister plants differing so greatly in a major physiological characteristic should have such similar growth rates is a very striking fact. This behavior of growth of these two types of plants is believed to be controlled by inherent factors that are not associated with chlorophyll and photosynthesis.

Some of the conditions that are classified as internal factors other than the inherent nature of the protoplasm are the concentration of the cell sap, the number and size of the growing points, and the accessibility of manufactured food. Although these factors are classified as internal, they may be altered to a marked degree by the external environment. The growth rate of a plant is thus a function of the internal factors, which at present are indefinitely defined, and the various conditions of the external environment. The external conditions merely accelerate or retard the manner in which the internal factors tend to express themselves (Briggs, 1928; Gregory, 1928). Changes of environment thus need not always have the same effect on the rate of growth. Even though two organisms have the same rate of growth under a given set of conditions, the internal conditions in the two cases may be different and may be differently affected by the changes in the environment.

In a discussion of the influence of internal factors upon growth, the effect of the size of the seed upon plant production could well be considered. The size of the seed, although governed by genetic tendencies, is influenced to some degree by its position on the parent plant and by the general environmental conditions. The reports of the experiments on the effect of the size of the seed upon the vigor and productivity of the plants derived from them are conflicting. This is apparently due to the fact that the comparative results vary greatly with the manner of testing (Kiesselbach, 1924). Kidd and West (1918) and Rotunno (1924) in reviewing the literature on the subject found that the evidence was in favor of the conclusion that larger seeds give rise to more vigorous plants and a better yield. Rotunno in his experiments with small, medium, and large seeds of the radish found that the percentage germination of all three sizes of seeds average the same within the limits of error. The medium seeds averaged the heaviest roots and produced with but one exception the highest percentage of salable roots.

Rudolfs (1923) working with bean seedlings from large and small seeds in the dark found that those from large seeds showed the better growth and that temperature acting as an accelerating factor increased this advantage. Thus at 10°C. large and small seeds required 9.8 days to attain a growth of 140 mm. At 15°C. the large seeds required 7.8 days and small seeds 8.8 days. At 20°C. the large seeds required only 3.1 days, while the small seeds require 8.5 days. In the large seeds a normal temperature coefficient for growth was shown, while in the small seeds the temperature coefficient was abnormally small. This seems to suggest that the amount of food stored in the seeds is of major importance for early growth.

Kiesselbach (1924) in reviewing the literature concerning the yields from seed grades of cereal crops summarized the published data as follows: (1) When the space planting was such as to permit maximum development, the individual plant grain yield was 19 per cent less for the small than for the large seed. This difference is apparently due to the immediate advantage of a greater reserve food supply in the larger seed, which results in a more vigorous initial growth. (2) When planted in equal number at a rate optimum for the large seeds, the small seeds yielded 11 per cent less grain per acre than the large. (3) When planted in equal weights at a rate optimum for the large seed, the small seed yielded 3 per cent less and the unselected seed 2 per cent more per acre than the large seed. The relatively low yield of plants from small seeds was largely overcome by planting a greater number of seeds. The results of experiments by Kiesselbach with winter and spring wheats and oats are very comparable to these reported results, and he concluded that there is no material or practical gain in the grain yield, under farm conditions, from the practice of grading small grain seed, provided it is free from inert matter.

Schmidt (1924) found with seeds of soybeans, buckwheat, lima beans, and corn that germination in the soil occurred more rapidly in light seeds than in heavy ones. Seeds of medium weight grade, or slightly heavier, were superior in germinating power by approximately 8 per cent over seeds of very heavy or very light weight. In Golden Bantam corn the number of ears, green weight of ears, and air-dry weight of stalks and husks varied in the same order as did the weight of the seeds that were planted. The superiority of the plants from heavier seeds over those grown from lighter ones decreased notably as the plants approached maturity and sometimes the difference had disappeared entirely by that time. Kotowski (1929) noted that the differences in plants of peas, broad beans, and cabbage due to seed size may be distinguished during the early growth of the plant but disappear before blooming and apparently have no influence on the yield. Mitchell (1934) found, however, that seed weight is an important factor in determining the size attained by 1-year-old seedlings of Scotch and White pines. According to Paton (1929) this effect is noticeable for 6 years. McComb (1934) also emphasized the effect of the size of seed upon the nature of the young seedling.

B. External Factors

Some of the external factors that may influence growth are temperature, light, moisture, electricity, and the amount and composition of the materials in the soil.

1. Temperature.—It has long been known that the growth of plants is closely correlated with temperature, since the rising temperature of the environment is attended within certain limits by an increased rate of growth.

a. Growth Curve in Relation to Temperature.—In general, the growth curve in relation to temperature shows three cardinal points: the minimum, the lowest temperature at which growth is exhibited; the optimum, the temperature at which growth is the greatest; and the maximum, the highest temperature at which growth will occur. The position of these three points, especially the optimum temperature, has varied rather widely in the reports of different investigators. Some of these differences may be attributed to the fact that the growth rate varies for different plants, for the different organs, and for different ages of the same plant or plant part (Newcombe, 1923). Thus for a 12-hr. period the optimum temperature for the growth of the coleoptile for oats was 30, for bluegrass 27, and for *Sorghum vulgare* 31°C. It has also been shown that the length of the time of exposure is an important factor in determining the position of the optimum, minimal, and maximal temperatures for growth.

Lehenbauer (1914) in corn seedlings found that the optimum temperature for growth for a 12-hr. period was 32°C., but that this was not the optimum when the length of the period of exposure was altered. At the temperatures above 31°C. the initial rate of growth was not maintained and there was a marked falling off of this rate during prolonged periods of exposure. Similar observations were made by Leitch (1916) in the growth of *Pisum sativum.* The minimum temperature for growth in this plant was −2°, the maximum 44.5°, and the optimum 28 to 30°C. In this instance, the relation of growth to temperature could be expressed as a uniform curve from −2 to 29°C., the curve showing a very close resemblance to that of respiration. Above 20°C., however, for each higher temperature a different curve had to be constructed to express the rate of growth in successive time intervals. This falling off of the growth rate with time is similar to the decrease in rate of enzyme action and photosynthesis under the influence of high temperatures during prolonged periods.

No definition of optimum temperature in regard to growth is thus of value unless it involves a consideration of time. The optimum temperature for growth has been defined by some as the highest temperature that will allow growth for some specified time period without decrease in rate. Lehenbauer believed it should be defined as the temperature which for a specified time period of exposure produces the greatest growth. He observed in corn seedlings that when the experiments were made over a 3-hr. period, the optimum temperature was 29°C. If the observations were over a 6-hr. period, the optimum temperature was 30°; and if the periods were from 9 to 27 hr., the optimum was 32°C. Above 27 hr. the optimum shifted to 31°C.

A definition of the maximal temperature for growth must also involve a consideration of time. The maximal temperature might thus be defined as the highest temperature at which growth ceases after a specific time exposure. With corn, Lehenbauer noted that a temperature of 40°C. allowed growth in all the plants tested for a period of at least 21 hr. With a temperature of 41°C., five of the six plants exhibited no growth after 15 hr., while the other plant ceased to grow after 18 hr. At 42°C., three plants ceased to grow after 15 hr. and three after 18 hr. At 43°C. three grew no more after 9 hr., two ceased to grow after 12 hr., and the sixth stopped growth after 15 hr. It may thus be stated that 43°C. is the maximal temperature for growth of these seedlings for an exposure period of 15 hr., and 42°C. for an 18-hr. period. It was observed that plants kept at 41 to 43°C. for 6 hr. after growth had ceased were not killed, since when the temperature was gradually lowered growth was again resumed.

Plitt (1935) noted that oat seedlings grown from seeds which were germinated at 25°C. exhibited a higher percentage of dry matter than those germinated at 5°C. The seedlings were more rapidly developed at the higher temperatures, and photosynthetic activity started earlier. Smith and Cochran (1935) noted that the maximum rate of growth of pollen tubes occurred at 70°F.

MacDougal (1917) noted that a retardation in the growth of wheat and corn seedlings occurred at more than one place in the temperature scale and at different times of the day. In some instances this was attributed to direct temperature effect, while in others this explanation was not deemed adequate, and the cause was attributed to varying imbibition capacity coincident with alternations of acidity or alkalinity.

Since growth is intimately associated with hydration and imbibition, it is of interest to note the effect of temperature upon these processes under different conditions of the medium. Sections of petioles of *Solanum* were placed in distilled water and in 4.2 and 2.6 per cent citric acid at temperatures ranging from 18 to 38°C. The swelling in distilled water was nearly three times as great at the higher temperature, while in the acid solution a retardation took place that limited the total at the higher temperature to something over a half that possible at the lower temperature. The total swelling in the acid at the lower temperature occupied an hour, while at the higher temperature it was a matter of 10 to 15 min. The total imbibitional capacity of the lower temperature was not reached for 8 to 10 hr., while at the higher temperature it was reached under 2 hr. The effect of temperature on growth is thus apparently determined to a marked degree by the chemical reactions of the tissues under observation.

Berkley and Berkley (1933) stated that cotton plants were least resistant to high temperatures at the higher relative humidities. They defined the "thermal death point" as that temperature at a given relative humidity that will kill the protoplasm immediately. A thermal death point is not valid without stating the age of the plant and the relative humidity at the time of its consideration.

It has been assumed by many in studying the effect of temperature on growth that the process is a unified one or a series of successive reactions. MacDougal (1920) emphasized the statement that growth is a group or "constellation" of activities and that the rate of one of these dependent on temperature may be the determining one when the particular process forms either the retarding or leading agency. On that account, he asserted that the relation of growth to temperature between 10 to 50°C. cannot be expressed by any simple formula. At times, however, the rates of metabolism, respiration, hydration, diffusion, and other processes may coincide in such a manner as to make possible the application of a simple formula for the effects of temperature within a limited range.

Kondo and Okamura (1930) reported that the best temperature for growth in length of the rice plant was 30 to 32°C.; for growth in weight, 34°C.; and for grain production, 30°C. The optimum temperature range for the growth of rice was thus 30 to 34°C. Shippy (1930) found that the range of temperature permitting the formation of callus from apple cuttings and grafts includes 0 to 40°C. For general callusing, temperatures below 20°C., rather than higher ones, have been the more satisfactory. Edwards, Pearl, and Gould (1934) and Pearl, Edwards, and Miner (1934) stated that the processes of growth in a relatively simple structure as a seedling are not homogeneous. Not all parts of the developing organism are growing at the same time or at the same rate. It appears that the processes are so interrelated that any excessive development of one function is counterbalanced by the reduced activity of another function. Nightingale (1935) grew peach and apple trees in sand at temperatures of 45, 50, 55, 60, and 95°F., and found that during the current season the maximal yield of roots and tops occurred at 65°F.

b. Resistance of Plants to Low Temperatures.—The large economic losses caused by the injury or death of plants during the autumn, winter, or spring by low temperatures have stimulated a considerable amount of research to determine the cause of this injury, the characteristics that retard it, and methods for its prevention, and to develop or discover plants that are inherently resistant to the effects of such temperatures. A thorough review of the literature pertaining to the effects of low temperatures on plants may be found in Abbe (1895), Salmon (1917), Rosa (1921), and West (1924), and the student is referred to these references for detailed information on the subject.

1. *Causes of Injury.*—The cause of injury by freezing has been assumed to be due primarily to the withdrawal of water from the cell. Upon freezing, water is usually withdrawn from the cell and freezes in the intercellular spaces. As the temperature falls, more water is frozen on the outer walls and more is drawn from the cell by imbibition. Since the force with which the remaining water is held increases rapidly with the progressive loss of water by virtue of the increasing concentration of the cell sap, the amount of water frozen at each successive degree is smaller and smaller.

According to Müller-Thurgau (1880, 1886) death from low temperatures is due to a denaturing of the protoplasm that is caused by the exosmosis of water to the intercellular spaces where it forms ice masses. Molisch (1897) considered it due to an injurious concentration of the cell sap caused by the removal of water in ice formation. Gorke (1906) considered that death results from this increased concentration together with the loss of some specific substances at their eutectic points. He considered that, under the ordinary conditions of life, these substances protect the protoplasm against the injurious effects of other components which remain in solution at freezing. Dexter

(1934) concluded from work with wheat crowns that the withdrawal of water by ice formation was not a fully reversible process, but that in hardier plants it was more nearly so. Thus when the plants are frozen beyond recovery, the water that has been removed by ice formation is not reabsorbed sufficiently to give the previous condition of turgidity.

Maximow (1912, 1914) believed that death from low temperatures is due to the rupturing of the protoplasm by masses of ice crystals that are formed in it. In experiments with pollen grains, Chandler and Hildreth (1935) believed that the theory of Maximow is correct. According to Stiles (1930), the death of a plant by freezing is caused by the formation of ice crystals in the protoplasm, which produces a disturbance in the relations between the dispersed and continuous phases, with an aggregation of the former. Such changes are usually irreversible, so that, upon thawing, the original colloidal system of the living protoplasm is not re-formed and is thus no longer living. Evidence in support of this view is derived from actual observations on the freezing of living and nonliving colloidal systems, and on the structural effect of varying the rate of freezing of such systems.

With very rapid freezing, it is possible to obtain the original system again upon thawing. By such treatment, cases are on record in which living plant and animal cells have remained alive through freezing and subsequent thawing. In frost-resistant plants, however, it is probable that water is bound to hydrophile colloids of the protoplasm and is nonfreezable, so that the formation of ice crystals does not occur. Chambers and Hale (1932) studied strips of the epidermal cells taken from onion bulbs that had been frozen at between −7 and −10°C. Ice readily appeared between the cell wall and the protoplasm, causing plasmolysis of the latter. The ice sometimes broke the cuter membrane of the protoplasm, but the tonoplast and vacuole remained intact. Ice was not definitely seen to form within the vacuole but the general impression was gained that it did form. Schaffnit and Lüdtke (1932) considered that death of a plant from frost or heat is a problem of metabolism. The entire process of metabolism need not be suspended. The cessation of activity of only one enzyme may be sufficient to cause death. Joslyn and Marsh (1933) stated that the physical changes which occur during freezing depend chiefly upon ice formation and osmotic relations; these involve changes in volume and texture. The chemical changes are those concerned with the hydrolysis of pectins and sucrose, and other changes are in color and flavor. Lutz (1935) believed that the recorded evidence indicated that at least in many cases the rapidity of thawing had no measurable influence on the injury resulting from exposure to a freezing temperature. In his experiments with onions and apples, the influence of temperature in thawing did not become pronounced until a temperature of 95°F. was applied. The killing by cold depends apparently in some cases on ice formation rather than on the direct effect of low temperature. The withdrawal of water from the cell by freezing may bring about death by the following causes: (1) the desiccation of the protoplasm, (2) the mechanical injury to the protoplasmic membrane, and (3) the precipitation of proteins by salting out due to the greater concentration of salts in the sap or to its increase in acidity as water is withdrawn. It should be noted here that chilling above the freezing point has an injurious effect upon some plants. Thus Sellschop and Salmon (1928) found that an exposure of 24 to 36 hr. at 0.5 to 5°C. was fatal to rice, velvet beans, and cotton, while with the same exposure cowpeas were completely defoliated. Peanuts and Sudan grass exposed to these temperatures for 48 hr. were apparently uninjured at first but died in about 2 weeks. Soybeans, potato, buckwheat, tomatoes, and flax were exceedingly hardy, showing no injury when exposed at these temperatures for 84 to 96 hr.

2. *Characteristics of Hardiness.*—Since certain varieties of plants are known to be hardy under conditions of low temperature, the question naturally arises as to what

characteristics enable them to survive while others perish. Strausbaugh (1921) stated that the dormant condition reached by hardy forms of plums appeared to involve fundamental changes in the colloidal condition of the protoplasm, whereby there was a marked retention of water against the force of dehydration. According to Newton (1924), the imbibition pressure of fresh leaves in the winter-hardened condition is in most cases directly related to hardiness. The imbibitional pressure of hardened leaves appears to depend on the physical state of the cell colloids of the living tissue, since this property was lost when the tissues were killed. The quantity of hydrophilic colloids contained in the pressed juice was found proportional to winter hardiness. Ackerman (1917) observed a high correlation between the sugar content of the hardened leaves and hardiness. Although the concentration of the cell sap may play some role in winter hardiness, it does not always show a relationship in that regard (Salmon and Fleming, 1918). According to Martin (1927), the hardy wheats are characterized by a low moisture content of the tissues, a high percentage of total solids in the cell sap, a high osmotic concentration of the cell sap, a high percentage of bound water, and a low rate of respiration at low temperature. Zacharowa (1925) in the freezing of roots found a positive correlation between cold resistance and alkaline reaction and a negative correlation between acid reaction and cold resistance. The water content of the soil may exert a marked influence on injury of plants by freezing. Thus Hill and Salmon (1927) noted that plants grown in a dry soil were injured more severely by freezing than similar plants in a wet soil. The specific heat of the water in the wet soil prevents a rapid change in temperature.

Recently much study has been given to the characteristics of plants that enable them to withstand low temperatures and other severe environmental conditions. Some of the observations in this regard are summarized in the table on pages 1067 and 1068.

3. *Hardening Plants.*—Plants in many instances may be made to endure low temperatures, without serious injury, by hardening. They may be hardened by being exposed to gradually decreasing temperatures, scantily supplied with water, severely pruned, or watered with various salt solutions. Apparently any treatment that checks the growth of the plant increases its resistance to cold. Harvey (1918) reported that cabbage plants became resistant to ½ hr. of freezing at −3°C. by exposure to 3°C. for 5 days. He considered that the principal hardening process in cabbage was to change the proteins to forms less easily precipitated, which was indicated by an increase in the amino acid content. Hardened plants contain a larger amount of bound water than the unhardened ones. The percentage of moisture frozen in hardened cabbage leaves at −3 and −4°C. was only two-thirds of that frozen in tender cabbage. That is, the actual amount of water remaining unfrozen at a given temperature was greater in hardened than in tender leaves, although the total moisture content was less.

Martin (1927) stated that during the hardening of wheats there is a decrease in the moisture content, an increase in the total solids in the sap, and an increased osmotic pressure and imbibition pressure of colloids as measured by the ability of the tissues to hold the sap against the forces of freezing and pressure. Hardened plants possess a larger amount of hydrophilic colloids, probably pentosans (Hooker, 1920), than the untreated ones. They lose less moisture by transpiration per unit of leaf surface than do tender plants (Rosa, 1921). Weimer (1929) found a correlation between the amount of bound water and cold resistance of hardened and unhardened alfalfa plants but could not show a similar correlation between varieties shown to differ in hardiness.

The term "bound water," as herein used, refers to that water which is associated with the colloidal material of the cell in such a manner that it will not freeze or at least it will not form crystals as does the free water under freezing conditions. Sayre

SUMMARY OF SOME RECENT OBSERVATIONS ON HARDINESS

Investigator	Plant or plant part	Characteristics of hardiness
DeLong (1929)...........	Wheat	Frost injury to standing wheat interrupts and prevents the translocation of materials from the stems and leaves to the grain.
Janssen (1929)...........	Wheat	The percentage of winterkilling varies with the date of seeding. Winterkilling was greatest in the seeding of Oct. 19, followed by Oct. 6, and Aug. 31. Temperature at which plant grows previous to freezing is important. Plants grown at 33°C. were killed when subjected to a temperature of −10°C., whereas plants grown at 5°C. were not injured when subjected to the same temperature.
Walter (1929)............	Evergreens	Concentration of cell sap increased as winter progressed. Resistance to frost increased with an increase of osmotic value. The more slowly the latter increases the more resistant the plant is to cold. Cold injury is due primarily to desiccation.
Hottes and Wilson (1930)..	Wheat	Resistance to high temperatures varies inversely with water content of seed.
Chandler (1931)...........	Pollen	Unwetted pollen remained viable after being subjected to temperatures of −15 to −18°C. for many hours. Moistened pollen was killed under similar conditions. Freezing kills cells by the physical effect of ice masses.
Newton, Brown, and Anderson (1931).	Winter wheat	Exposure of the expressed sap of unhardened plants to frost decreased the coagulable protein and increased amino acids. This splitting of proteins in hardened plants may be a result of frost instead of an adaptation against it and the value of sugar as a protection to winter plants may lie partly in its ability to delay this action.
Newton and Brown (1931).	Winter wheat	Maximal precipitation of proteins in expressed sap of leaves of unhardened plants when they were subjected for 5 hours to a temperature of −7°C. The maximum precipitation occurred at pH 5.1 and 7.3. Dialysis of electrolytes increased subsequent to precipitation of proteins by frost. This precipitation is irreversible.
Peltier and Tysdal (1931)..	Alfalfa	Hardy alfalfa varieties become dormant earlier and harden more rapidly. They retain their hardiness longer in the spring.
Friedberg (1932)..........	Wheat	Cold-resistant varieties tiller deeper than sensitive wheats. The capacity to develop a tiller from the coleoptile is a varietal characteristic by which the wheat renews growth when the growing tip is killed by cold.
Martin (1932)............	Spring wheat	Late heading of wheats of the Pacific coast indicates resistance to cold.
Timmons and Salmon (1932).	Alfalfa	The resistance to artificial low temperatures was correlated with the severity of the winters of the regions to which the various strains had been adapted.
Waldron (1932)............	Wheat	Plants injured by frosts in April and May showed decreased yields.
Dunn (1933)...............	Cabbage, Brussels sprouts, and alfalfa	Pectic substances may be partially responsible for cold resistance. Proteins are apparently converted into other compounds in the hardening process, and the protoplasmic constituents are probably of primary importance in the hardening of the plants to cold.

SUMMARY OF SOME RECENT OBSERVATIONS ON HARDINESS.—(*Continued*)

Investigator	Plant or plant part	Characteristics of hardiness
Salmon (1933)	Wheat	Plants frozen artificially at night were injured less than similar plants frozen during the day. Plants in dry soil injured more than those in wet soil. If soil frozen previous to the test, no difference observed in injury to plants. The less injury in wet soil due to lag in temperature.
Schaffnit and Wilhelm (1933).	Potatoes and tomatoes	Application of potassium fertilizer lessens freezing injury by lowering the freezing point.
Collison and Harlan (1934).	Baldwin apple trees	Trees given nutritional treatment conducive to growth and yield showed less winter injury than those not receiving this favorable treatment.
Dexter (1934, 1935)	Wheat	Hardiness associated with a higher percentage of dry matter and with low concentration of soluble salts. At low temperatures there is an increase in soluble organic nitrogen, and a decrease in respiratory rate that is not correlated with a change in sugar content or enzyme activity. Usually an increase in sugar that may or may not be accompanied by increased hardiness.
Greathouse and Stuart (1934).	Red clover	Winter hardiness associated with greater concentration of carbohydrate and nitrogen, lower content of moisture, a large amount of unfreezable water, a lower specific conductivity, and a slightly higher pH value.
Megee (1935)	Alfalfa roots	The higher the conductivity of expressed juice the less the hardiness. No direct relationship existed between winter hardiness and heat of swelling, moisture equivalent, freezing point, chemical composition, respiration, and amount of moi ture. Heredity is the most plausible explanation of winter hardiness.
Worzella (1935)	Winter wheat	Cold resistance inherited in the same manner as other quantitative characters.
Woods (1935)	Raspberries	Increased nutrients produce more vigorous growth and such plants are more susceptible to winter injury than the less vigorous canes.
Laude (1937)	Winter wheat, rye, barley, and oats	Water content and amount of expressed sap increased as active growth began after dormancy. The total solids in the sap decreased. The loss of cold resistance was usually more rapid in the varieties possessing greater midwinter hardiness.

(1932) considered that no satisfactory definition of bound water can be made. He stated that all the water which does not show some of the common properties of liquid water may be considered as bound water. He described three methods for determining bound water.

Harvey (1930) found that the threshold for producing hardiness in cabbage was about 5°C. Alternate exposures for 12 hr. at 0°C. and 12 hr. at 10°C. produced greater hardiness than continuous exposure at the average of these two temperatures.

Walton and Fort (1931) stated that mannitol and dextrin were found in frozen sugar cane in Louisiana. Acetic acid fermentation also appeared under such conditions. Bourne (1935), however, did not notice these facts in frozen cane in Florida.

[Sch]affnit and Lüdtke (1932) reported that with winter wheat, winter vetch, and cabbage the lowering of the temperature altered the entire relationship of the nitrogen compounds. Meyer (1932) found that the threshold temperature for the hardening of evergreen leaves was about 6°C. A few hours' exposure daily to this or a lower temperature suffices to keep them in a hardened condition. The basis for cold resistance lies in some physicochemical properties of the protoplasm which as yet are not well understood and cannot be discerned by the gross measurements that at present are generally employed.

Dexter (1933) stated that the hardening of plants is favored by conditions which tend toward the accumulation or conservation of carbohydrates and other reserve foods, which favor photosynthesis, and which reduce the respiratory rate and the extension of vegetative parts. Tysdal (1933) found that the hardier varieties of alfalfa hardened more rapidly under short days than under normal days than did the less hardy ones. The short day reduced growth, a process that is conducive to hardening. Plants hardened under white light could withstand cold better than those hardened under red and blue light. Peltier and Kiesselbach (1934) and Suneson and Peltier (1934) noted that the seedlings of winter wheat which were yet dependent on the endosperm excelled in hardiness those which had emerged 1 or 2 weeks previously and were independent of it.

Calvert (1935) found in wheat that the total water and the free water per 100 g. of expressed sap were significantly higher in the morning than in the afternoon, while the content of bound water was the reverse of this. The total water, the free water, and the bound water per gram of dry matter in the expressed sap were all significantly greater in the forenoon than in the afternoon. The free water exceeded the bound water in the morning, while the bound water exceeded the free water in the afternoon. Dexter (1935) found indications that there is no loss of electrolytes in any of the varieties of alfalfa during the process of hardening.

4. *Measurement.*—It was stated by Dexter, Tottingham, and Graber (1930, 1932) that the degree of resistance of plants to injury by cold weather may be measured by the diffusion of electrolytes and other substances from chilled or frozen tissues into water, when such tissues are thawed. Tottingham, Shands, and Delwiche (1931) considered that Chibnall's method of extraction for the separation of the vacuolar sap and cytoplasmic fluid in plant cells can be used for investigating the winter hardiness of wheat. Peltier and Tysdal (1932) reported a method for growing and hardening alfalfa in the greenhouse.

Dexter (1932) described a method by which leaves may be tested for hardiness by absorbing the expressed juice on ashless filter pads and, after the extraction of the pad with distilled water, determining the quality of sap by measurements of the electrical conductivity. In every case, extracts of tender varieties have higher electrical conductivities than those from hardier plants. Extractions by freezing or by pressure served to differentiate the varieties on a hardiness basis, while the extraction of plants injured by heat, ether, or grinding did not. A considerable part of freezing injury may be due to the mechanical effect of the pressure of the ice that is formed.

5. *Hardiness and Drought Resistance.*—According to Maximow (1929) an increase in the accumulation of sugars, other soluble carbohydrates, and hydrophilic colloids gave a greater resistance to frost. The same factors also aid a plant in enduring drought. Waldron (1931) in a study involving five varieties of wheat under semiarid conditions found that there was a positive relationship between frost resistance and drought resistance. Tysdal (1933) found that plants which were kept severely wilted for 10 to 14 days were much more resistant to cold than those which had been watered regularly. As reported by Schröder (1909), Savage and Jacobson (1935), and Vassiliev and Vassiliev (1936), plants resistant to drought have the same general characteristics as those resistant to cold.

2. Light.—In addition to its influence on the formation of chlorophyll and the process of photosynthesis, light exerts a formative effect upon plants.

a. Growth in Darkness.—When a plant is grown in darkness, striking structural and morphological changes result. It is greatly elongated, the tip of the stem has a characteristic hooked form, and the leaves remain very narrow and scalelike. A plant grown under such conditions and with such an appearance is said to be "etiolated." According to Priestley (1926), the main morphological and structural features of etiolation are determined by a redistribution of meristematic growth at the shoot apex, which results from the greater difficulty experienced by the meristem in drawing nourishment from the vascular supply, because when grown in the dark the walls between the vascular strands and meristem are rendered relatively impermeable by the retention in them of the protein and fatty substances that form the surface of the protoplast. The preponderance of growth in length over superficial growth in the shoot in etiolated plants suggests that the superficial layers of meristem are making less growth, while new tissues are being added to the stem as the result of the activity of the more deeply situated meristematic cells. Priestley obtained evidence from micro- and macrochemical data which indicated that the fatty substances released from the differentiating cells are only slowly removed from the walls between these protoplasts under conditions of etiolation. This slow removal of fatty substances from the walls is obviously in close correlation with the deep-seated meristematic activities. Trumpf (1924) found that exposing etiolated plants to light for only 1 to 2 min. and then returning them to darkness produced marked changes in the appearance of the plant. The most striking results were the marked expansion of the leaf lamina and the disappearance of the plumular hook. Priestley (1925) by the exposure to light of the etiolated plants of *Vicia faba* and *Pisum sativum* produced the changes observed by Trumpf and also greatly shortened the stems, so that the etiolated plants had much the same appearance as normal ones, except that they lacked the green color. It appears that the light action in these cases is photocatalytic.

b. Rate of Growth in Light and Darkness.—If stem elongation is taken as the measure of growth, it will be found that in most instances the rate of growth during the night will exceed that during the day, provided that the temperature does not fall too low during the night. Thus Osmaston (1918) noted that the increase in height of the bamboo was 9.65 ft. in 14 days of which 3.7 ft. occurred by day and 5.95 ft. by night. Porterfield (1928) reported that the ratio of the night growth of bamboo to that of daylight was 1.8 to 2.8 as reported by investigators in Java, Ceylon, and India. In Algiers, however, the growth of this plant was greater during the day than during the night. Porterfield found the ratio of day growth to night growth to be 1.6. Brown and Trelease (1918) found that the shoots of *Cestrum nocturnum*, which is one of the most rapidly growing plants around Manila, P. I., wilted during the day and decreased in length, but that later in the afternoon they returned to their original length and increased it during the night. Prescott (1921), according to Hanna (1925), found that the growth rate of corn in Egypt as measured by increase in height showed two maxima during the 24 hr., one a short time after sunrise and the other in the late afternoon about sunset. The total growth, however, during the 12 hr. of night from 8 P.M. to 8 A.M. was generally greater than that during the remaining 12 hr. of the daytime. Miller (1925) observed at Manhattan, Kans., that corn plants which were about 4 ft. high increased in height 6 cm. from 7 A.M. to 7 P.M. and 8 cm. from 7 P.M. to 7 A.M.

Mason (1925) reported that in the date palm, normal growth, which is manifested by the pushing up of the leaves from the growth center, is made chiefly in the time between sunset and sunrise but also at a reduced rate in the daytime when direct sunlight is cut off by the clouds. In full sunlight, elongation of the date palm ceases entirely. Leaf growth in this plant, however, may be induced in darkness obtained by enclosing the plant in a dark chamber at any hour of the day. Partial growth may also be obtained by screening the plant from direct sunlight, exposing it to reflected light, or placing it with a north exposure.

Coster (1927) in a study of 19 species of plants found that the majority of them grew more slowly on hot days than during the nights, but these diurnal differences were decreased or absent on rainy days. Some species showed equal growth rates during the day and night. Darrow (1929) noted in the brambles that there was a periodic, diurnal march of elongation with great regularity of two maxima and two minima. Thus the growth in length was most rapid from 1 to 2 P.M. until 11 P.M., and from daylight until 10 or 11 A.M. The slowest rate of growth in length was from 11 P.M. until daylight and from 10 to 11 A.M. to 1 to 2 P.M. Elongation was slightly greater during the day than at night.

Porterfield (1930) noted in observations on the growth rate of *Phyllostachys nigra* that 87.8 per cent of the cases exhibited greater day growth, 9.6 per cent greater night growth, and 2.5 per cent the same rate of growth during each period. Stone (1933) reported that, of the total growth of potato leaves in the greenhouse, 54.1 per cent occurred during the night and 45.9 per cent during the day. Karling (1934) found that the trunk of *Achras zapota* reached its maximum diameter between 6 and 7 A.M., and gradually decreased to a minimum at approximately 5 P.M. However, rain, humidity, and changes in temperature altered the rhythm to a considerable degree.

The question arises as to what is the cause of the greater elongation of stems during the night. In many cases it is apparently due to a better water supply for growth. The demands for water by transpiration are so great during the day that there is not a sufficient amount available in the growing regions for the maximal elongation rate of the cells. Brown and Trelease (1918) believed that the absence of growth and even shrinkage of *Cestrum nocturnum* during the day is connected with excessive transpiration. This plant grows better when exposed to the sun than it does in the shade. During the daytime a sufficient amount of water is apparently available in the leaf for photosynthesis, and the plant accumulates food to be used in elongation at night. Mason (1925) believed that in the desert it is the action of the rays of a wave length of 570mμ in the yellow to 405mμ in the violet and probably, also, the ultraviolet rays that inhibit the growth of the palm leaf. In the absence of direct sunlight, the growth of these leaves was apparently synchronous with the closing of the stomata, the checking of transpiration, and increased turgescence in the meristematic tissue. In some succulents the rate of growth varies during the day and night according to the acidity of the tissue. Thus, according to Long (1915), MacDougal noted in *Opuntia blakeana* that the growth increase in diffuse light was 6.9 per cent, while in bright light it was 12.4 per cent. In the daylight in this instance the acidity decreased, while at night acids were accumulated.

Increased rate of growth is not always due to an increase in the osmotic pressure, since Borowikow (1912) has found in the sunflower that the most rapidly growing seedling is not always the most turgescent one. Vogt (1915) in observations on the growth of the coleoptile of barley obtained evidence that the greater growth in darkness is due largely to the stimulating effect of the previous illumination. He believed that a knowledge of the effects of light on the hydration of colloids will be necessary to explain the behavior of growth in light and darkness.

c. Different Rays.—Under this heading is reviewed some of the work that has been done on the effects of the rays of the ordinary spectrum, X rays, and polarized light on the growth of plants.

1. *Rays of the Spectrum.*—Schanz (1919) mentioned the fact that in Holland many growers prefer crude glass to regular window glass for forcing. The former cuts out more of all rays than the latter and is especially effective in screening out the ultraviolet. On this account, Schanz carried on some experiments to determine the effect of the different rays of the spectrum upon plant growth. He stated that cucumber, *Fuchsia, Begonia,* beet, and potato became taller the more the short rays of light were eliminated. The maximal height was obtained under red light and the minimal under the blue-violet light. He concluded, in general, that light of short wave lengths, particularly the ultraviolet, was detrimental to the growth of plants and recommended the use of glass for greenhouses which eliminates these rays. He believed

also that the ultraviolet rays have an important relation to the development of anthocyanin in the epidermal layer, since this pigment failed to develop in red-leaved lettuce and other plants when these rays were excluded.

When plants were grown in daylight from which all wave lengths shorter than 529mμ were eliminated, Popp (1926) found that they developed the following characteristics as compared with plants grown in the entire spectrum of daylight: a greater final height in soybeans, tomatoes, and *Coleus* but a decided decrease in height in sunflowers, petunia, buckwheat, and Sudan grass. There was a considerable decrease in the thickness of stems and a reduction in the number of branches and side shoots. There was a good development of chlorophyll but a reduction in the amount of anthocyanin of the leaves and flowers. The number of flowers was reduced and there was a delay in their time of appearance. There was also a very weak development of seeds, fruits, and general storage organs. The decrease in starch and total carbohydrates was considerable, but usually there was an increase in total nitrogen. When all wave lengths shorter than 472mμ were removed, the same effects were produced but to a somewhat lesser degree. When only ultraviolet rays were eliminated, none of the foregoing results was obtained with any of the plants used, although there was a small increase in the length of stems in all the plants except buckwheat. In general, there was very little difference between plants that received all the rays of the spectrum of daylight and those from which all ultraviolet rays were eliminated. The results of the work of Popp indicate that, although the ultraviolet rays are not indispensable, the blue-violet end of the spectrum is necessary for the normal vigorous growth of plants. This conclusion is substantiated by the work of Shirley (1928), who grew plants under five different light qualities with three intensities and found that the plants produced more dry matter per unit of light intensity under the complete solar spectrum than under any portion of it.

Eltinge (1928) found that raying with an unscreened quartz mercury light caused injury to all the plants used. This ultraviolet radiation had little if any effect upon the decomposition of the chlorophyll or upon the hydrogen-ion concentration of the plants. Raying with a lamp screened by Vitaglass, which allowed the passage of the rays 578 to 289mμ, was beneficial to some plants, while it produced little visible effects in others. Raying with a lamp screened by quartzlite glass (578 to 313mμ) injured none and benefited many of the plants. Eltinge believed that her results emphasize the fact that each plant has its own ultraviolet requirement for its best growth, which can be determined only by experiment. Tincker (1930), in using various types of commercial glass which allow the ultraviolet rays to pass, found that a heavier crop of radishes and carrots was produced under them than under ordinary glass but that the rate of growth of lettuce was not affected.

Colla (1927) subjected young wheat seedlings in the dark to the action of light rays of a length of 3300 to 3900 Å. After 8 hr. of continuous illumination, these plants curved toward the light and the new leaves developed sufficient chlorophyll to give a pallid-green color to the plants. Fuller (1930) showed that plants of tomato and cucumber that had been treated with ultraviolet irradiation increased in growth approximately 34 per cent over the controls. The ash content in the treated plants was greater by 13 per cent than that of the controls. It was observed by Sheard and Johnson (1930) that marked changes in the difference of potential between the base and tip of intact leaves can be produced by ultra- and infraviolet radiations. In darkness or subdued light the base of the leaf is positive with respect to the tip. Under intense artificial illumination, behind ordinary glass, the difference in potential decreases so that the tip of the leaf may become positive relative to the base.

Shull and Lemon (1931) stated that in general two physiologically different regions of ultraviolet radiation have been recognized—the abiotic, or destructive, region, and the biotic, or biologically valuable, region. The abiotic are the shorter and the biotic the longer rays (Fuller, 1931). The ultraviolet rays of sunlight are not harmful to ordinary plants, nor do they seem to be useful, since plants can be grown very successfully without them. Seed coats allow only the longer ultraviolet radiations to penetrate. The lowest limit of rays transmitted by any of the seed coats tested was at 3020 Å, and the penetration of those shorter than 3630 Å was always very slight. In the corn seed, different portions of the same seed coat are not alike in permeability to radiations. Bucholtz (1931) reported that the expanded leaves of *Mnium* and the stamen hairs of *Tradescantia* were not injured by strong intensities of the rays, 2378 to 3654 Å. Wynd and Fuller (1931) found that tomato and cucumber plants which had been stimulated by ultraviolet light showed a definite increase in calcium and a decrease in phosphorus, in percentage on a dry basis.

Masure (1932) subjected the dry seeds of *Pisum sativum* to a radiation of 3650 Å, and found that this wave length exerted a stimulative action on the subsequent rate of growth of the hypocotyls. Singh, Kapoor, and Choudhri (1936) found that the treatment of seeds with ultraviolet radiations for 10 min., or less, accelerated germination and induced luxuriant growth, while an exposure for periods of 20 min. retarded both germination and development.

Fuller (1932) found that radiations from an unscreened quartz mercury lamp produced decided injury in the tissues of plants of tomato and bean. The major part of this injury was due to the infrared radiations. Meier (1934) found that the wave lengths of ultraviolet ranging from 2652 to 3022 Å killed *Chlorella vulgaris*. It is stated by Popp and Brown (1933) in their comprehensive review of the literature on the effect of ultraviolet radiation upon plants that the influence of these rays upon seed plants, other than their destructive action, is yet to be ascertained.

Arthur and Stewart (1935) grew plants of buckwheat under mazda, neon, mercury-vapor, and sodium-vapor lamps, and found that if the dry weight produced under the mazda lamp were taken as one, the values for the other lamps were neon, 1.10; sodium vapor, 0.90; and the mercury vapor, 0.66. Calculations of the dry weights that might have been produced, if equal amounts of energy in the visible region had been used, were mazda, 1.00; sodium vapor, 1.41; neon, 1.21; and mercury vapor, 0.62. Johnston (1932) and Brackett and Johnston (1932) found that the infrared region of the spectrum was of little or no benefit in chlorophyll formation by the tomato. The near infrared region of the spectrum is of considerable biological importance. Haig (1934, 1935) noted in the response of the seedlings of oats to light that if the illuminations were confined to the extreme tip of the coleoptiles only the low-energy portion of the response curve was obtained, but when it was confined to the basal portion the high-energy part appeared. The tip had a distinct maximum at 4800 Å, while the base curve exhibited no maximum in the visible range. The high portions of the base curve may represent a broad maximum somewhere between 4000 and 4700 Å. The reactive time for positive phototropic responses for short exposures to white light decreased with increasing intensity to 1,000 meter-candles.

Johnston (1934, 1935) found that the sensitivity of the oat plant increased rapidly from 4100 to 4400 Å, fell to 4575 Å, then again rose to a second maximum at 4750 Å. From this point the sensitivity decreased rapidly to 5000 Å, from which point it gradually decreased to 5461 Å—the threshold of sensitivity on the long-wave-length side. Growth was retarded to the greatest extent in blue light.

Burkholder and Pratt (1936) noted that the leaflets of *Mimosa pudica* responded readily in blue to long ultraviolet and in long-red radiations, but showed little or no response in the orange, yellow-green, or infrared rays. Ball (1936) found that the shorter red rays were as effective as white light in producing inhibition of anthesis in the flowers of *Turnea ulmifolia*. There was a decrease in efficiency of the red rays beyond 700mμ. The yellow and violet regions were much less effective than the red in this regard. The infrared rays produced little or no effect.

Atkins (1936) found that the maximal phototropic response of *Avena* to the rays of the spectrum was in the region of 4480 Å with diminution in the ultraviolet and in the red. With *Lepidium sativum*, the initial maximum was about 4640 Å in the blue with a second maximum in the red. Plants containing red anthocyanin pigments in the epidermis gave a sluggish response, the pigments absorbing light without inducing curvatures.

2. *X Rays.*—Johnson (1926) found that the catalase activity was depressed in the young seedlings of *Helianthus annuus* whose seeds had been exposed to medium doses of X rays, but that in older seedlings little change could be noted. When stunted growth accompanied heavy irradiation, the rate of respiration was decreased. Cells from the root tip of seedlings whose growth had been inhibited by irradiation showed elongation, large vacuoles, and the absence of a nucleus. Potatoes irradiated before sprouting with a slight dose of X rays produced 27 per cent more tubers per hill, but the average weight was 18 per cent less than the controls, thus making the average total weight per hill practically the same (Johnson, 1928). When the seeds of sunflower were irradiated with medium doses of X rays, an inhibiting effect proportional to the dosage was noted in the growth of the seedlings for about 3 weeks. After that, however, the plants grew rapidly until at maturity, when there was little difference in height or time of blossoming between the controls and the plants from treated seeds. Williams (1923, 1925) noted that the cells of *Saxifraga umbrosa* and *Elodea* after a short exposure to the β and γ rays from radium showed an increase in the rate of circulation of the protoplasm and an increase in the permeability of the protoplasmic membrane as indicated by the exosmosis of solutes. Large doses produced shrinkage and vacuolation effects that were irreversible. Extensive experiments have been conducted by Stadler (1930) on the genetic effect of X rays on barley and corn. A discussion of these effects, however, belongs primarily to the province of genetics and will not be considered here.

Some of the more recent work on the effect of X rays on plant activities is shown in the table on page 1076.

Johnson (1936) reported that of the 70 species of plants from 35 families that have been studied in regard to reaction to X rays, 15 species were apparently unaffected, 15 were slightly affected, especially during their early growth, while the remaining 40 species were noticeably affected. Members of the *Chenopodiaceae*, *Umbelliferae*, and *Brassicaceae* frequently gave indications of injury soon after treatment, but, by the time maturity was reached, showed little effect. Those plants noticeably injured by radiation were marked by decrease in total height, increased branching in *Caryophyllaceae*, and frequent occurrence of dichotomous branching particularly in the *Carduaceae*. Other effects were irregularities in shape, margins, and chlorophyll development of the leaves and delayed and reduced flowering. The members of the *Solanaceae*, *Scrophulariaceae*, and *Carduaceae* were noticeably X-ray susceptible, as were the genera *Ambrosia*, *Ricinus*, *Lavatera*, *Cobaea*, and *Ipomoea*.

3. *Polarized Light.*—The effect of polarized light on growth was noted by Macht (1926, 1928). Plants of *Lupinus albus* exposed to polarized light grew much better than did the controls. The growth of wheat and sunflowers was also hastened, although not to such a marked degree. It was also found, as mentioned in Chap. XI, that polarized light hastened the conversion of starch to sugar.

RESULTS OF TREATMENTS WITH X RAYS

Investigator	Parts treated	Effects of treatment
Reinhard and Tucker (1928)	Seedlings (*Vicia faba*)	Showed their greatest sensitivity to irradiation at 10 and 11 A.M., and 1 and 5 P.M.
Sprague and Lenz (1929)...	Potatoes	Definite lesions on leaves, may reduce number of, but increase size of, tubers.
Wigoder and Patten (1929).	Dry and soaked seeds of *Vicia faba*, barley, white mustard, timothy, and peas	Failure to develop roots and shoots. The maximal effect occurred with actively dividing cells in the first 4 mm. of the root tip. Mitotic rate greatly decreased 3 hr. after irradiation and after 3 days was absent entirely. Cell division resumed after 5 to 8 days, when many abnormal, multinucleate cells were observed.
McKay and Goodspeed (1930).	Pollen of cotton	Seeds obtained from crossing normal pistils with X-ray treated pollen were planted. Many of the seeds were infertile.
Cattell (1931).............	Wheat seedlings	Growth of lateral roots reduced one-half in length. The effects detected in a few hours after treatment.
Johnson (1931)...........	Tomato and various other seedlings, tulip bulbs, and wild potato tubers	Tulip bulbs and wild potato tubers grown from irradiated material showed slight increase over controls. Marked leaf anomalies in tomato leaves. Leaflets often absent, badly twisted, or joined together. Older leaves spotted or variegated. More branches than control and abnormalities of floral parts.
Moore and Haskins (1932).	Seeds of grapefruit	Plants from these seeds flowered prematurely.
Johnson (1933)...........	Plants of *Atriplex hortensis*	Red forms showed greatest decrease in height. Green varieties showed the greatest number of leaf abnormalities.
Shull and Mitchell (1933)..	Seeds of corn, wheat, oats, and sunflower	Treatment of seeds produced stimulative effects upon seedlings. The conditions for the best results are metallic screens, high voltage, low amperage, and short exposures.
Benedict and Kersten (1934).	Seeds of wheat	Increase in diastatic activity and reducing sugars with treatment for five seconds. Both decreased if irradiated longer. Irradiated seeds absorb water less rapidly than controls.
Francis (1934)............	Seeds of wheat	Retarded growth of seedlings. Depressed respiratory rate. Linear growth retarded. Lateral roots most sensitive and coleoptiles least sensitive.
Skoog (1934)............	*Avena* coleoptiles	Great decrease in amount of growth substance diffusing from them. Does not decrease transport of growth substances.
Haskins and Moore (1935).	Seeds of *Citrus*	High percentage of seedlings showed a deficiency of chlorophyll. Many grades of albinism noted. Many monstrosities.
Havas (1935)............	Various seeds and seedlings	Radiations from a radioactive mud obtained from Hungary showed stimulating effects on some plants and depressing effects upon others at an early stage of growth.
Nogüchi (1935)...........	Seeds of sunflower	Seedlings from treated seeds showed abnormal development of vegetative parts. Abnormality of leaves the most common symptom. The minimum time for inducing leaf abnormality is irradiation for 3 to 4 min.
Reynolds (1935)..........	Tomato plants	Injury by irradiation from an unscreened mercury-vapor arc is more difficult to show on those plants which have developed during the summer months than those which have grown during the winter months. In summer, 1 per cent, and, in winter, 0.2 per cent of solar radiation is ultraviolet.
Collins and Maxwell (1936).	Dry seeds of corn	Found a range of treatment that caused all the plants to die in the seedling stage without reducing the rate of germination.
Meier (1936).............	*Chlorella vulgaris*	The maximal lethal sensitivity was at 2600 Å. The wave length of 3130 Å, which is but slightly longer than the shortest wave length of solar radiation reaching the surface of the earth, had no lethal effect.

Semmens (1930) found that the leaves of alfalfa, nasturtium, and grape grown in polarized light with alternating periods of light and darkness showed temporary negative photoperiodism, leaf fall, disappearance of starch, and signs of starvation. He believed that plants require both polarized and unpolarized light for the best growth.

d. Intensity of Light.—According to Withrow (1936), the processes involving visible light are of two types—those which require light intensities above 100 foot-candles and those which require lower light intensities. Photosynthesis falls within the first class and, aside from the length of day factor, is apparently the limiting factor that prevents the normal growth of plants in greenhouses during cloudy weather. In other respects, plants in greenhouses appear to grow quite normally in light intensities of 100 foot-candles or less. Light intensities of 20 foot-candles, or less, are apparently sufficient for the synthesis of chlorophyll, for preventing etiolated characteristics, for proper permeability, and for phototropic response. To sustain vigorous growth of plants in windowless greenhouses without the aid of sunlight would probably require intensities of 1,000 foot-candles for a period of 12 hr. per day. Low light intensities prevailing during the winter months frequently involve intensities of as low as 100 to 300 foot-candles during late morning and early afternoon, and plants grow very slowly under such conditions. It would probably be necessary to double at least these lower values with artificial radiation to satisfactorily accelerate growth.

Under artificial light, Harvey (1922) found that some plants grew and bloomed over a wide range of light intensities, while others bloomed only in a very limited range. Thus, with the light intensity measured in foot-candles, the range of corn was 500 to 3,000; tobacco, 800 to 10,000; *Amaranthus retroflexus*, 500 to 10,000; squash, 500 to 750; and wheat, 650 to 800. Popp (1926) grew four varieties of soybeans for 7 days at a temperature of 19 to 23°C. under light intensities of 4,285, 1,536, 560, 390, 250, and 26 foot-candles, respectively. He observed that the lower the light intensity the more rapid was the rate of stem elongation during the period of initial growth. The greatest general height was attained by plants under a light intensity averaging 560 foot-candles and the lowest under 26 foot-candles. The thickness of the stem was directly proportional to the light intensity, and there was a gradual decrease in vigor with decreasing light intensity. Adams (1924) believed that light of lower intensity but more prolonged will produce the same result as light of higher intensity but of shorter duration. Observations as regards height, weight, and time of flowering on numerous plants grown in the greenhouse and in the open showed that the results were as satisfactory under 2 hr. exposure to light at midday as under 3 hr. exposure during the morning and afternoon.

Shirley (1931) noted that a light intensity of approximately 66 per cent was sufficient for the establishment of the seedlings of Norway pine, while light values below 17 per cent resulted in uncertain growth. It was observed by Arthur, Guthrie, and Newell (1930) that the tomato will be damaged by an intensity of light that causes little or no injury to other plants. Arthur and Stewart (1931) reported that plants of sunflower, buckwheat, dahlia, and tobacco grown during June and July produced greater dry weight of tissue when shaded. The tobacco plant reached a maximal dry weight in a light intensity of 35 per cent, sunflower in 78 per cent, and dahlia and buckwheat in 58 per cent. In August and September, however, the increase in dry weight increased with higher transmission, and were highest in open sunlight. Steinbauer (1932) found with tree seedlings that the greatest response to an increased concentration of the nutrient solution occurred at the higher light intensities. Ashby and Oxley (1935) found that the dry weight of *Lemna minor* in light intensities of 80 to 1,600 foot-candles increased linearly with light intensity. Panchaud (1935) found that the ratio of water to the dry substance in the radish tended to decrease as the light intensity was increased.

e. Length of Exposure to Light.—Although it has been known for a long time that the duration of exposure to light markedly affects the growth of plants, it is only within the past two decades that the subject has been extensively investigated. The striking results obtained by Garner and Allard (1920) in their extensive investigations on the effect of the relative length of day and night upon growth and reproduction in plants created a great amount of interest and stimulated research in this field. These investigators suggested the term "photoperiodism" to designate the response of organisms to the relative length of day and night and "photoperiod," to designate the favorable length of day for each organism.

1. *Continuous Light.*—In general, it may be stated that the plants which are exposed longest to daylight attain the greatest production of dry matter. This subject has been especially studied by Adams (1920 to 1925). In Lat. 45.5° N. during June and July, plants of corn, wheat, soybean, white mustard, wax bean, flax, tomato, and sunflower were darkened from 1 to 5 hr. each day. In almost all cases, the plants exposed longest to the action of light gave the greatest average weight and the greatest average height. Wheat, rye, flax, hemp, soybeans, tomatoes, buckwheat, and sunflowers were grown in darkness, in different periods of daylight ranging from 3 to 15 hr., and in artificial light in addition to full daylight, bringing the total exposure of light up to 18 to 20 hr. In the case of the tomato, soybean, buckwheat, and hemp, there appears to be an upper limit to the amount of light which a plant can utilize and above which it makes no additional growth.

Harvey (1922) grew a large number of plants under continuous artificial illumination with electric lights. These lights were mounted with ordinary enameled reflectors at a distance of 5 ft. from the plants and gave about 457 lumens per square foot. Under these conditions, wheat, oats, barley, rye, flax, buckwheat. white clover, peas, beans, and a number of weeds grew from seed to maturity and set good seed. Potatoes, tomatoes, red and alsike clover, and squash bloomed but set no seed. Adams (1925) grew the castor bean from seed to seed under artificial light, while Tjebbes and Upholf (1921) and Hostermann (1922) obtained increased yields from greenhouse crops by the use of artificial light. Harrington (1926) grew cereals in the greenhouse during the winter through the use of artificial light to supplement the natural daylight. By this means he was able to grow to maturity three successive generations of wheat in one year.

Redington (1929) reported that *Zea mays, Gossypium herbaceum, Cucurbita pepo, Pisum sativum,* and *Linum usitatissimum* showed more growth with 16 hr. of light per day than in continuous light. Flowering, however, generally occurred earlier in continuous light. Reid (1929) found that the exposure of the plants of corn, cowpea, soybean, muskmelon, and sunflower to the normal length of day in May and June produced an inhibitory effect on the growth in length of the stem and hypocotyl but a stimulatory effect on the growth of leaves and foliaceous cotyledons. Arthur, Guthrie, and Newell (1930) stated that the tomato is sensitive to continuous light. In cabbage plants the total carbohydrate was doubled in a 19-hr. day as compared to a 5-hr. day. Eaton (1931) found in the soybean that the amount of growth and nodule development was in direct proportion to the length of day. Ramaley (1931) noted that numerous garden vegetables, grains, weeds, native herbs, and ornamentals grown in the greenhouse in continuous light were more elongated than the controls. Flowering was hastened in some cases and inhibited in others. In continuous light, the root systems were less extensive and the stems showed a thinner cortex, less vascular tissue, and more pith than the controls.

Darrow (1933) stated that in Alaska the production of strawberries and tomatoes occurs under continuous light of 6 weeks' duration. Burge, Wickwire, and Fuller (1936) found that, in continuous illumination, irritability in *Mimosa pudica* did not fall during the night but remained at high day level. In continuous darkness it remains at the low night level.

2. *Short and Long Days.*—The work of Garner and Allard (1920, 1923) and of Tincker (1925, 1928, 1929) showed that plants, in their behavior toward the length of light exposure, arrange themselves into three general groups according to the effect of the treatment upon the time of flowering: (*a*) Those species and varieties which are apparently little affected in

their time of blooming by the duration of the exposure to light. (*b*) Those plants which are caused to flower by the action of short days. Such plants are commonly called "short-day" plants. (*c*) Those plants which are forced into flowering through the action of long days. These have been designated "long-day" plants.

(*a*) *Flowering.*—Some of the plants that have been found to require short days to produce flowering are Maryland Mammoth tobacco, Biloxi soybeans, ragweed, *Aster linariifolius*, *Phaseolus vulgaris*, *P. multiflorus*, *Cosmos*, and *Chrysanthemum*. Some of those which require a long day for the production of flowers are Mandarin soybeans, *Hibiscus*, goldenrod, climbing hempweed, timothy, radish, and perennial rye grass. The plants that are placed in the short-day group are those which can bloom only under a daily light period of 12 hr. or less, while those which are placed in the long-day group require a length of daylight of more than 12 hr. (Kellerman, 1926).

The influence of the relative length of day upon the date of flowering can be illustrated by considering certain varieties of soybeans. If the Biloxi variety is planted about the middle of April in the latitude of Washington, D. C., approximately 125 days elapse between germination and the time of blooming. If, however, these plants are exposed to daylight for only 7 to 12 hr. the time elapsing between germination and blooming is only 28 days. The fact that certain varieties of plants are early or late maturing is apparently connected with the different responses of these plants to the relative length of day and night. Thus for plantings in the field through the month of May, the average number of days from germination to blossoming for the Mandarin, Peking, Tokyo, and Biloxi varieties of soybeans was approximately 27, 56, 70, and 105 days, respectively. When these plants are placed in the daylight for only 7 to 12 hr. daily, they all bloom in approximately 28 days and thus all become early maturing. On the other hand, the late-maturing varieties may be maintained in the active vegetative condition for a prolonged period of time by continued exposure to a relatively long daily period of illumination.

It is impossible to review here all the work that has been done on photoperiodism. Some of the more important studies have been by Darrow and Waldo (1930, 1933) on the strawberry; Garner and Allard (1930) and Austin (1935) on soybeans; Forster, Tincker, Vasey, and Wadham (1932), Tincker (1932) and Hurd-Karrer (1933) on wheat; Kondo, Okamura, Isshiki, and Kasahara (1932) and Pan (1936) on rice; Starkey (1932), Laurie and Poesch (1932), and Post (1934) on chrysanthemums; Greene, Withrow, and Richman (1932) and Withrow and Richman (1933) on greenhouse flowers; Evans and Allard (1934) on timothy; Ramaley (1934) and Withrow and Benedict (1936) on numerous greenhouse plants; Robinson (1933) on flax; Allard (1935) and Kramer (1936) on various woody plants. Johnston (1936) reviews some of the work in this regard.

According to Withrow (1936) the intensity of the light necessary for the lengthening of the day varies widely for different plants. For the stock this value is between 5 and 10 foot-candles. For Chinese aster ⅓ foot-candle is as effective as 100 foot-candles in inducing earlier blooming. Even ⅒ foot-candle, or about twice the intensity of bright moonlight, is sufficient to markedly influence the flowering of the "Heart of France" aster. Commercially, 1 to 25 foot-candles are now used. The extra light can be applied at any time during the night for not less than 4 hr. or more than 6 hr. Garner and Allard (1931) reported that breaking the continuity of the daily illumination period of plants by darkening them in the middle of the day for periods of ½ to 5 hr. materially affected the general nutrition and amount of growth but as a rule failed to influence reproduction to a degree comparable with that produced by excluding the early morning or late afternoon light of the long summer days. Thus far the chrysanthemum is the only plant that is commercially treated with the short day.

The red rays of the visible spectrum are the ones that are largely effective in bringing long-day plants into flower and in delaying the blooming of short-day plants. The blue rays are much less effective than the red rays. The lengthening of the day with green and yellow light has practically no effect.

The work of Gilbert (1926) and Adams (1925) suggests that the phenomena of response to relative length of day may be influenced by the temperature and humidity conditions to which the plant may be subjected during the growth period. Gilbert observed in soybeans and cotton that there was a definite retardation of flowering with the lower temperatures and higher humidity, while *Cosmos* flowered much earlier and more normally under lower temperatures and higher humidity. *Salvia* and buckwheat exhibited no reaction to temperature and humidity conditions. In the case of *Xanthium pennsylvanicum* grown under known conditions of temperature and relative daylight, it was found that temperature was the determining factor in influencing the time of flower primordia formation but was associated with the response to the length of day. Adams (1925) observed in numerous experiments carried out at different times of the year that a large number of plants showed as good growth under an average exposure to daylight of 569 hr. at a mean temperature of 60.8°F. as they did with an average exposure to light of 500 hr. at a temperature of 68.2°F. The temperature must always be taken into account in experiments dealing with the relation of plants to light.

Plitt (1932) showed that in some cases temperature had an influence on the response of radish to the length of day. Berkley (1931) believed that temperature differences may be substituted for length of day in certain combinations. Thus fruiting may occur in the cotton plant under any length of day, of 8 hr. or more, provided that the temperature and other factors are favorably adjusted.

The work of Schaffner (1923, 1927) on the influence of the relative length of day upon the reversal of sex in hemp and corn should be noted here. When hemp is planted in the latitude of Colorado or Ohio between May 10 and Aug. 10 in the greenhouse or in the open, it develops the typical dioecious state. If, however, the planting is done in the greenhouse between Aug. 20 and May 1, sex reversal is almost sure to take place in inverse proportion to the length of daylight and the time of planting and the subsequent period of growth. This reversal of the sexual state is in both directions. In the staminate plants the reversal is from maleness to femaleness and in the carpellate plants from femaleness to maleness. In corn planted on Nov. 1 in the greenhouse, all individuals showed some degree of female expression in the tassel, whereas if planted in the spring or summer, they showed only purely staminate tassels.

Richey and Sprague (1932) showed that the reversal of sex expression in maize is influenced by environment and heredity, and that shorter periods of daylight and lower temperature tend to cause silks to develop in the tassel. It was noted by Allard (1932) that the hog peanut (*Facata comosa*) developed blue, aerial, perfect flowers only when the days were not less than 13.5 hr. in length. The cleistogamic flowers were developed under lengths of day ranging from 5 to 13.5 hr.

(*b*) *Development of Storage Organs.*—The duration of the daily illumination period not only influences the quantity of photosynthetic material that is formed but also may determine the use that the plant can make of it. The changes in the character of vegetative growth due to the alteration of the period of light exposure are frequently as clear and striking as the change from the vegetative to the flower behavior. The Globe spring radish when exposed to a 7-hr. day enlarged its root for at least a year. The Irish potato and the Jerusalem artichoke developed tubers only under relatively short days. Carrots did not produce a bulbous root under a very short day but under a 14-hr. day developed a typical carrot root. In no case did the light period best adapted to the formation of bulbs, tubers, or storage roots coincide with the best daylight period adapted for the upward or top growth of the particular plant under consideration (Kellerman, 1926).

In Puerto Rico, McClelland (1928) found that onions were very sensitive to the duration of the daily light period, and that different varieties varied in their response. Short days favored leaf growth but inhibited bulb formation, for which longer days are necessary. The variety Bermuda white showed itself to be adapted for growing in such latitudes as have a maximal day length only slightly in excess of 13 hr. The Prize Taker and other varieties were found to be wholly unsuited for growing under such day lengths, the plants remaining in the spring-onion stage rather than forming bulbs. Under daily light exposures of a little more than 15 hr., normal bulb formation was rapid for all the varieties tried. The longer exposures favored growth of tops in potatoes, while the shorter exposures favored tuberization.

It was observed by Zimmerman and Hitchcock (1929) that the length of day markedly affects the type of root system formed by dahlia cuttings. The long-day exposure gave rise to a fibrous root system, while short-day illumination resulted in a heavy root storage. Weaver and Himmel (1929) found in investigating both short- and long-day plants that the development of the root system for different exposures was in all cases in direct correlation with the development of tops.

(*c*) *Chemical Changes.*—According to Garner, Bacon, and Allard (1924), the light period influences the acidity relations, the form of carbohydrates present in the plant, and probably the water content of the tissues.

In the case of short-day plants, a relatively long daily illumination period was associated with a progressive increase in the actual acidity, particularly in the regions of the growing points. This increase continued until the upper portion of the plant became more acid than the lower portion. Exposure to a relatively short daily light period was followed by only a moderate increase until a level was approached at which flowering was initiated. Under the short-day exposure the upper portions of the plant were less acid than the lower portions. After flowering had been initiated, there was a progressive increase in acidity in the vegetative parts of the plant. An abrupt transfer from a long day to a short day caused a sudden and sharp decrease in acidity in the region of the growing point, which usually occurred about 3 to 5 days after the transfer had been made. This drop in acidity, which was believed to indicate definite transition from the vegetative to the flowering condition, was only temporary and rapidly rose to the original level.

In the case of *Cosmos* a transfer from a long to a short day resulted in a material increase in reducing sugar in the upper portion of the stem within 48 hr. after the transfer had been made. Two days later this increase in sugar content was found to be in the form of polysaccharides and was accompanied by a slight decrease in the water content. Twelve days later after flowering buds had appeared, the increased carbohydrate was in the form of monosaccharoses, while a considerable increase in the water content had occurred. In the summer radish, the elongation of the stem resulting from an exposure to a long day was associated with an increased content of reducing sugar in the tissues with the maximal concentration in the upper portion of the stem.

Deats (1925) found that the cell-sap concentration in the leaves was greatest in those plants exposed to a long day, while it was more concentrated in the fruit of the tomato which had been exposed to a short day. He suggested that the differences in the relative length of day and night might influence the form of plant development by a change in the nitrogen/carbohydrate ratio.

It was noted by Pfeiffer (1926, 1928) that, in plants which were subjected to short exposures to light, there was less total growth, a lower production of differentiated tissue, and lower carbohydrate and protein reserves than in those which had been longer in the light. Hurd-Karrer and Dickson (1934) reported that young wheat plants in the tillering stage had the highest carbohydrate and the lowest total nitrogen percentage in the long day, which accelerated the elongation of the culm and flowering. The lowest percentage of carbohydrate and the highest of nitrogen were produced by the short day, which retarded heading and resulted in large, vegetative plants with some sterility and reduced yields. The pH values of the cell sap were generally highest in the short day with a low temperature, and lowest in the long day with a high temperature. Biddulph (1935) reported that the plants of *Cosmos sulphureus* reacting to short-day treatment required 7 days for the initiation of flower primordia. The change of the foliar primordium to a floral one was accompanied by a marked accumulation of carbohydrate and protein in the stem tip. These were hydrolyzed at the base of the stem and then translocated to the tip. Krassinsky, Kondrashova, and Vinogradova (1936) reported that the shortening of the day caused a considerable and regular increase of activity of peroxidase and catalase.

(*d*) *Localization of Response.*—Garner and Allard (1925) found that, when two coordinate branches of *C. bipinnatus* were exposed to light for daily periods of different length, each branch responded in a characteristic manner to its particular light period, more or less independently of the other. With *C. sulphureus* when the upper portion of the plant was exposed to the full length of the day of summer and the lower portion received only 10 hr. of light daily, the latter promptly flowered, while the former remained vegetative. Knott (1926, 1927) noted that the catalase activity of the apical portion of the stem of spinach decreased as the spinach plant changed from a vegetative to a reproductive type of growth. If elongation ceased and vegetative growth was resumed, the high catalase activity returned. In the bud leaves of the *Cosmos* there was the same marked change in the catalase activity as was found in the stem apex of the spinach. No such change, however, could be noticed in the leaves immediately adjacent to the apex or bud leaves. Knott believed that these facts suggest a highly localized response in the plant to the changed length of day.

3. Electricity.—This topic can best be discussed under the headings of the effects of electricity on crop production, and of the occurrence of electrical phenomena in the tissues of the plant.

a. Effects on Crop Production.—Blackman, Legg, and Gregory (1923) exposed the coleoptile of the barley seedling for periods of 1 to 3 hr. to an electric discharge from a point charged positively to about 10,000 volts and placed 2 cm. above the coleoptile so that a current of the order of 0.5×10^{-10} amp. passed through it.

The increase in the rate of growth reached 4 per cent of the normal rate in the first hour, and if the current was continued for another 2 hr., the effect went on increasing, so that at the end of the third hour an increase of 5 per cent of the normal occurred. When the current was stopped there was a striking aftereffect, the rate of growth continuing to rise for at least 4 hr. after the cessation of the discharge. The aftereffect was greater than the direct effect and resulted in the fifth hour in an increase of 12.6 per cent of the actual rate of the control plants. This aftereffect, as measured by the increase in the rate of growth in the fifth hour from the start of the discharge, was greater with the short discharge of 1 hr. than with the longer discharge period of 3 hr.

When the point was negatively charged and a current of the same intensity passed through the seedling for 3 hr., the rate of growth increased during the first hour, but the increase instead of becoming greater with time as with a current in the other direction became less, so that at the end of the third hour the rate was little if at all above the normal. When after 3 hr. the current was stopped, a stimulating aftereffect occurred, but not so marked as with the positively charged point. These results were treated statistically by Gregory and Batten (1926) and were found to provide significant evidence of the physiological effect of the discharge.

The significance of electrical discharges on field crops was pointed out by Jorgensen and Priestley (1914) and by Jorgensen and Stiles (1917). The most extensive field work in this regard, however, was performed by Blackman (1924) in experiments with spring-sown oats and barley, winter wheat, and clover. A discharge was given at the rate of 0.5 to 1.0 milliamp. per acre from three insulated wires stretched above the crop at a height of about 7 ft. and charged to a voltage of 40,000 to 80,000. The discharge was usually given for 6 hr. a day in two periods, 3 hr. in the morning and 3 hr. in the afternoon. In the case of the 18 experiments with oats and barley, 14 gave positive results in favor of the electrified plots, while 4 showed decreased yields compared with the controls. The increased yield ranged from 10 to 50 per cent, while none of the negative results was so much as 10 per cent. The beneficial effects on winter wheat and clover, however, were uncertain. Blackman and Legg (1924) ran pot cultures of young cereals concurrently with the above field experiments. Networks charged to a high voltage were suspended over the plants by insulating supports and so arranged that they could be raised or lowered and thus vary the strength of the current passing to the plant. The plants showed marked sensitiveness to currents of very low intensity, definite increases in dry weight being obtained with currents as low as 1×10^{-10} amp. per plant. The optimum intensity of current appeared to be below 1×10^{-8} amp.

Briggs, Campbell, Heald, and Flint (1926) in electrocultural experiments extending over 8 years at the Arlington Experimental Farms found no well-defined increase in yield due to electrical treatment, as the results obtained were well within the limit of experimental error. Collins, Flint, and McLane (1929) found that treatment with a current intensity of 10^{-9} amp. was slightly positive for corn seedlings under laboratory conditions.

Shibusawa and Shibata (1927) found that an electric current of very low intensity was induced in plants in the greenhouse by applying high voltage to a thin, wire network suspended at a distance of 15 to 20 cm. above the plants. Under such treatment the yield of buckwheat was increased 12.6 per cent and that of tobacco 21.7 per cent. Singh (1932) subjected seedlings in the soil to an electric current by means of electrodes. After 13 days this treatment increased the root growth as much as 400 per cent but did not visibly affect the growth of the tops. Musso (1933) found that oats headed earlier, formed a greater number of stalks, and yielded 20 per cent more dry weight than the controls when a strong, positive potential was applied to the soil. Fraser and Pidgeon (1933), however, could find no increase in the yield of wheat by treating the seeds electrically.

The knowledge concerning the proper conditions under which electrical treatments should be given is very meager, and much more experimental work is needed before any definite statements can be made concerning the effects of electricity on growth and crop production.

The mode of the action of an electrical current in those cases where increased yields have been reported is not known. Blackman (1924) noted that in some cases the treated crops had a deeper green than those of the controls. The effect of the discharge is apparently of the nature of a stimulus, since the additional energy in most of the experiments was less than one-thousandth of the energy that the plant was receiving from the sun.

It should be mentioned here that the atmosphere becomes heavily charged with electricity during the dust storms that occur in the Great Plains area, and it is believed by some that this causes the marked injury to cereals which has been observed after such storms.

b. Electrical Phenomena in Tissues.—Waller (1900) and Waller (1924, 1925) reported that, when light falls upon a plant tissue, electricity is produced by the tissue under certain conditions. This photoelectric current is of interest on account of the distinct significance that it appears to possess in reference to plant metabolism. It was believed that the nature of the photochemical reactions in the plant may be indicated by this electrical response. This photoelectric response is shown mainly by the green parts of plants and is apparently dependent upon the metabolic activity of the chlorophyll, although it may occur in the absence of chlorophyll, as shown in the cases of *Hydrangea* petals and in blanched celery. The direction of the current is apparently determined by the kind of leaf and by the conditions to which it is subjected. Dixon and Bennet-Clark (1927) found that a moderate electrical current in a tissue caused a fall in electrical resistance which became less and less rapid. After a few minutes the resistance began to rise and reached the normal in about an hour. They suggested that the ability of a current to stimulate a cell is determined by the potential difference across the membrane of the cell.

The electrical phenomena in plant tissues have been studied more recently by Marsh (1928), Vysotskii (1928), Umrath (1928), Lund (1929, 1931, 1932), Lund and Bush (1930), Glass (1933), Rehm (1936), and Seifriz (1936).

It was observed by Marsh (1928) that the e.m.f. of a given length of onion root was the algebraic sum of the e.m.fs. of the cells of such length. He considered that the electric polarity of the root or a part of it is a fluctuating quantity which may be increased, diminished, or reversed by changes in the environment or the internal processes of cells. Lund (1929) reported the existence of an easily measurable, inherent electric polarity in the main stems of the Douglas and white firs. This e.m.f. varies from 30 to 200 mv. in different trees. The apical growing point is generally positive with respect to the more basal parts of the stem. This e.m.f. is equal to the algebraic sum of the e.m.fs. of the individual segments of the stem. He reported that the cells of the onion root, of the leaf of *Bryophyllum*, and of Douglas fir generate electric currents continuously. The orientation of the electric polarities is such that they act in series and parallel and sometimes may be observed to oppose one another's electromotive forces like electric batteries. In the Douglas fir the unstimulated stem and each lateral branch exhibit an electric polarity. The apex is electropositive to the basal parts of the stem, while the unstimulated apex of the main stem is electropositive to the apex of each lateral branch. This positive electric dominance of the main apex corresponds to its dominance in growth.

Lund and Bush (1930) reported that the distal part of the petiole of *Bryophyllum* was positive to the growing points and selected points in the lamina. Lund (1931) found that the cortex of the Douglas fir is the origin of a characteristic e.m.f. The orientation of this radial e.m.f. in the cortex is opposite that in the wood. This e.m.f. increases toward the apex, while that in the wood decreases. In 1932 he reported that the longitudinal and radial e.m.fs. in the intact stem below the apex are both greatly decreased by lowering the temperature and increased by raising it. The radial e.m.f. of the isolated wood axis is affected in a similar manner. The e.m.fs. of dead stems is not affected by temperature changes. Lund believed that electric currents continually flow along certain circuits in the tree, thus correlating the living parts of the whole tree and making it a definite and continuous electrified system.

Rehm (1936) reported that in the scarlet runner bean (*Phaseolus multiflora*) certain electrical polarities in the region of the axillary buds possessed characteristics that suggest the possibility of their playing a role in bud inhibition. They possessed a relatively constant orientation in both young and old tissues which was maintained under varied conditions. He stated that these potential differences are due to the activities of living cells.

Seifriz (1936) in studies of the nature of protoplasm reported that the tissues of apple, pear, banana, onion, carrot, and potato, among others, are transformed into a soft, usually dark-colored mass when they are the cathode of a 110-volt, direct electric current. When they were used as the anode, little or no change was produced. The primary cause of the degradation is thought to be proteolysis catalyzed by intracellular proteolytic enzymes that are rendered active by the reducing conditions existing at the cathode.

4. Hydrogen-ion Concentration.—The effects of the hydrogen-ion concentration on the various activities of the plant have been stated in the discussion of these activities in the text, so that little remains to be said in regard to its direct effect upon growth. McCall (1923) emphasized the fact that the intensity of the acidity of the soil in many instances is of greater significance in biochemical processes than is the quantity of acid present. Kurz (1923) reported that the hydrogen-ion concentration *per se* was not the main factor in determining the distribution of the species which he considered. Reed and Haas (1924) noted that the injury to walnut seedlings in the absence of calcium was due to the lack of this element rather than to the high concentration of hydroxyl ions. Knudson (1925) stated that the hydrogen-ion concentration may affect many factors in plants other than the alteration of their growth.

Numerous workers have studied the effects of the reaction of the medium upon plant development. A summary of these observations is tabulated in the following table:

Investigator	Plants	Optimum pH	Investigator	Plants	Optimum pH
Wherry (1924)...	Timothy	9.0	Katchioni-Walther (1930).	Buckwheat	5.0 to 6.5
	Alfalfa, barley, and orchard grass	8.0			
	Alsike clover and beet	7.5	Meyer (1930).....	*Nelumbo lutea*	4.5 to 9.0
	Radish, rutabaga, and wheat	7.0	Wilson (1930)....	Onion	6.0 to 7.0
	Lupine, peas, and red clover	6.0	Emmert (1931)....	Tomatoes	8.4
				Lettuce	7.5
	Oats	5.0			
	Flax and turnip	4.0	Waltman (1931)...	Everbearing strawberry	5.3 to 5.5
Sideris (1926)....	Pineapple	5.0	Burkholder (1932).	Bean	Could not determine
Powers (1927)....	Alfalfa and alsike	5.6 to 6.0			
	Hungarian vetch	5.3			
	Spearmint	6.0	Gourley (1932).....	Iris	7.0 to 7.4
Wann (1927).....	*Chlorella*	3.4	Hicks (1932)......	*Lemna trisulca*	4.9 to 7.3
Arrhenius (1929).	Rye, oats, potato, and timothy	5.0	Wessels (1932)....	Cauliflower	5.5 to 6.6
				Potato	4.8 to 5.4
	Beets, barley, and alfalfa	7.0 to 7.2			

Aslander (1929) believed that plants in a fertile soil can tolerate a higher acidity than those growing in an infertile soil. Tait and Knott (1933) concluded from a study of the muck soils of New York that it is impossible to designate any particular soil reaction as the optimum for a given crop. The effect of the soil reaction in plant growth is to some extent an expression of the effects of such soil reaction on the availability of materials. Cooper (1932) suggested that a well-balanced supply of available cationic and anionic nutrient materials, with relatively high oxidation-reduction potentials, is much more effective than the hydrogen-ion concentration of the soil in determining the dominance of various pasture plants. Howell (1932) stated that the seedlings of western yellow pine grow best in an acid medium, but under field conditions other factors may have more influence on growth than the soil reaction.

The relation of the hydrogen-ion concentration to the growth of the cell was discussed by Pearsall (1925). He suggested that one of the conditions for the development of dividing cells in normal nongrowing tissue is the development of a gradient of hydrogen-ion concentration such that the protoplasm or its principal proteins can be reduced to an isoelectric condition. It is at this point that the proteins show minimum swelling, minimum osmotic pressure, maximum precipitation, and least viscosity. If the protoplasm and principal proteins were made approximately isoelectric by an alteration in the hydrogen-ion concentration, they would tend to lose water to adjacent media of higher osmotic potential, and as a result the condition in the protoplasmic or protein aggregate would favor synthesis. In stems there is a superficial layer of growing cells next to the cuticle that is usually markedly acid (pH 3.5 to 4.5), while the ground tissue usually has a pH value between 5 and 6. When lateral roots develop, they do so outside the xylem with its markedly acid reaction and not opposite the phloem, which is relatively alkaline. The surroundings of the dividing cells usually have a reaction of about pH 4.5 to 5.0 as against the normal tissue reaction of pH 5.5 to 6.5. Small (1929) found that the tissue of the xylem is more acid than that of the phloem. The various tissues of the broad bean, potato, and sunflower never showed a reaction less than pH 7.0.

5. Gases.—In 1908 Crocker and Knight found that the flowers of carnation are extremely sensitive to traces of illuminating gas. They observed that concentrations of 1 part in 40,000 parts of air killed the young buds and prevented the opening of those already showing the petals. A concentration of 1 part of gas to 80,000 parts of air caused the closing of the open flowers after 12 hr. exposure. Doubt (1917) reported that the following plants were found to be well adapted for use in detecting illuminating gas in greenhouses: *Lycopersicum esculentum, Salvia splendens, Mimosa pudica, Ricinus communis,* and *Datura stramonium.* Traces of gas 50 p.p.m. caused epinastic growth of the petals of all of these plants. This amount of gas is far below the limit of odor. The following plants were not injured by gas unless there was enough present to be detected by odor: *Caladium esculentum, Lupinus perennis, Eriobotrya japonica, Phoenix canariensis,* and others.

The effects of illuminating gas on plants have been reported by Hitchcock, Crocker, and Zimmerman (1930, 1931, 1932, 1934), Deuber (1932, 1933, 1934), and Zinkernagel (1932). This gas hastens the abscission of leaves, fruits, and flowers of numerous plants. Apparently the abscission layer of these parts is stimulated to produce many new cells, which results in the separation of the leaf from the stem. The dormancy of the buds of young oak and catalpa trees may be overcome by treatment with illuminating gas. The epinasty of leaves is one of the universal effects of this gas upon leafy plants.

Hitchcock, Crocker, and Zimmerman (1934) reported that the toxicity of gassed soils is due primarily to the constituents of illuminating gas that combine with the soil particles. After this combination, the gas constituents are not readily removed by aeration or leaching. According to these investigators (1932), the injury by gas to plants was lessened when it was first passed through water, sodium hydroxide, soil or moss peat before reaching the plant. Zinkernagel (1932) found that the treatment of onion roots for a few hours with illuminating gas caused the spindle fibers of cells in the early stages of division to disappear. This interfered with the movement of the chromosomes to the poles. Later in the absence of this gas the spindle reformed and cell division was completed. When in gas, the roots did not respond to gravity but grew upward.

Crocker, Zimmerman, and Hitchcock (1932) tested 202 species and varieties of plants with 38 different gases. Approximately 44 per cent of these plants showed induced epinasty but only in the presence of five of these gases: ethylene, acetylene, propylene, carbon monoxide, and butylene. If the concentration of ethylene were taken as one, then the concentration required for a response was 500 for acetylene and propylene, 5,000 for carbon monoxide, and 500,000 for butylene. All the effective gases were carbon gases with unsaturated bonds. Their effectiveness is not correlated with their solubilities in water, but may be related to their ability to combine with one or more constituents of the protoplasm.

According to Rodriguez (1932) the use of smoke on pineapple plants in the field in Puerto Rico resulted in the general flowering of all plants and in early fruit production. The quantity of the smoke was apparently unimportant. One or more constituents of the smoke, and not the increased temperature, was responsible for the flowering.

Deuber (1934) placed the roots of tomato plants that had been exposed to illuminating and ethylene gases into nutrient solutions containing methylene blue in concentrations of 1 to 10 p.p.m. The results indicated that this pigment thus presented to the roots helped these plants to recover from the effects of various gases. Zimmerman and Crocker (1934) found that the plants of 65 different genera were susceptible to injury from the vapors emanating from soil or tankage moistened with a solution of mercuric chloride.

The influence of sulphur dioxide on plants has been studied by various workers. Zimmerman and Crocker (1930, 1934) found that the leaves of tomato, salvia, *Coleus*, geranium, castor bean, buckwheat, roses, and several other plants were injured after treatment for 1 hr. in 3 to 4 p.p.m. of this gas. The age of the tissues was a factor in the susceptibility of leaves to injury. The middle-aged leaves were the least resistant and the younger ones the most resistant. Interveinal portions of the leaves were more readily injured than the tissues along the veins. The wilted plants were more resistant than the turgid ones. Plants treated during the night were more resistant than those treated during the day. Buckwheat was the most susceptible, being injured by a fumigation of 0.46 p.p.m. for 7 hr., while orchids were the most resistant, withstanding 60 p.p.m. for several hours.

Zimmerman, Crocker, and Hitchcock (1933) treated plants from 108 species with carbon monoxide, and, of these, 45 showed epinastic growth of leaves, while several showed hyponasty.

The effects of ethylene on plants have been more extensively studied than those of any other gas. Some of the workers who have reported in this regard are Harvey (1913), Knight and Crocker (1913), Harvey and Rose (1915), Harvey (1915), Zimmerman, Hitchcock, and Crocker (1931, 1932), and Deuber (1936).

Harvey (1913) found that the castor bean is a very delicate indicator of the presence of ethylene, since it responded by a nastic drooping of 15 to 30 deg. when exposed to as little as 1 part ethylene to 10,000,000 parts of air. Knight and Crocker (1913) showed that tobacco smoke of a concentration of 1,000 p.p.m. caused a reduction in the rate of elongation of the sweet pea, while 5,000 p.p.m. completely stopped elongation. They determined that ethylene was the component that thus affected growth, and that the sweet pea was a more delicate indicator of the presence of this gas than any chemical test.

Harvey and Rose (1915) believed that ethylene is the most harmful constituent of illuminating gas except in extremely high concentrations of gas when other substances may play a part. Harvey (1915) investigated the changes in the metabolism of the sweet pea which were caused by the action of ethylene. In the treated plants the simple soluble substances increased at the expense of the higher soluble and insoluble forms. Thus the sugars, amino acids, amides, polypeptids, and lipoids increased by 8 per cent, while the insoluble substances as proteins, starch, and cellulose correspondingly diminished. The water content of the treated and untreated plants, however, remained the same. The osmotic pressure of the treated plants increased as did also the permeability of the protoplasmic membrane.

Ethylene apparently always causes epinasty in leaves, and may also cause abscission of leaves and flowers, depending upon the environmental conditions and the kind of plant. The effects of illuminating gas on plants are due in a large degree to the ethylene in it. According to Crocker, Zimmerman and Hitchcock (1932) this gas does not act directly by inducing epinasty of peticles but probably acts indirectly by modifying the equilibrium position of the petiole with gravity, as shown by Neljubow (1901, 1911).

Priestley (1922) suggested that the effect of the injurious components of illuminating gas on the stems and roots of etiolated plants is due to their inhibition of the formation of an unbroken primary endodermis normally present in these structures. The unsaturated hydrocarbons prevent the formation of a functional endodermis by preventing the normal accumulation of unsaturated acids in the region of the future Casparian strip. Woffenden and Priestley (1924) traced the effect of coal gas upon the stem of *Sambucus* to its action upon cork and lenticel formation. When the cork cells are cut off from the phellogen, they are very susceptible to the presence of coal gas. The fatty acids diffusing from the protoplasm fail to remain in and on the walls, so that no suberin lamella can be formed. This is due apparently to the greater mobility of unsaturated fatty acids in the presence of the gaseous unsaturated hydrocarbons present in the coal gas. They believed that coal gas will not be toxic to cork-enveloped tissue except in the case of leakage from buried gas mains around the roots.

6. Other Factors.—It was reported by Thornton (1930) that treatment with carbon dioxide in concentrations of 5 to 30 per cent was effective in prolonging the life of cut roses when stored at 38 to 50°F. He stated (1931) that the percentage of carbon dioxide necessary to cause injury to fruits and vegetables in storage is related directly to the firmness of the tissue and inversely to the amount of moisture upon their surfaces. Miller and Brooks (1932) could not find any change in the percentage of the various carbohydrates of peaches and cherries when they were treated in storage with carbon dioxide. Pearl, Edwards, Winsor, and Winsor (1934) stated that the seedlings of cucumber showed a diminished growth rate and less efficient translocation of food materials as the ventilation was progressively lessened.

It was shown by Sampson and McCarty (1930) in California that the rate of growth of the needle grass (*Stipa pulchra*) in winter is controlled by atmospheric temperatures; in spring and early summer it is partly related to the internal factors, among which the food supply and growth habits are predominant. Root growth occurred in the autumn after the top growth had practically ceased. Bailey (1933) found that soaking the seeds of dwarf bean in distilled water for varying periods of time resulted in a progressive decrease in the growth rate of the plants produced and in the time necessary for them to reach maturity. Livingston (1934), in a comprehensive paper, emphasized that the general organization of the organism and its environment must be studied together if we are to understand the former. Brain (1935) found that plants which were rotated on a horizontal clinostat showed an increase in root growth and a decrease in shoot growth. He believed that this fact indicated that a different growth mechanism is involved in the opposite tropisms of shoot and root.

Brink (1924) stated that the growth of the tubes of cultured pollen grains required sugar. There was an increase of growth in length of 60 to 142 per cent by the addition of sterile yeast to the sugar medium. The growth-promoting substance contained in the yeast was water soluble, heat stable, and active in small amounts. The addition of extracts of the potato and of pistils of flowers to the artificial medium regularly increased the growth of pollen tubes. Reid (1930) believed that the seedlings of the Hubbard squash utilized their nitrogenous reserves in a different manner at opposite seasons of the year. Thus the leaves of plants grown in December contained much more of the total nitrogen than those grown during May. The plants grown in May contained much more of their nitrogen in the roots. Gustafson and Laing (1931) considered that the decrease in the productivity of the later developed flower clusters in tomatoes under ordinary conditions is due to a decrease in available nutrient material, which has been utilized by the first-formed fruit. Hence a lack of nutrient material and not the presence of inhibiting substances is the cause of the smallness of the later formed fruit. Harrison (1931) noted that the shorter the grass was cut and the greater its leaf area was reduced, the smaller was the quantity of roots produced, and that the application of mineral fertilizers did not compensate for the lack of top growth. Gould, Pearl, and Edwards (1934) noted that when a considerable portion of the cotyledonary tissues of the canteloupe was removed before planting, the growth in length of the hypocotyl was in excess of that expected from the available food present. They stated that this illustrates "the physiological factor of safety" in that the seedling does not use all the available stored material.

Walter (1923) noted that the rootlets of germinating seeds of water cress, nasturtium, and peas would grow only in an atmosphere with a relative humidity of not less than 97 to 98 per cent. Nightingale and Mitchell (1934) grew two sets of tomato plants under identical conditions for growth, except that one set was grown under a relative humidity of 35 per cent and the other under a relative humidity of 95 per cent. Those grown under the latter condition were far better plants when judged by every criterion of development and appearance.

Lloyd and Barnes (1932) considered that the chemical activity of water appears to be associated with the colloidal form, trihydrol, which is present in large quantities even at relatively high temperatures. The growth and photosynthesis of *Spirogyra* were accelerated to a greater degree in water cultures rich in trihydrol than in those rich in monohydrol and dihydrol.

The effects of the tillers of corn upon the yield of grain have been studied by Lyon (1905), Montgomery (1909), Williams (1912), Williams and Etheridge (1912), Thompson (1926), McClelland (1928), and Dungan (1931). Their observations indicated that the removal of tillers reduced the yield of grain to a greater or lesser degree in practically all cases.

C. Vernalization

1. General Consideration.—The terms "Iarovization," "Jarovization," "Yarovization," and "Iarovtsii" have been used by the Russians to indicate a certain treatment of seeds. Each means literally "making springlike." The English have used "Springification," "Springization," and "Vernalization" to describe this treatment. The last named has become widely used and will be used in this discussion.

According to Thomson (1936) vernalization is the name applied to a treatment given to seeds before sowing to hasten the time of flowering of the plants that will develop from them. Martin (1933) and Laude (1933) stated that it is a method of treating the seeds of winter cereals so that they may mature from spring sowing, and of treating the seeds of spring cereals and other spring crops to advance the time of reproduction and maturity. Thomson (1936) stated that the essence of the vernalization treatment of seeds is that the stages incident to the developing plant are passed before the seeds are sown and yet the seeds are in a condition to be sown by the ordinary methods. Briefly this consists of starting germination and then holding it in check by the proper moisture, temperature, and light relations, as experience has indicated. The seeds are germinated until the radicles are just beginning to break the seed coats and then restraining treatments are applied. These treatments cannot be successfully applied to dormant seeds, but the embryo must have started development and growth. After the seeds have been treated in this manner, they may be planted immediately or dried and kept for a later planting.

According to McKinney and Sando (1933), Martin (1933), Laude (1933), and McKee (1935), in winter cereals the vernalization process consists of adding water to the seeds in an amount that will barely produce visible germination. This requires a period of 1 to 2 days in the processing chamber with the temperature at 10 to 12°C. The seeds are then transferred to a temperature of 3 to 5°C., stirred frequently, and their moisture content kept constant by the additions of water. The time required in the cool room will vary, depending upon the temperature and variety of the seeds, but the average time is from 35 to 45 days. Lysenko (1932) stated that vernalization, at least for some plants, should be completed in darkness. He also stated that with plants requiring high temperatures for germination, such as corn, foxtail, millet, soybeans, Sudan grass, and other sorghums, the treatment during the germination stage is much the same as with the cereal grains. After germination has begun, however, the seeds are kept at 20 to 30°C. for 5 to 10 days.

Vernalization is really an old and well-known seed treatment under a new name. The fact that the low temperature requirement of winter cereals can be satisfied in the early stages of germination has been known since 1837. A process very similar to what is now called "vernalization" was well described in 1849 in the "New American Farm Book," a standard reference book on agriculture at that time. In 1857 Klippart wrote in the annual report of the Ohio State Board of Agriculture, "To convert winter wheat into spring wheat, nothing more is necessary than that the winter wheat should be allowed to germinate slightly in the fall or winter but kept from vegetating by a low temperature of freezing until it can be sown in the spring. This is usually done by soaking and sprouting the seed and freezing it while in this state and keeping it frozen until the season for spring sowing has arrived. Only two things seem requisite, germination and freezing. The experiment of converting winter wheat into a spring wheat has met with great success. It retains many of its primitive winter wheat qualities and is inferior in no respect to the best varieties of spring wheat and produces at the rate of 28 bu. per acre." Klebs (1913), Gassner (1918), Maximow and Pojarkova (1925), and Maximow and Krotkina (1930) have contributed to this work.

2. Economic Value.—Vernalization has been used in field practice to a considerable extent in Russia during the past 6 years and has attracted attention in western Europe and in North America. McKinney and Sando (1930, 1933); Gfeller, Derrick, and Fraser (1933); Sprague (1934); Peltier and Kiesselbach (1934); McKinney, Sando, and others (1934); Burr and Turner (1935); McKee (1935); and Bell (1936) have studied the practicability of vernalization. The available evidence indicates that the process does not offer great possibilities in commercial wheat production in the United States. There are available in this country so many spring varieties well adapted to the spring-wheat regions that there appears to be no need for the vernalization of winter varieties. It also seems doubtful if vernalization of spring varieties will return yields proportional to the cost of treating. It has been suggested that it might be used for spring seeding in winter-wheat areas where the fall sowing has perished. In many cases, however, the need for spring sowing cannot be determined in sufficient time to permit vernalization. In England, winter-sown wheat always yields best, and its time for planting relieves the congestion of work. Vernalized winter wheat yields better there than spring wheat but produces less than winter-sown wheat. In Russia, however, there is a great need for wheat that will head before the dry season occurs, and it appears that vernalization will supply the remedy. The drought and long days of Russia are against growing such crops as maize and millet, but if the vernalization of short-day plants can be accomplished there seems to be no doubt but that they will grow.

Sprague (1934) reported that the vernalization of certain corn hybrids resulted in statistically significant hastening of sexual maturity, but the difference was so slight as to be of no agronomic importance, and a general reduction in germination and vigor was associated with the treatment.

According to Laude (1933), the obvious difficulties, such as the necessity for accurate control of temperature, moldy seed, low germination, poor stands, and those difficulties inherent in drying the seeds or in planting moist, partly germinated seeds, are such as to indicate that vernalization offers little of immediate value to the practical farmer. Vernalization, however, has its value in experimental culture in the greenhouse as shown by McKinney and Sando (1930). It advances the time of growth and reproduction in winter cereals, so that two or three generations may be grown in the same time that is required for one generation under ordinary treatment.

3. Theory of the Process.—According to Maximow (1934, 1935), Purvis (1934), and Thomson (1936), Lysenko, the Russian authority on vernalization, stated that the results obtained from this process are based upon the following concepts of plant behavior: (*a*) growth and development are not identical phenomena but are independent of each other; (*b*) the entire process of the development of an annual seed plant consists of a series of individual steps and stages; (*c*) these stages always proceed in a strict sequence, and a subsequent stage cannot occur until the preceding stage has been completed; and (*d*) in the same plant different stages of development require, for their completion, different external conditions.

According to Lysenko there are at least two outstanding stages in the development of a plant:

a. The Thermo-stage.—The indications are that this stage must be completed before the initials of the reproductive organs can be formed. The conditions required for the completion of this stage are low temperatures of from 0 to 20°C., suitable moisture, and adequate aeration. The time required for the completion of this stage varies with the type of plant and the prevailing environmental conditions. The effect on the thermo-stage is incurred only when the dormant period is broken and the embryo is induced to start growth without being allowed to penetrate the seed coat. This indicates that seeds thus treated have ceased to be seeds in the physiological sense and have become equivalent to growing plants. To all external appearances, however, these may differ in no respect from resting seeds. Many varieties of winter wheat fail to head when sown in the spring because the temperatures are too high for the accomplishment of this thermo-stage. Once this stage has been completed, the other stages in the cycle of development can normally occur. Thomson (1936) stated that it has long been known that if the seedlings of beet or members of the genus *Brassica* get slightly frosted the plants will bolt, although normally they do not flower until after winter. Full-grown cabbages were transplanted to a warm greenhouse in October and 2 years later were yet growing. Those that were transplanted at the same time into a cool greenhouse flowered in 22 weeks, while cabbages left in the open until December and then transplanted into a warm greenhouse flowered in 6 weeks. It has been noted in American agricultural practice that if winter wheat is sown late and germination starts, but is not completed, before cold weather begins, so that the seedling plants do not appear until spring, the plants develop and head in a normal manner. However, if the seeds remain dormant until spring, and then germinate, the resulting plants will not head. This suggests that the winter cold influences germinating seeds but not dormant ones.

Draghetti (1933) used the term "crypto-vegetation" to signify the state of vegetation maintained by certain plants under the low-temperature conditions of winter. As a result of climate this state is characterized by the cessation or intermittent functioning of photosynthesis. He found that the wheat plants continued to absorb nutrient salts, especially the nitrogenous ones, to perform metabolic activities, and to increase in weight during the winter. These facts show that the plants do not have a definite winter rest period.

b. The Photo-stage.—The changes involved in the thermo-stage are not capable of initiating reproduction. A photo-stage is necessary, and it can only become effective after the thermo-stage is completed. Light and darkness play no part in the thermo-stage. The photo-stage requires high temperature and can be effected only under conditions of a long day or under continuous illumination. This long-day requirement does not hold for the entire cycle of development of plants but only during this one particular stage through which they must pass immediately after the thermo-stage. The photo-stage can be most rapidly effected in wheat in continuous illumination and less rapidly under long days, but it is delayed indefinitely under short-day conditions.

McKinney and Sando (1935) believed that sexual reproduction in spring and winter wheat is not dependent upon a critical temperature or a photoperiod since this process occurs over a wide range of such factors. The time, however, when sexual reproduction occurs is greatly influenced by the temperature and the photoperiod. Most spring wheats complete their life cycle quickly when given a long day and temperatures at 70°F. or above, throughout the life cycle. On the other hand, winter wheats complete their life cycle most rapidly when given a short day and low temperatures during the earlier stages of growth and a long day and high temperatures during later stages of development.

Bibliography

Abbe, C., 1895: Influence of cold on plants,—a résumé, *Exp. Sta. Rec.*, **6**: 777–781.

Abbott, C. E., 1935: Blossom-bud differentiation in citrus trees, *Am. J. Bot.*, **22**: 476–485.

Abbott, O., 1923: Chemical changes at beginning and ending of rest period in apple and peach, *Bot. Gaz.*, **76**: 167–184.

Ackerman, A., 1917: Beiträge zur Kenntniss der Kältresistanz der Winterweizen, *Z. Pflanzenzucht.*, **5**: 349–359.

Adams, J., 1920: Relation of flax to varying amounts of light, *Bot. Gaz.*, **70**: 153–156.

———, 1923: The effect on certain plants of altering the daily period of light, *Ann. Bot.*, **37**: 72–94.

———, 1924: Does light determine the date of heading out in winter wheat and winter rye? *Am. J. Bot.*, **11**: 535–539.

———, 1924: Duration of light and growth, *Ann. Bot.*, **38**: 510–523.

———, 1924: The effect on tomato, soybean, and other plants of altering the daily period of light, *Am. J. Bot.*, **11**: 229–233.

———, 1925: Some further experiments on the relation of light to growth, *Am. J. Bot.*, **12**: 398–413.

ALLARD, H. A., 1932: Length of day in relation to the natural and artificial distribution of plants, *Ecology*, **13**: 221–234.

———, 1932: Flowering behavior of the hog peanut in response to length of day, *J. Agr. Res.*, **44**: 127–137.

———, 1935: Response of the woody plants *Hibiscus syriacus*, *Malvaviscus conzattii* and *Buginvillea glabra* to length of day, *J. Agr. Res.*, **51**: 27–34.

ALLEN, F. W., 1932: Physical and chemical changes in the ripening of deciduous fruits, *Hilgardia*, **6**: 381–441.

ALLEN, R. L., 1849: "American Farm Book," C. M. Saxton, New York, XII + 325 pp.

ALMOSLECHNER, E., 1934: Die Hefe als Indikator für Wuchsstoffe, *Planta*, **22**: 515–542. Quoted by Avery, 1935.

ANDERSEN, A. M., 1931: The use of dilute nitric acid on the germination of seeds of *Poa compressa*, abst., *Am. J. Bot.*, **18**: 889.

ANDREWS, F. M., 1932: A measurement of growth, *Proc. Ind. Acad. Sci.*, **41**: 77–80.

ANTHONY, R. D., and W. S. CLARKE, JR., 1932: Growth record on fertilized apple trees grown in metal cylinders, *J. Agr. Res.*, **44**: 245–266.

APPLEMAN, C. O., 1911: Physiological behavior of enzymes and carbohydrate transformations in after-ripening of the potato tuber, *Bot. Gaz.*, **52**: 306–315.

———, 1912: Changes in Irish potatoes during storage, *Md. Sta. Bull.* 167.

———, 1914: Biochemical and physiological study of the rest period in the tubers of *Solanum tuberosum*, *Md. Sta. Bull.* 183.

———, 1916: Biochemical and physiological study of the rest period in the tubers of *Solanum tuberosum*, *Bot. Gaz.*, **61**: 265–294.

———, 1918: Physiological basis for the preparation of potatoes for seed, *Md. Sta. Bull.* 212.

———, 1918: Special growth-promoting substances and correlation, *Sci.*, **48**: 319–320.

——— and E. V. MILLER, 1926: A chemical and physiological study of maturity in potatoes, *J. Agr. Res.*, **33**: 569–579.

ARMSTRONG, H. E., and E. F. ARMSTRONG, 1911: The function of hormones in regulating metabolism, *Ann. Bot.*, **25**: 507–519.

ARMSTRONG, J. M., and W. J. WHITE, 1935: Factors influencing seed setting in alfalfa, *J. Agr. Sci.*, **25**: 161–179.

ARRHENIUS, O., 1929: Soil acidity. Plant growth and its practical application, *Proc. Int. Cong. Plant Sci., Ithaca*, **1**: 53–54.

ARTHUR, J. M., J. D. GUTHRIE, and J. M. NEWELL, 1930: Some effects of artificial climates on the growth and chemical composition of plants, *Am. J. Bot.*, **17**: 416–482.

——— and W. D. STEWART, 1931: Plant growth under shading cloth, abst., *Am. J. Bot.*, **18**: 897.

———, 1932: Artificial light and plant growth, *Agr. Eng.*, **13**: 288–291.

——— and W. D. STEWART, 1935: Relative growth and dry weight production of plant tissue under mazda, neon, sodium and mercury vapor lamps, *Contr. Boyce Thompson Inst.*, **7**: 119–130.

ASHBY, E., and T. A. OXLEY, 1935: The interaction of factors in the growth of *Lemna*. VI, An analysis of the influence of light intensity and temperature on the assimilation rate and the rate of frond multiplication, *Ann. Bot.*, **49**: 309–336.

ASLANDER, A., 1929: Concentration of the nutrient medium *versus* its hydrogen-ion concentration as manifest by plant growth, *Svensk Bot. Tidskr.*, **23**: 96–140; *Exp. Sta. Rec.*, **67**: 22.

Atkins, G. A., 1936: The effect of pigment on phototropic response: A comparative study of reactions to monochromatic light, *Ann. Bot.*, **50**: 197–218.

Auchter, E. C., and C. P. Harley, 1924: Effect of various lengths of day on development and chemical composition of some horticultural plants, *Proc. Am. Soc. Hort. Sci.*, **21**: 199–214.

Austin, J. P., 1935: Some effects of the photoperiod on development of *Impatiens balsamea*, *Plant Physiol.*, **10**: 545–552.

Austin, S., 1933: Vegetation and reproduction in the soy bean, *Sci.*, **78**: 363–364.

———, 1935: Effects of exfloration on plant metabolism, *Plant Physiol.*, **10**: 225–243.

Avery, G. S., 1935: Differential distribution of a phytohormone in the developing leaf of *Nicotiana*, and its relation to polarized growth, *Bull. Torrey Bot. Club*, **62**: 313–330.

——— and P. R. Burkholder, 1936: Polarized growth and cell studies on the *Avena* coleoptile, phytohormone test object, *Bull. Torrey Bot. Club*, **63**: 1–16.

Axentjev, B. N., 1930: Über die Rolle der Schalen von Samen und Früchten, die bei der Keimung auf Licht reagieren, *Beih. bot. Cent.*, **46**: 119–202; *Biol. Abst.*, **8**: 1911.

Babb, M. F., 1928: The effect of ethylene upon the vitamin B content of celery, *Sci.*, **68**: 231.

Bachmann, F., and F. Bergann, 1930: Über die Werkigkeit von Strahlen verschiedener Wellenlänge für die phototropische Reizung von *Avena sativa*, *Planta*, **10**: 744–755.

Bailey, L. H., 1891: Some preliminary studies of the influence of the electric arc lamp upon greenhouse plants, *N. Y. Sta.* (*Cornell*) *Bull.* 30.

Bailey, W. M., 1933: Structural and metabolic after-effects on soaking seeds of *Phaseolus*, *Bot. Gaz.*, **94**: 688–713.

Bakke, A. L., 1913: The effect of smoke and gases on vegetation, *Contr. Bot. Dept. Iowa State Col.*, No. 54; *Proc. Iowa Acad. Sci.*, **20**: 169–188.

Baldwin, H. I., 1932: Alcohol separation of empty seed, and its effect on the germination of red spruce, *Am. J. Bot.*, **19**: 1–12.

———, 1934: Effect of after-ripening treatment on germination of white pine seeds of different ages, *Bot. Gaz.*, **96**: 372–376.

Ball, N. G., 1936: The effect of nocturnal illumination by different regions of the spectrum on the subsequent opening of flower buds, *New Phytol.*, **35**: 101–116.

Barnes, T. C., 1932: The physiological effect of trihydrol water, *Proc. Nat. Acad. Sci.*, **18**: 136–137.

Barton, L. V., 1932: Effect of storage on the vitality of *Delphinium* seeds, *Contr. Boyce Thompson Inst.*, **4**: 141–153.

———, 1934: Dormancy in *Tilia* seeds, *Contr. Boyce Thompson Inst.*, **6**: 69–90.

———, 1935: Storage of vegetable seeds, *Contr. Boyce Thompson Inst.*, **7**: 323–331.

Bell, G. D. H., 1936: Experiments on vernalization, *J. Agr. Sci.*, **26**: 155–171.

———, 1936: Vernalization: its meaning and practical application, *Agr. Prog.*, **13**: 76–82.

Benedict, H. M., 1915: Senile changes in leaves of *Vitis vulpina* and certain other plants, *N. Y. Sta.* (*Cornell*) *Mem.* 7.

——— and H. Kersten, 1934: Effect of soft X-rays on germination of wheat seeds, *Plant Physiol.*, **9**: 173–178.

Berkley, D. M., and E. E. Berkley, 1933: Super optimal and thermal death temperatures of the cotton plant as affected by variations in relative humidity, *Ann. Mo. Bot. Gard.*, **20**: 583–603.

BERKLEY, E. E., 1931: Studies of the effects of different wave lengths of day, with variations in temperature on vegetative growth and reproduction in cotton, *Ann. Mo. Bot. Gard.*, **18**: 573–601.

BIDDULPH, O., 1935: Histological variations in *Cosmos* in relation to photoperiodism, *Bot. Gaz.*, **97**: 139–155.

BLACKMAN, V. H., 1919: The compound interest law and plant growth, *Ann. Bot.*, **33**: 353–360.

———, 1920: The significance of the efficiency index of plant growth, *New Phytol.*, **19**: 97–100.

———, A. T. LEGG, and F. G. GREGORY, 1923: The effect of a direct electric current of very low intensity on the rate of growth of the coleoptile of barley, *Proc. Roy. Soc.*, B, **95**: 214–228.

———, 1924: Field experiments in electroculture, *J. Agr. Sci.*, **14**: 240–267.

——— and A. T. LEGG, 1924: Pot culture experiments with an electric charge, *J. Agr. Sci.*, **14**: 268–287.

BLAKE, A. K., 1935: Viability and germination of seeds and early life history of prairie plants, *Ecol. Monographs*, **5**: 405–460.

BONNER, J. F., 1932: The production of growth substance by *Rhizopus suinus*, *Biol. Zent.*, **52**: 565–582.

———, 1933: Studies on the growth hormones of plants. IV, On the mechanism of the action, *Proc. Nat. Acad. Sci.*, **19**: 717–719.

———, 1933: The action of the plant growth hormone, *J. Gen. Physiol.*, **17**: 63–76.

———, 1934: The relation of hydrogen ions on the growth rate of the *Avena* coleoptile, *Protoplasma*, **21**: 406–423.

———, 1934: Studies on the growth hormone of plants. V, The relation of cell elongation to cell wall formation, *Proc. Nat. Acad. Sci.*, **20**: 393–397.

——— and K. V. THIMANN, 1935: Studies on the growth hormones of plants. VII, The fate of growth substance in the plant and the nature of the growth process, *J. Gen. Physiol.*, **18**: 649–658.

BOOTH, E. G., 1929: Daily growth of the oat kernel and effect on germination of immaturity and controlled low temperatures, *Minn. Sta. Tech. Bull.* 62.

BORODIN, D. N., 1933: Yarovization formula for winter oats and barleys, abst., *Am. J. Bot.*, **20**: 677.

BOROWIKOW, G. A., 1912: Über die Ursachen des Wachstums der Pflanzen, *Biochem. Z.*, **48**: 230–246; **50**: 119–128; *Plant World*, **16**: 347–349 (1913).

BORTHWICK, H. A., and W. W. ROBBINS, 1928: Lettuce seed and its germination, *Hilgardia*, **3**: 275–304.

BOTJES, J. O., 1933: Aethyleen als vermoedelijke oorzaak van de groeiremmende werking van rijpe appels, *Tijdschr. Plantenziek.*, **39**: 207–211.

BOURNE, B. A., 1935: Effects of freezing temperatures on sugar cane in the Florida everglades, *Fla. Sta. Tech. Bull.* 278.

BOYSEN-JENSEN, P., 1910: Über die Leitung des phototropischen Reizes in Avenakeimpflanzen, *Ber. bot. Ges.*, **28**: 118–120.

———, 1931: Über Bildung eines Wachstumsregulators durch *Aspergillus niger*, *Biochem. Z.*, **239**: 243–249.

———, 1932: Über die Bildung und biologische Bedeutung des Wachstumsregulators bei *Aspergillus niger*, *Biochem. Z.*, **250**: 270–280.

———, 1933: Über den Nachweis von Wuchsstoff in Wurzeln, *Planta*, **19**: 345–350; *Biol. Abst.*, **9**: 746.

———, 1933: Significance of growth substance for the growth and the geotropic curvature of the roots of *Vicia faba*, *Planta*, **20**: 688–698; *Chem. Abst.*, **28**: 1380.

———, 1936: "Growth Hormones in Plants." Translated by G. S. Avery and P. R. Burkholder. McGraw-Hill Book Company, Inc., New York and London, XIV + 268 pp.

BRACKETT, F. S., and E. S. JOHNSTON, 1932: The functions of radiation in the physiology of plants. I, General methods and apparatus, *Smithsonian Misc. Coll.*, **87**, No. 13: 1–10.

BRAIN, E. D., 1935: Studies in the effects of prolonged rotation of plants on a horizontal klinostat. I, Growth rate, *New Phytol.*, **34**: 97–112.

BRAMBLE, W. C., 1932: Breaking the dormancy of tree seedlings by chemical treatment, *Sci.*, **75**: 193–194.

BREAZEALE, J. F., 1927: Vitamin-like substances in plant nutrition, *Ariz. Sta. Tech. Bull.* 16.

BRENCHLEY, W. E., 1919: Some factors in plant competition, *Ann. App. Biol.*, **6**: 142–170.

———, 1920: On the relations between growth and the environmental conditions of temperature and bright sunshine, *Ann. App. Biol.*, **6**: 211–244.

BRIGGS, G. E., F. KIDD, and C. WEST, 1920: A quantitative analysis of plant growth. I, II, *Ann. App. Biol.*, **7**: 103–123, 202–223.

———, ———, and ———, 1920: Methods in the quantitative analysis of plant growth—a reply to criticism, *Ann. App. Biol.*, **7**: 403–406.

———, 1925: Plant yield and the intensity of external factors.—Mitscherlich's Wirkungsgesetz, *Ann. Bot.*, **39**: 475–502.

———, 1928: A consideration of some attempts to analyse growth curves, *Proc. Roy. Soc.*, B, **102**: 280–285.

BRIGGS, L. J., A. B. CAMPBELL, R. H. HEALD, and L. H. FLINT, 1926: Electroculture, *U. S. Dept. Agr. Dept. Bull.* 1379.

BRINK, R. A., 1924: The physiology of pollen. I, The requirements for growth, *Am. J. Bot.*, **11**: 218–229.

———, 1924: The physiology of pollen. II, Further consideration regarding the requirements for growth, *Am. J. Bot.*, **11**: 283–295.

———, 1924: The physiology of pollen. III, Growth *in vitro* and *in vivo*, *Am. J. Bot.*, **11**: 351–365.

———, 1924: The physiology of pollen. IV, Chemotropism: effects on growth of grouping grains, formation and function of callose plugs; Summary and conclusions, *Am. J. Bot.*, **11**: 417–436.

BROWN, A. B., 1935: Cambial activity, root habit and sucker shoot development in two species of poplar, *New Phytol.*, **34**: 163–179.

BROWN, A. H., 1933: Effects of sulphuric-acid delinting on cotton seeds, *Bot. Gaz.*, **94**: 755–770.

BROWN, W. H., and S. F. TRELEASE, 1918: Alternate shrinkage and elongation of growing stems of *Cestrum nocturnum*, *Philippine J. Sci.*, C, **13**: 353–360.

BRUMAN, A. J., 1932: "Yarovization" a new factor in plant breeding technique, *J. Heredity*, **23**: 465–466.

BUCHOLTZ, A. F., 1931: The effect of monochromatic ultra-violet light of measured intensities on behavior of plant cells, *Ann. Mo. Bot. Gard.*, **18**: 489–508.

BURGE, W. E., G. C. WICKWIRE, and H. J. FULLER, 1936: Daily variation in the sensitivity of *Mimosa* with special reference to the action of light, *Bot. Gaz.*, **97**: 674–677.

BURKHOLDER, P. R., and R. PRATT, 1936: Leaf-movements of *Mimosa pudica* in relation to light, *Am. J. Bot.*, **23**: 46–52.

——— and ———, 1936: Leaf-movements of *Mimosa pudica* in relation to the intensity and wave length of the incident radiation, *Am. J. Bot.*, **23**: 212–219.

———, 1936: The rôle of light in the life of plants. I, Light and physiological processes. II, The influence of light upon growth and differentiation, *Bot. Rev.* **2**: 1–52, 97–168.

Burkholder, W. H., 1932: Effect of the hydrogen-ion concentration of the soil on the growth of the bean and its susceptibility to dry root rot, *J. Agr. Res.*, **44**: 175–181.

Burnett, L. C., and A. L. Bakke, 1930: The effect of delayed harvest upon the yield of grain, *Iowa Sta. Res. Bull.* 130.

Burr, S., and D. M. Turner, 1935: Vernalisation of tomatoes, *Gard. Chron.* III, **98**: 288.

Busse, W. F., 1930: Effect of low temperatures on germination of impermeable seeds, *Bot. Gaz.*, **89**: 169–179.

——— and C. R. Burnham, 1930: Some effects of low temperature on seeds, *Bot. Gaz.*, **90**: 399–412.

Caldwell, J. S., 1934: Hydrion concentration changes in relation to growth and ripening in fruits, *U. S. Dept. Agr. Tech. Bull.* 403.

Calvert, J., 1935: Drought resistance in wheat: the bound and free water of expressed sap from wheat leaves in relation to time and soil moisture, *Protoplasma*, **24**: 505–524.

———, 1935: Diurnal variation in "bound and free water" and other factors in the sap expressed from the leaves of *Phalaris tuberosa*, *Protoplasma*, **24**: 525–530.

Cattell, W., 1931: The effects of X-rays on the growth of wheat seedlings, *Sci.*, **73**: 531–532.

Chace, E. M., and F. E. Denny, 1924: Use of ethylene in the coloring of citrus fruit, *Ind. and Eng. Chem.*, **16**: 339–340.

——— and E. G. Church, 1927: Effect of ethylene on the composition and color of fruits, *Ind. and Eng. Chem.*, **19**: 1135–1139.

Chambers, R., and H. P. Hale, 1932: The formation of ice in protoplasm, *Proc. Roy. Soc.*, B, **110**: 336–352.

Chambers, W. H., 1923: Cultures of plant cells, *Proc. Soc. Exp. Biol. and Med.*, **21**: 71–72.

Chance, H. L., 1934: The influence of various types of defoliation and leaf wounding upon the growth and yield of beans, *Am. J. Bot.*, **21**: 85–108.

Chandler, W. H., 1931: Freezing of pollen: Evidence as to how freezing kills plant cells, abst., *Am. J. Bot.*, **18**: 892.

——— and A. C. Hildreth, 1935: Evidence as to how freezing kills plant tissue, *Proc. Am. Soc. Hort. Sci.*, **33**: 27–35.

Child, C. M., and A. W. Bellamy, 1920: Physiological isolation by low temperature in *Bryophyllum*, *Bot. Gaz.*, **70**: 249–267.

Cholodny, N., 1924: Ueber die hormonale Wirkung der Organspitze bei der geotropischen Krümmung, *Ber. bot. Ges.*, **42**: 356–362.

———, 1926: Beiträge zur Analyse der geotropischen Reaktion, *Jahrb. wiss. Bot.*, **65**: 447–459.

———, 1927: Wuchshormone und Tropismen bei den Pflanzen, *Biol. Zent.*, **47**: 604–626.

———, 1928: Beiträge zur hormonalen Theorie von Tropismen, *Planta*, **6**: 118–134.

———, 1934: Über die Bildung und Leitung des Wuchshormons bei den Wurzeln, *Planta*, **21**: 517–530; *Biol. Abst.*, **10**: 37.

Clark, N. A., 1926: Plant growth promoting substances, hydrogen-ion concentration and the reproduction of *Lemna*, *Plant Physiol.*, **1**: 273–279.

———, 1929: Vitamins and the world of plants, *Sci. Mo.*, **28**: 343–348.

——— and B. H. Thomas, 1934: Microorganisms and vitamin production in green plants, *Sci.*, **79**: 571–572.

CLARK, W. G., 1935: Note on the effect of light on the bioelectric potentials in the *Avena* coleoptile, *Proc. Nat. Acad. Sci.*, **21**: 681–684.

COGGESHALL, M., 1931: Influence of acetic, propionic, normal butyric and sulphuric acids and potassium acetate on elongation of primary roots of seedlings of white lupine, *Plant Physiol.*, **6**: 389–447.

COLIN, M. H., 1922: L'inuline dans les plantes greffées. La greffe Soleil annuel-Topinambour, *Rev. Gén. Bot.*, **34**: 145–155, 202–213.

COLLA, S., 1927: L'azione dei raggi ultra violetti sulle piante etiolate, *Boll. Soc. Ital. Biol. Sperim.*, **2**: 724–726; *Biol. Abst.*, **4**: 354.

COLLINS, G. N., L. H. FLINT, and J. W. MCLANE, 1929: Electrical stimulation of plant growth, *J. Agr. Res.*, **38**: 585–600.

——— and L. R. MAXWELL, 1936: Delayed killing of maize seedlings with X-rays, *Sci.*, **83**: 375–376.

COLLISON, R. C., and J. D. HARLAN, 1934: Winter injury of Baldwin apple trees and its relation to previous tree performance and nutritional treatment, *N. Y. Sta. (Geneva) Bull.* 647.

COOPER, H. P., 1932: Relation of hydrogen-ion concentration of soils to the growth of certain pasture plants, *Plant Physiol.*, **7**: 527–532.

COOPER, W. C., 1935: Hormones in relation to root formation on stem cuttings, *Plant Physiol.*, **10**: 787–794.

COPEMAN, P. R., 1928: Autocatalysis and growth, *Ann. App. Biol.*, **15**: 613–622.

COSTER, C. H., 1927: Die täglichen Schwankungen des Längenzuwachses in den Tropen, *Rec. trav. bot. Néerl.*, **24**: 257–306; *Biol. Abst.*, **2**: 1512; *Exp. Sta. Rec.*, **61**: 817.

COVILLE, F. V., 1920: The influence of cold in stimulating the growth of plants, *J. Agr. Res.*, **20**: 151–160.

CROCKER, W., and L. I. KNIGHT, 1908: Effect of illuminating gas and ethylene upon flowering carnations, *Bot. Gaz.*, **46**: 259–276.

——— and W. E. DAVIS, 1914: Delayed germination in seeds of *Alisma plantago*, *Bot. Gaz.*, **57**: 285–322.

———, 1916: Mechanics of dormancy in seeds, *Am. J. Bot.*, **3**: 99–121.

——— and L. V. BARTON, 1931: After-ripening, germination, and storage of certain Rosaceous seeds, *Contr. Boyce Thompson Inst.*, **3**: 385–404.

———, P. W. ZIMMERMAN, and A. E. HITCHCOCK, 1932: Ethylene-induced epinasty of leaves and the relation of gravity to it, *Contr. Boyce Thompson Inst.*, **4**: 177–218.

———, A. E. HITCHCOCK, and P. W. ZIMMERMAN, 1935: Similarities in the effects of ethylene and the plant auxins, *Contr. Boyce Thompson Inst.*, **7**: 231–248.

CURTIS, O. F., and H. I. CHANG, 1930: The relative effectiveness of the temperature of the crown as contrasted with that of the rest of the plant upon the flowering of celery, *Am. J. Bot.*, **17**: 1047–1048.

DANIEL, L., 1921: À propos des greffes de Soleil sur Topinambour, *C. R. Acad. Sci. Paris*, **172**: 610–612.

DARLINGTON, H. T., 1931: The 50-year period for Dr. Beal's viability experiment, *Am. J. Bot.*, **18**: 262–266.

DARROW, G. M., 1929: Effect of light, temperature, and transpiration on elongation of canes of raspberry and other brambles, *Proc. Am. Soc. Hort. Sci.*, **26**: 308–311.

———, 1930: Experimental studies on the growth and development of strawberry plants, *J. Agr. Res.*, **41**: 307–325.

——— and G. F. WALDO, 1930: Fruit-bud formation in the strawberry in spring in southeastern states, *Sci.*, **72**: 349–350.

——— and ———, 1933: Photoperiodism as a cause of the rest period in strawberries, *Sci.*, **77**: 353–354.

———, 1933: Tomatoes, berries and other crops under continuous light in Alaska, *Sci.*, **78**: 370.

DASTUR, R. H., and G. A. KAPADIA, 1931: Mechanism of curvature in the tendrils of *Cucurbitaceae*, *Ann. Bot.*, **45**: 279–301.

DAVIES, P. A., 1930: The relation of total nitrogen to regeneration in the willow, *Sci.*, **72**: 275–276.

DAVIES, W., G. A. ATKINS, and P. C. B. HUDSON, 1936: Effect of ascorbic acid and indolyl acetic acid on regeneration of willow branches and germination, *Nature*, **137**: 618.

DAVIS, E. E., 1934: Influence of thyroxin on the growth of plants, *Plant Physiol.*, **9**: 377–384.

DAVIS, W. B., and C. G. CHURCH, 1931: The effect of ethylene on the chemical composition and the respiration of the ripening Japanese persimmon, *J. Agr. Res.*, **42**: 165–182.

———, 1935: Detection and measurement of freezing injury in Valencia oranges, *Am. J. Bot.*, **22**: 557–566.

DEATS, M. E., 1925: The effect on plants of the increase and decrease of the period of illumination over that of the normal day period, *Am. J. Bot.*, **12**: 384–393.

DEBEER, G. R., 1924: "Growth," Edward Arnold & Company, London, 120 pp.

DEGMAN, E. S., and E. C. AUCHTER, 1934: Metaxenia studies in apples, *Proc. Am. Soc. Hort. Sci.*, **32**: 213–220.

DELONG, G. E., 1929: The effect of cutting garnet wheat at different stages of maturity and on consecutive dates after the occurrence of frost, *Sci. Agr.*, **9**: 566–574.

DENNY, F. E., 1924: Hastening the coloration of lemons, *J. Agr. Res.*, **27**: 757–771.

———, 1926: Hastening the sprouting of dormant potato tubers, *Am. J. Bot.*, **13**: 118–126.

———, 1926: Second report on the use of chemicals for hastening the sprouting of dormant potato tubers, *Am. J. Bot.*, **13**: 386–396.

———, 1928: Chemical treatments for controlling the growth of buds of plants, *Ind. and Eng. Chem.*, **20**: 578–581.

——— and E. N. STANTON, 1928: Chemical treatments for shortening the rest period of pot-grown woody plants, *Am. J. Bot.*, **15**: 327–337.

——— and ———, 1928: Localization of response of woody tissues to chemical treatments that break the rest period, *Am. J. Bot.*, **15**: 337–344.

———, 1928: The importance of temperature in the use of chemicals for hastening the sprouting of dormant potato tubers, *Am. J. Bot.*, **15**: 395–404.

———, 1929: Role of mother tuber in growth of potato plant, *Bot. Gaz.*, **87**: 157–194.

———, 1930: Shortening the rest period of *Gladiolus* by treatment with chemicals, *Am. J. Bot.*, **17**: 602–613.

———, 1930: Sucrose and starch changes in potatoes treated with chemicals that break the rest period, *Am. J. Bot.*, **17**: 806–818.

———, 1930: Direct *vs.* indirect effects in the treatment of potato tubers with chemicals, abst., *Am. J. Bot.*, **17**: 1049–1050.

——— and L. P. MILLER, 1932: Effect of ethylene chlorhydrin vapors upon dormant lilac tissues, *Contr. Boyce Thompson Inst.*, **4**: 513–528.

———, 1933: Effect of ethylene chlorohydrin vapors upon the chemical composition of *Gladiolus* corms, *Contr. Boyce Thompson Inst.*, **5**: 435–440.

——— and L. P. MILLER, 1934: Hastening the germination of dormant *Gladiolus* cormels with vapors of ethylene chlorohydrin, *Contr. Boyce Thompson Inst.*, **6**: 31–38.

——— and———, 1935: Production of ethylene by plant tissue as indicated by the epinastic response of leaves, *Contr. Boyce Thompson Inst.*, **7**: 97–102.

——— and ———, 1935: Further experiments on shortening the rest period of potato tubers, *Contr. Boyce Thompson Inst.*, **7**: 157–181.

———, 1935: Testing plant tissue for emanations causing leaf epinasty, *Contr. Boyce Thompson Inst.*, **7**: 341–347.

———, 1936: Gravity-position of tomato stems and their production of the emanation causing leaf epinasty, *Contr. Boyce Thompson Inst.*, **8**: 99–104.

DEUBER, C. G., 1931: Chemical treatments to shorten the rest period of tree seeds, *Sci.*, **73**: 320–321.

———, 1932: Effects of illuminating gas on young oak trees, abst., *Am. J. Bot.*, **19**: 845.

———, 1932: Stimulative effects of illuminating gas on trees, *Sci.*, **75**: 496–497.

———, 1933: Accelerating seed germination with gas, *Am. Gas Assoc. Month.*, **15**: 313–315.

———, 1933: Stimulative effects of gas on small trees, *Am. Gas. Assoc. Month.*, **15**: 380–383.

———, 1933: Observations of shade trees in the vicinity of gas leaks, *Am. Gas. Assoc. Month.*, **15**: 465–466.

———, 1934: Methylene blue and gassed plants, *Sci.*, **79**: 389–390.

———, 1936: Effects on trees of an illuminating gas in the soil, *Plant Physiol.*, **11**: 401–412.

DEXTER, S. T., 1930: A method for determining hardiness in plants, *Sci.*, **71**: 220–221.

———, W. E. TOTTINGHAM, and L. F. GRABER, 1930: Preliminary results in measuring the hardiness of plants, *Plant Physiol.*, **5**: 215–223.

———, ——— and ———, 1932: Investigations of the hardiness of plants by measurement of electrical conductivity, *Plant Physiol.*, **7**: 63–78.

———, 1932: Studies of the hardiness of plants: A modification of the Newton pressure method for small samples, *Plant Physiol.*, **7**: 721–726.

———, 1933: Effect of several environmental factors on the hardening of plants, *Plant Physiol.*, **8**: 123–139.

———, 1933: Decreasing hardiness of winter wheat in relation to photosynthesis, defoliation and winter injury, *Plant Physiol.*, **8**: 297–304.

———, 1934: Salt concentration and reversibility of ice-formation as related to the hardiness of winter wheat, *Plant Physiol.*, **9**: 601–618.

———, 1935: Growth, organic nitrogen fractions, and buffer capacity in relation to hardiness of plants, *Plant Physiol.*, **10**: 149–158.

———, 1935: Salt concentration and reversibility of ice formation as related to the hardiness of alfalfa, *Plant Physiol.*, **10**: 403–406.

DICKSON, H., 1932: Polarity and the production of adventitious growing points in *Marchantia polymorpha*, *Ann. Bot.*, **46**: 683–702.

DIXON, H. H., and T. A. BENNET-CLARK, 1927: Response of plant tissue to electric currents, *Notes, Bot. School Trinity Col. Dublin*, **4**: 33–54.

DOLK, H. E., and K. V. THIMANN, 1932: Studies on the growth hormone of plants. I, *Proc. Nat. Acad. Sci.*, **18**: 30–46.

DOUBT, S. L., 1917: The response of plants to illuminating gas, *Bot. Gaz.*, **63**: 209–224.

DOVE, W. F., and E. MURPHY, 1936: The vitamin C content of apples and its relation to human welfare, *Sci.*, **83**: 325–327.

DRAGHETTI, A., 1933: Importance of winter cryptovegetation in the forced culture of wheat, *Ann. Tech. Agr.*, **6**: 484–501; *Exp. Sta. Rec.*, **72**: 596.

DUBUY, H. G., 1936: The change in the response of *Avena* coleoptiles to growth regulators produced by aging, *Proc. Nat. Acad. Sci.*, **22**: 272–275.

DUGGAR, B. M., 1937: "Biological Effects of Radiation," McGraw-Hill Book Company, Inc., New York and London, 2 vol., 1342 pp.

DUNGAN, G. H., 1931: An indication that corn tillers may nourish the main stalk under some conditions, *J. Am. Soc. Agron.*, **23**: 662–670.

DUNN, S., 1933: Relation of hydrophilic colloids to hardiness in cabbage, brussels sprouts, and alfalfa plants as shown by the dye adsorption test, *Plant Physiol.*, **8**: 275–286.

DUSTMAN, R. B., 1934: Effect of ethylene, ethylene chlorohydrin, and ultra-violet light on carbohydrate content of stored apples, *Plant Physiol.*, **9**: 637–643.

DUVEL, J., 1905: The vitality of buried seeds, *U. S. Bur. Plant. Ind. Bull.* 83.

EASTHAM, A., 1914: The longevity of some common seeds, *Agr. Gaz. Canada*, **1**: 544–546.

EATON, F. M., 1924. Assimilation-respiration balance as related to length of day reactions of soybeans, *Bot. Gaz.*, **77**: 311–321.

EATON, S. V., 1931: Effects of variation in dry-length and clipping of plants on nodule development and growth of soy bean, *Bot. Gaz.*, **91**: 113–144.

ECKERSON, S. H., 1913: A physiological and chemical study of after-ripening, *Bot. Gaz.*, **55**: 286–299.

EDWARDS, T. I., 1932: Temperature relations of seed germination, *Quart. Rev. Biol.*, **7**: 428–443.

———, 1934: Relations of germinating soy beans to temperature and length of incubation time, *Plant Physiol.*, **9**: 1–30.

———, R. PEARL, and S. A. GOULD, 1934: The growth and duration of life of *Celosia cristata* seedlings at different temperatures, *J. Gen. Physiol.* **17**: 763–781.

———, ——— and ———, 1934: Influence of temperature and nutrition on the growth and duration of life of *Cucumis melo* seedlings, *Bot. Gaz.*, **96**: 118–135.

EGOROV, M. A., 1915: À propos du rôle des éléments de cendre dans la vie des plantes, *Zhur. Opytn. Agron.*, **16**: 270–280. Quoted by Murneek, 1932.

ELMER, O. H., 1932: Growth inhibition of potato sprouts by the volatile products of apples, *Sci.*, **75**: 194.

———, 1936: Growth inhibition in the potato caused by a gas emanating from apples, *J. Agr. Res.*, **52**: 609–626.

ELTINGE, E. T., 1928: The effect of ultra-violet radiation upon higher plants, *Ann. Mo. Bot. Gard.*, **15**: 169–240.

EMMERT, E. M., 1931: The effect of soil reaction on the growth of tomatoes and lettuce and on the nitrogen, phosphorus and manganese content of the soil and plant, *Ky. Sta. Res. Bull.* 314.

EVANS, C. R., 1933: Germination behavior of *Magnolia grandiflora*, *Bot. Gaz.*, **94**: 729–754.

EVANS, M. W., and H. A. ALLARD, 1934: Relation of length of day to growth of timothy, *J. Agr. Res.*, **48**: 571–586.

EWART, A. J., 1908: On the longevity of seeds, *Proc. Roy. Soc. Victoria*, **21**: 1–210.

FARR, W. K., 1931: Cotton fibers. I, Origin and early stages of elongation, *Contr. Boyce Thompson Inst.*, **3**: 441–458.

———, 1933: Cotton fibers. III, Cell divisions in the epidermal layer of the ovule subsequent to fertilization, *Contr. Boyce Thompson Inst.*, **5**: 167–172.

FERNALD, E. I., 1925: The inhibition of bud development as correlated with the osmotic concentration of sap, *Am. J. Bot.*, **12**: 287–305.

———, 1931: Freezing point depressions of asparagus shoots determined by a thermoelectric method, abst., *Am. J. Bot.*, **18**: 892.

FISHER, R. A., 1920: Some remarks on the methods formulated in a recent article on the quantitative analysis of plant growth, *Ann. App. Biol.*, **7**: 367–372.

FITTING, H., 1905, 1906: Die Reizleitungsvorgänge bei den Pflanzen, *Ergebn. Physiol.*, **4**: 684–763; **5**: 155–249.

———, 1909: Die Beeinflussung der Orchideenblüten durch die Bestäubung und durch andere Umstände, *Z. Bot.*, **1**: 1–86.

FIVAZ, A. E., 1931: Longevity and germination of seeds of *Ribes*, particularly *R. rotundifolium*, under laboratory and natural conditions, *U. S. Dept. Agr. Tech. Bull.* 261.

FLEMION, F., 1931: After-ripening, germination and vitality of seeds of *Sorbus aucuparia*, *Contr. Boyce Thompson Inst.*, **3**: 413–439.

———, 1933: Physiological and chemical studies of after-ripening of *Rhodotypos kerrioides* seeds, *Contr. Boyce Thompson Inst.*, **5**: 143–159.

———, 1934: Physiological and chemical changes preceding and during the after-ripening of *Symphoricarpos racemosus* seeds, *Contr. Boyce Thompson Inst.*, **6**: 91–102.

———, 1934: Dwarf seedlings from non after-ripened embryos of peach, apple and hawthorn, *Contr. Boyce Thompson Inst.*, **6**: 205–209.

FLINT, L. H., 1934: Light in relation to dormancy and germination in lettuce seed, *Sci.*, **80**: 38–40.

——— and E. D. MCALISTER, 1935: Wave lengths of radiation in the visible spectrum inhibiting the germination of light sensitive lettuce seed, *Smithsonian Misc. Coll.*, **94**: No. 5.

FLORELL, V. H., and B. FAULKNER, 1934: Growth of wheat during the heading period, *J. Am. Soc. Agron.*, **26**: 954–965.

FORSTER, H. C., M. A. H. TINCKER, A. J. VASEY, and S. M. WADHAM, 1932: Experiments in England, Wales and Australia on the effect of the length of day on various cultivated varieties of wheat, *Ann. App. Biol.*, **19**: 378–412.

FRANCIS, D. S., 1934: The effect of X-rays on growth and respiration of wheat seedlings, *Bull. Torrey Bot. Club*, **61**: 119–153.

FRASER, J. G. C., and L. M. PIDGEON, 1933: Electrolysis of seed of cereals, *Sci. Agr.*, **14**: 141–148.

FREELAND, R. O., 1933: Some morphological and physico-chemical changes accompanying proliferation of *Bryophyllum* leaves, *Am. J. Bot.*, **20**: 467–480.

FRIEDBERG, R., 1932: The tillering of wheat: Relation between the depth of the tillering node and cold resistance, *Ann. Agron.* (*Paris*), **2**: 215–228; *Exp. Sta. Rec.*, **72**: 44.

FRIESNER, R. C., 1920: Daily rhythms of elongation and cell division in certain roots, *Am. J. Bot.*, **7**: 380–406.

FULLER, H. J., 1930: Stimulatory effects of ultra-violet radiation upon higher plants, *Sci.*, **72**: 535–536.

———, 1931: Stimulatory effects of radiation from a quartz mercury vapor arc upon higher plants, *Ann. Mo. Bot. Gard.*, **18**: 17–39.

———, 1932: The injurious effects of ultra-violet and infra red radiations on plants, *Ann. Mo. Bot. Gard.*, **19**: 79–84.

GANE, R., 1934: Production of ethylene by some ripening fruits, *Nature*, **134**: 1008.

———, 1934: Identification of ethylene among the volatile products of ripe apples, *Dept. Sci. and Ind. Res., Food Invest. Bd. Rept.* (*Gr. Brit.*), **1934**: 122–123; *Exp. Sta. Rec.*, **74**: 40.

———, 1935: The formation of ethylene by plant tissues, and its significance in the ripening of fruits, *J. Pom. and Hort. Sci.*, **13**: 351–358.

GARDNER, F. E., 1929: Composition and growth initiation of dormant pear shoots as influenced by temperatures, *Plant Physiol.*, **4**: 405–434.

GARDNER, W. A., 1921: Effect of light on germination of light-sensitive seeds, *Bot. Gaz.*, **71**: 249–288.

Garner, W. W., and H. A. Allard, 1920: Effect of the relative length of day and night and other factors of the environment on growth and reproduction in plants, *J. Agr. Res.*, **18**: 553–607.

——— and ———, 1923: Further studies in photoperiodism, the response of the plant to relative length of day and night, *J. Agr. Res.*, **23**: 871–921.

———, C. W. Bacon, and H. A. Allard, 1924: Photoperiodism in relation to hydrogen-ion concentration of the cell sap and the carbohydrate content of the plant, *J. Agr. Res.*, **27**: 119–157.

——— and H. A. Allard, 1925: Localization of the response in plants to relative length of day and night, *J. Agr. Res.*, **31**: 555–567.

——— and ———, 1930: Photoperiodic response of soy beans in relation to temperature and other environmental factors, *J. Agr. Res.*, **41**: 719–737.

——— and ———, 1931: Effect of abnormally long and short alternations of light and darkness on growth and development of plants, *J. Agr. Res.*, **42**: 629–639.

——— and ———, 1931: Duration of the flowerless condition of some plants in response to unfavorable lengths of day, *J. Agr. Res.*, **43**: 439–445.

———, 1933: Comparative responses of long-day and short-day plants to relative lengths of day and night, *Plant Physiol.*, **8**: 347–356.

Gassner, G., 1918: Beiträge zur physiologischer Characteristik Sommer-und Winter-annueller Gewächse, insbesondere der Getriedepflanzen, *Z. Bot.*, **10**: 417–480.

Gautheret, R., 1933: Cultures de méristèmes de racines de *Zea mays*, *C. R. Acad. Sci. Paris*, **197**: 85–87.

Gericke, W. F., 1918: Effects of rest and no-rest periods upon growth of *Solanum*, *Bot. Gaz.*, **65**: 344–353.

———, 1924: Growth-inhibiting and growth-stimulating substances, *Bot. Gaz.*, **78**: 440–455.

Gfeller, F., R. A. Derrick, and J. G. C. Fraser, 1933: Yarovization or vernalization of cereal crops, *Sci. Agr.*, **13**: 728–729.

Giersbach, J., 1934: After-ripening and germination of *Cotoneaster* seeds, *Contr. Boyce Thompson Inst.*, **6**: 323–338.

Gilbert, B. E., 1926: Interrelation of relative day length and temperature, *Bot. Gaz.*, **81**: 1–25.

———, 1926: The response of certain photoperiodic plants to differing temperature and humidity conditions, *Ann. Bot.*, **40**: 315–321.

Gill, N., 1933: The relation of flowering and cambial activity, *New Phytol.*, **32**: 1–12.

Glass, H. B., 1933: Effect of light on the bioelectric potentials of isolated *Elodea* leaves, *Plant Physiol.*, **8**: 263–274.

Goodspeed, T. H., 1929: The effect of X-rays and radium on species of the genus *Nicotiana*, *J. Heredity*, **20**: 243–259.

Gorke, H., 1906: Über chemische Vorgänge beim Erfrieren der Pflanzen, *Landw. Versuchs-Stat.*, **65**: 149–160.

Goss, W. L., 1924: The vitality of buried seeds, *J. Agr. Res.*, **29**: 349–362.

Gould, S. A., R. Pearl, T. I. Edwards, and J. R. Miner, 1933: Available food, relative growth and duration of life in seedlings of *Cucumis melo*, *Proc. Nat. Acad. Sci.*, **19**: 228–233.

———, ——— and ———, 1934: On the effects of partial removal of the cotyledons upon the growth and duration of life of canteloup seedlings without exogenous food, *Ann. Bot.*, **48**: 575–599.

Gourley, J. H., 1932: Response of iris to soil reaction, *Plant Physiol.*, **7**: 739–742.

Gračanin, M., 1928: Orthophosphorsäure als Stimulator der Keimungsenergie und Aktivator der Keimfähigkeit der Samen, *Biochem. Z.*, **195**: 457–468.

GREATHOUSE, G. A., and N. W. STUART, 1934: A study of physical and chemical properties of red clover roots in the cold hardened and unhardened condition, *Md. Sta. Bull.* 370.

GREENE, L., R. B. WITHROW, and M. F. RICHMAN, 1932: The response of greenhouse crops to electric light supplementing daylight, *Ind. Sta. Bull.* 366.

GREGORY, F. G., 1926: The effect of climatic conditions on the growth of barley, *Ann. Bot.*, **40**: 1–27.

——— and L. BATTEN, 1926: A critical statistical study of experimental data on the effect of minute electric currents on the growth rate of the coleoptile of barley, *Proc. Roy. Soc.*, B, **99**: 122–130.

———, 1928: Studies in the energy relations of plants. I, The effect of temperature on the increase in area of leaf surface. II, The effect of temperature on increase in area of leaf surface and in dry weight of *Cucumis sativus*, *Ann. Bot.*, **42**: 469–507.

———, 1928: The analysis of growth curves.—a reply to criticism, *Ann. Bot.*, **42**: 531–541.

GROVES, J. F., 1917: Temperature and life duration of seeds, *Bot. Gaz.*, **63**: 169–189.

GUNDERSON, M. F., and C. E. SKINNER, 1934: Production of vitamins by a pure culture of *Chlorococcum* grown in darkness on a synthetic medium, *Plant Physiol.*, **9**: 807–815.

GUSTAFSON, F. G., 1926: Growth studies on fruits, *Plant Physiol.*, **1**: 265–272.

———, 1927: Growth studies on fruits: an explanation of the shape of the growth curve, *Plant Physiol.*, **2**: 153–161.

——— and H. E. LAING, 1931: Growth studies of fruits. The effect of the removal of alternate flower clusters on the development of fruits in the tomato, *Mich. Acad. Sci., Arts, and Letters*, **13**: 95–99.

GUTHRIE, J. D., 1933: Effect of chemical treatments of dormant potato tubers on the conductivity of the tissue and on the leaching of electrolytes from the tissue, *Contr. Boyce Thompson Inst.*, **5**: 83–94.

———, 1933: Changes in the glutathione content of potato tubers treated with chemicals that break the rest period, *Contr. Boyce Thompson Inst.*, **5**: 331–350.

———, 1934: Metabolism of citric, sulphuric and nitric acid in the potato tuber. An explanation for the high pH of the juice of tubers treated with ethylene chlorhydrin, *Contr. Boyce Thompson Inst.*, **6**: 247–268.

HABER, E. S., P. M. NELSON, and P. P. SWANSON, 1931: Association of vitamins with nutritional conditions in plants, *Iowa Sta. Ann. Rept.*, **1931**: 105.

HABERLANDT, G., 1922: Cell-division hormones and their relation to wound repair, fertilization, parthenogenesis, and adventive embryogeny, *Biol. Zent.*, **42**: 145–172; *Exp. Sta. Rec.*, **49**: 219.

HACKENBERG, H., 1909: Ueber die Substanzquotienten von *Cannabis sativa* und *Cannabis gigantea*, *Beih. bot. Cent.*, I, **24**: 45–67.

HAIG, C., 1934: The spectral sensibility of *Avena*, *Proc. Nat. Acad. Sci.*, **20**: 476–479.

———, 1935: The phototropic responses of *Avena* in relation to intensity and wave length, *Biol. Bull.*, **69**: 305–324.

HAMMETT, F. S., 1929: The chemical stimulus essential for growth by increase in cell number, *Proc. Am. Phil. Soc.*, **68**: 151–161.

HANNA, W. F., 1924: Growth of corn and sunflowers in relation to climatic conditions, *Bot. Gaz.*, **78**: 200–215.

———, 1925: The nature of the growth rate in plants, *Sci. Agr.*, **5**: 133–138.

HARRINGTON, G. T., 1916: Agricultural value of impermeable seeds, *J. Agr. Res.*, **6**: 761–796.

———, 1923: Use of alternating temperatures in the germination of seeds, *J. Agr. Res.*, **23**: 295–333.

HARRINGTON, J. B., 1926: Growing wheat and barley hybrids in winter by means of artificial light, *Sci. Agr.*, **7**: 125–130.

HARRISON, C. M., 1931: Effect of cutting and fertilizer applications on grass development, *Plant Physiol.*, **6**: 669–684.

HARRISON, G. J., 1931: Metaxenia in cotton, *J. Agr. Res.*, **42**: 521–544.

HARTSHORN, R., 1931: Some effects of acetylene on the ripening processes of bananas, *Plant Physiol.*, **6**: 467–484.

HARVEY, E. M., 1913: The castor bean plant and laboratory air, *Bot. Gaz.*, **56**: 439–442.

———, 1915: Some effects of ethylene on the metabolism of plants, *Bot. Gaz.*, **60**: 193–214.

——— and R. C. ROSE, 1915: The effects of illuminating gas on root systems, *Bot. Gaz.*, **60**: 27–45.

———, 1931: A method for comparing growth rates by means of a protractor, *Sci.*, **74**: 46–47.

HARVEY, R. B., 1918: Hardening process in plants and developments from frost injury, *J. Agr. Res.*, **15**: 83–113.

———, 1922: Growth of plants in artificial light, *Bot. Gaz.*, **74**: 447–451.

———, 1922: Growth of plants in artificial light from seed to seed, *Sci.*, **56**: 366–367.

———, 1925: A new method of blanching celery, *Minn. Hort.*, **53**: 41.

———, 1926: The ripening of fruits by ethylene gas, *Minn. Hort.*, **54**: 140.

———, 1928: Artificial ripening of fruits and vegetables, *Minn. Sta. Bull.* 247.

———, 1928: Ethylene as a ripener of fruits and vegetables, *Sci.*, **67**: 421–422.

———, 1930: Time and temperature factors in hardening plants, *Am. J. Bot.*, **17**: 212–217.

———, 1931: The action of toxic agents used in the eradication of noxious plants, *J. Am. Soc. Agron.*, **23**: 481–489.

HASKINS, C. P., and C. N. MOORE, 1935: Growth modification in citrus seedlings grown from X-rayed seeds, *Plant Physiol.*, **10**: 179–185.

HASSELBRING, H., 1932: Effect of hot water treatment on the carbohydrate changes in narcissus bulbs during storage, *Plant Physiol.*, **7**: 145–152.

HAUSEN, S., 1935: Effect of vitamin C (ascorbic acid) on the growth of plants, *Nature*, **136**: 516.

HAVAS, L., 1935: Ascorbic acid (vitamin C) and the germination and growth of seedlings, *Nature*, **136**: 435.

———, 1935: Follicular (Oestrus) hormone and plant tumours, *Nature*, **136**: 516.

——— and J. CALDWELL, 1935: Some experiments on the effects of animal hormones on plants, *Ann. Bot.*, **49**: 727–748.

———, 1935: Some effects of radioactive mud upon germination of seeds and growth of seedlings, *J. Agr. Sci.*, **25**: 198–216.

HAWKER, L. E., 1932: Experiments on the perception of gravity by roots, *New Phytol.*, **31**: 321–328.

———, 1933: The effect of temperature on the geotropism of seedlings of *Lathyrus odoratus*, *Ann. Bot.*, **47**: 501–515.

HAYNES, D., and H. K. ARCHBOLD, 1928: Chemical studies in the physiology of apples. X, A quantitative study of chemical changes in stored apples, *Ann. Bot.*, **42**: 965–1017.

HEDRICKS, E., and R. B. HARVEY, 1924: Growth of plants in artificial light. II, Intensities of continuous light required for blooming, *Bot. Gaz.*, **77**: 330–334.

HEYN, A. N. J., 1931: Der Mechanismus der Zellstreckung, *Rec. trav. bot. Néerl.*, **28**: 113–244.

——— and J. VAN OVERBEEK, 1931: Weiteres Versuchsmaterial zur plastischen und elastischen Dehnbarkeit der Zellmembran, *Proc. Kon. Akad. Wet. Amsterdam*, **34**: 1190–1195.

———, 1934: Weitere Untersuchungen über den Mechanismus der Zellstreckung und die Eigenschaften der Zellmembran. III, Die Änderungen der Plastizität der Zellwand bei verschiedenen Organen, *Jahrb. wiss. Bot.*, **79**: 753–789.

HIBBARD, A. D., 1933: Metaxenia in apple and squash, *Proc. Am. Soc. Hort. Sci.*, **30**: 140–142.

HIBBARD, R. P., 1930: The physiological effect of ethylene gas upon celery, tomatoes and certain fruits, *Mich. Sta. Tech. Bull.* 104.

HICKS, L. E., 1932: Ranges of pH tolerance of the *Lemnaceae*, *Ohio J. Sci.*, **32**: 237–244.

HILL, D. D., and S. C. SALMON, 1927: The resistance of certain varieties of winter wheat to artificially produced low temperatures, *J. Agr. Res.*, **35**: 933–939.

HITCHCOCK, A. E., W. CROCKER, and P. W. ZIMMERMAN, 1932: Effect of illuminating gas on the lily, narcissus, tulip and hyacinth, *Contr. Boyce Thompson Inst.*, **4**: 155–176.

———, ———, and ———, 1932: The effect on plants of passing illuminating gas through the soil, abst., *Am. J. Bot.*, **19**: 846.

———, ———, and ———, 1934: Toxic action in soil of illuminating gas containing hydrocyanic acid, *Contr. Boyce Thompson Inst.*, **6**: 1–30.

———, 1935: Indole-3-*n* propionic acid as a growth hormone and the quantitative measurement of plant response, *Contr. Boyce Thompson Inst.*, **7**: 87–95.

———, 1935: Tobacco as a test plant for comparing the effectiveness of preparations containing growth substances, *Contr. Boyce Thompson Inst.*, **7**: 349–364.

——— and P. W. ZIMMERMAN, 1935: Absorption and movement of synthetic growth substances from soil as indicated by the response of aerial parts, *Contr. Boyce Thompson Inst.*, **7**: 447–476.

——— and ———, 1936: Effect of growth substances on the rooting response of cuttings, *Contr. Boyce Thompson Inst.*, **8**: 63–80.

HOOKER, H. D., 1920: Pentosan content in relation to winter hardiness, *Proc. Am. Soc. Hort. Sci.*, **17**: 204–207.

———, 1925: Plant growth, *Proc. Nat. Acad. Sci.*, **11**: 710–713.

HOSTERMANN, G., 1922: Kulturversuche mit elektrischem Licht, *Gartenwelt*, **26**: 74–75; *Bot. Abst.*, **11**: 383.

HOTTES, C. F., 1929: Studies in experimental cytology, *Plant Physiol.*, **4**: 1–30.

——— and H. K. WILSON, 1930: Resistance of wheat to high temperatures under different methods of heating, *J. Am. Soc. Agron.*, **22**: 108–113.

HOWARD, W. L., 1910: An experimental study of the rest period in plants. The winter rest, *Mo. Sta. Res. Bull.* 1.

———, 1915: An experimental study of the rest period in plants. The summer rest of bulbs and herbaceous perennials, *Mo. Sta. Res. Bull.* 15.

———, 1915: An experimental study of the rest period in plants. Pot grown woody plants, *Mo. Sta. Res. Bull.* 16.

———, 1915: An experimental study of the rest period in plants. Seeds, *Mo. Sta. Res. Bull.* 17.

———, 1915: An experimental study of the rest period in plants. Physiological changes accompanying breaking of the rest period, *Mo. Sta. Res. Bull.* 21.

HOWELL, J., 1932: Relation of western yellow pine seedlings to the reaction of the culture solution, *Plant Physiol.*, **7**: 657–671.

HUELIN, F. E., 1933: Effects of ethylene and of apple vapors on the sprouting of potatoes, *Dept. Sci. and Ind. Res., Food Invest. Bd. Rept. (Gr. Brit.)*, **1932**: 51–53.

HURD-KARRER, A. M., and A. D. DICKSON, 1934: Carbohydrates and nitrogen relations in wheat plants with reference to type of growth under different environmental conditions, *Plant Physiol.*, **9**: 533–565.

———, 1933: Comparative responses of a spring and winter wheat to day length and temperature, *J. Agr. Res.*, **46**: 867–888.

HUTCHINGS, S. S., 1932: Light in relation to seed germination of *Mimulus ringens*, *Am. J. Bot.*, **19**: 632–643.

INAMDAR, R. S., S. B. SINGH, and T. D. PANDE, 1925: The growth of the cotton plant in India. I, The relative growth rates during successive periods of growth and the relation between growth rate and respiratory index throughout the life cycle, *Ann. Bot.*, **39**: 281–311.

ISBELL, C. L., 1931: Regenerative capacities of leaf and leaflet cuttings of tomato and of leaf and shoot cuttings of potato, *Bot. Gaz.*, **92**: 192–201.

JANSSEN, G., 1929: Physical measurements of the winter wheat plant at various stages in its development, *Plant Physiol.*, **4**: 477–491.

JENSEN, I. J., 1925: Winter wheat studies in Montana with special reference to winter killing, *J. Am. Soc. Agron.*, **17**: 630–631.

JOHNSON, E. L., 1926: Effect of X-rays upon growth, development, and oxidizing enzymes of *Helianthus annuus*, *Bot. Gaz.*, **82**: 373–403.

———, 1928: Growth and germination of sunflowers as influenced by X-rays, *Am. J. Bot.*, **15**: 65–77.

———, 1928: Tuberization of potatoes increased by X-rays, *Sci.* **68**: 231

———, 1931: On the alleged stimulating action of X-rays upon plants, *Am. J. Bot.*, **18**: 603–614.

———, 1931: Effect of X-irradiation upon growth and reproduction of tomato, *Plant Physiol.*, **6**: 685–694.

———, 1933: The influence of X-radiation on *Atriplex hortensis*, *New Phytol.*, **32**: 297–307.

———, 1936: Susceptibility of seventy species of flowering plants to X-radiation, *Plant Physiol.*, **11**: 319–342.

JOHNSON, L. P. V., 1935: General preliminary studies on the physiology of delayed germination in *Avena fatua*, *Can. J. Res.*, **13**: 283–300.

JOHNSTON, E. S., F. S. BRACKETT, and W. H. HOOVER, 1931: Relation of phototropism to the wave length of light, *Plant Physiol.*, **6**: 307–313.

———, 1932: The functions of radiation in the physiology of plants. II, Some effects of near infra red radiation on plants, *Smithsonian Misc. Coll.*, **87**, No. 14.

———, 1934: Phototropic sensitivity in relation to wave length, *Smithsonian Misc. Coll.*, **92**, No. 11.

———, 1935: Photoperiodism: a specific growth response to light, *Smithsonian Rept.*, **1934**: 313–323.

———, 1936: Sun rays and plant life, *Smithsonian Rept.* **1936**: 353–371.

JONES, J. A., 1920: Physiological study of maple seeds, *Bot. Gaz.*, **69**: 127–152.

———, 1928: Overcoming delayed germination of *Nelumbo lutea*, *Bot. Gaz.*, **85**: 341–344.

JONES, J. P., 1928: A physiological study of dormancy in vetch seed, *N. Y. Sta. (Cornell) Mem.* 120.

JONES, W. N., 1925: Polarity phenomena in sea kale roots, *Ann. Bot.*, **39**: 359–372.

JORGENSEN, I., and J. H. PRIESTLEY, 1914: The distribution of the overhead electrical discharge employed in recent agricultural experiments, *J. Agr. Sci.*, **6**: 337–349.

——— and W. STILES, 1917: Atmospheric electricity as an environmental factor, *J. Ecology*, **5**: 203–209.

JOSLYN, M. A., and G. L. MARSH, 1933: Changes occurring during freezing and storage and thawing of fruits and vegetables, *Calif. Sta. Bull.* 551.

JOZEFOWICZ, M., 1930: The effect of certain treatments on the germination of tomato seeds, *Ann. App. Biol.*, **17**: 504–513.

———, 1930: Some observations on tomato plants from seed submitted to high temperatures, *Ann. App. Biol.*, **17**: 514–521.

KARLING, J. S., 1924: A preliminary account of the influence of light and temperature on growth and reproduction in *Chara fragilis*, *Bull. Torrey Bot. Club*, **51**: 469–488.

———, 1934: Dendrograph studies on *Achras zapota*, *Am. J. Bot.*, **21**: 161–193.

KATCHIONI-WALTHER, L. S., 1930: Influence of H-ion concentration on the development and growth of buckwheat, *Exp. Sta. Rec.*, **62**: 318–319.

KATER, J. M., 1927: A cytological study of dormancy in the seed of *Phaseolus vulgaris*, *Ann. Bot.*, **41**: 629–643.

KEARNEY, T. H., and G. J. HARRISON, 1932: Pollen antagonism in cotton, *J. Agr. Res.*, **44**: 191–226.

KEEBLE, F., M. G. NELSON, and R. SNOW, 1929: Integration of plant behavior. I, Separate geotropic stimulations of tip and stump in roots, *Proc. Roy. Soc.*, B, **105**: 493–498.

———, ———, and ———, 1930: The integration of plant behavior. II, The influence of the shoot on the growth of roots in seedlings, *Proc. Roy. Soc.*, B, **106**: 182–188.

———, ———, and ———, 1930: A wound substance retarding growth in roots, *New Phytol.*, **29**: 289–293.

———, ———, and ———, 1931: The integration of plant behavior. III, The effect of gravity on the growth of roots, *Proc. Roy. Soc.*, B, **108**: 360–365.

———, ———, and ———, 1931: The integration of plant behavior. IV, Geotropism and growth substance, *Proc. Roy. Soc.*, B, **108**: 537–545.

——— and ———, 1935: The integration of plant behavior. V, Growth substance and traumatic curvature of the root, *Proc. Roy. Soc.*, B, **117**: 92–119.

KEINHOLTZ, R., 1924: The effect of high temperatures on the germination and subsequent growth of corn, *Philippine J. Sci.*, **25**: 311–347.

KELLERMAN, K. F., 1926: A review of the discovery of photoperiodism: the influence of the length of daily light periods upon the growth of plants, *Quart. Rev. Biol.*, **1**: 85–94.

KELLICOTT, W. E., 1904: The daily periodicity of cell division and of elongation in the root of *Allium*, *Bull. Torrey Bot. Club*, **31**: 529–550.

KEMPTON, J. H., 1924: The rate of growth of green and albino maize seedlings, *J. Agr. Res.*, **29**: 311–312.

KERTESZ, Z. I., 1930: The chemical changes in peas after picking, *Plant Physiol.*, **5**: 399–412.

KIDD, F., and C. WEST, 1918: Physiological pre-determination: The influence of the physiological condition of the seed upon the course of subsequent growth and upon the yield. II, Rev. Lit., Chap. I, "Parental conditions affecting the potentiality" of the seed, *Ann. App. Biol.*, **5**: 112–142.

———, ———, and G. E. BRIGGS, 1920: What is the significance of the efficiency index of plant growth? *New Phytol.*, **19**: 88–96.

——— and ———, 1932: Effects of ethylene and of apple vapors on the ripening of fruits, *Dept. Sci. and Ind. Res., Food Invest. Bd. Rept. (Gr. Brit.)* **1932**: 55.

——— and ———, 1934: The influence of the composition of the atmosphere upon the incidence of the climacteric in apples, *Dept. Sci. and Ind. Res., Food Invest. Bd. Rept. (Gr. Brit.)*, **1933**: 51–57.

KIESSELBACH, T. A., 1924: Relation of seed size to yield of small grain crops, *J. Am. Soc. Agron.*, **16**: 670–682.

KLEBS, G., 1913: Physiologie der Fortpflanzung. Handwörterb., *Naturwiss.*, **4**: 276–296.

KLIPPART, J. H., 1858: An essay on the origin, growth, diseases, varieties, etc., of the wheat plant, *Ann. Rept. Ohio State Bd. Agr.*, **12**: 562–816.

KNIGHT, L. I., and W. CROCKER, 1913: Toxicity of smoke, *Bot. Gaz.*, **55**: 337–371.

KNOTT, J. E., 1926: Further localization of the response in plant tissues to relative length of day and night, *Proc. Am. Soc. Hort. Sci.*, **23**: 67–70.

———, 1927: Catalase in relation to growth and to other changes in plant tissue, *N. Y. Sta. (Cornell) Mem.* 106.

———, 1932: Rapidity of response of spinach to change in photoperiod, *Plant Physiol.*, **7**: 123–130.

KNUDSON, L., 1925: Hydrogen-ion concentration and plant growth, *J. Am. Soc. Agron.*, **17**: 711–716.

KOCH, K., 1934: Investigations on the transverse and longitudinal transport of the growth substance in plant organs, *Planta*, **22**: 190–220.

KÖGL, F., and A. J. HAAGEN-SMIT, 1931: I, Mitteilung über pflanzliche Wachstumsstoffe. Über die Chemie des Wuchsstoffs, *Proc. K. Akad. Wetensch. Amsterdam*, **34**: 1411–1416.

———, ———, and H. ERXLEBEN, 1933: Über ein Phytohormon der Zellstreckung. Reindarstellung des Auxins aus menschlichem Harn, *Z. physiol. Chem.*, **214**: 241–262.

———, H. ERXLEBEN, and A. J. HAAGEN-SMIT, 1934: Über die Isolierung der Auxine *a* und *b* aus pflanzlichen Materialien, *Z. physiol. Chem.*, **225**: 215–229.

——— and ———, 1934: Über die Konstitution der Auxine *a* und *b*, *Z. physiol. Chem.*, **227**: 51–73.

———, A. J. HAAGEN-SMIT, and H. ERXLEBEN, 1934: Über den Einfluss der Auxine auf das Wurzelwachstum und über die chemische Natur des Auxins der Graskoleoptilen, *Z. physiol. Chem.*, **228**: 104–112; *Chem. Abst.*, **29**: 195.

———, 1935: Über Wuchsstoffe der Auxin- und der Bios-Gruppe, *Ber. chem. Ges.*, **68**: 16–28.

KONDŌ, M., and T. OKAMURA, 1930: On the influence of various temperatures in duration of storage and various moistures of rice upon the preservation of germination power of hulled rice, *Ber. Ōhara-Inst. Landw. Forsch., Kurischiki, Japan*, **4**: 315–341.

——— and ———, 1930: Beziehung zwischen der Wassertemperatur und dem Wachstum der Reispflanzen, *Ber. Ōhara-Inst. Landw. Forsch., Kurischiki, Japan*, **4**: 395–411.

——— and ———, 1931: Beziehung zwischen Wassertemperatur und Wachstum der Reispflanzen, *Ber. Ōhara-Inst. Landw. Forsch., Kurischiki, Japan*, **5**: 67–85.

——— and ———, 1931: Storage of rice. V, On the influence of a desiccating material upon the preservation of the germinating power of hulled rice having different moisture contents and stored at different temperatures, *Ber. Ōhara-Inst. Landw. Forsch., Kurischiki, Japan*, **5**: 221–242.

———, ———, S. ISSHIKI, and Y. KASAHARA, 1932: Untersuchungen über "Photoperiodismus" der Reispflanzen, *Ber. Ōhara-Inst. Landw. Forsch., Kurischiki, Japan*, **5**: 243–280.

KORSTIAN, C. F., 1927: Factors controlling germination and early survival in oaks, *Yale Forestry Bull.* 19.

KOTOWSKI, F., 1929: Effect of size of seed on plant production, *Proc. Int. Cong. Plant Sci.*, **2**: 974–987.

KOTTE, W., 1922: Kulturversuche mit isolierten Wurzelspitzen, *Beitr. allgem. Bot.*, **2**: 413–434.

———, 1922: Wurzelmeristem in Gewebekultur, *Ber. bot. Ges.*, **40**: 269–272.

KRAMER, P. J., 1934: Methods of breaking dormancy in certain forest trees, *J. Forestry*, **32**: No. 7.

———, 1936: Effect of variation in length of day on growth and dormancy of trees, *Plant Physiol.*, **11**: 127–137.

KRASSINSKY, N., A. A. KONDRASHOVA, and N. I. VINOGRADOVA, 1936: On photoperiodism and changes in the enzymatic system, *Ann. Bot.*, **50**: 293–304.

KURZ, H., 1923: Hydrogen ion concentration in relation to ecological factors, *Bot. Gaz.*, **76**: 1–30.

LA RUE, C. D., 1933: Intumescences on poplar leaves. II, Physiological considerations, *Am. J. Bot.*, **20**: 159–175.

———, 1935: Regeneration in monocotyledonous seedlings, *Am. J. Bot.*, **22**: 486–492.

———, 1936: Tissue cultures of spermatophytes, *Proc. Nat. Acad. Sci.* **22**: 201–209.

———, 1936: The effect of auxin on the abscission of petioles, *Proc. Nat. Acad. Sci.*, **22**: 254–259.

LAUDE, H. H., 1933: Review of the literature regarding vernalization, *Rept. Agr. Seminar, Kans. State Col.*, Manhattan, Kans.

———, 1937: Cold resistance of winter wheat, rye, barley and oats in transition from dormancy to active growth, *J. Agr. Res.*, **54**: 899–817.

———, 1937: Comparison of the cold resistance of several varieties of winter wheat in transition from dormancy to active growth, *J. Agr. Res.*, **54**: 919–926.

LAURIE, A., and G. H. POESCH, 1932: Photoperiodism. The value of supplementary illumination and reduction of light on flowering plants in the greenhouse, *Ohio Sta. Bull.* 512.

LEGGATT, C. W., 1929: The comparative value of scarified and unscarified sweet clover seed, *Sci. Agr.*, **9**: 611–614.

———, 1929: The agricultural value of hard seeds of sweet clover in Alberta, *Sci. Agr.*, **9**: 683–688.

LEHENBAUER, P. A., 1914: The growth of maize seedlings in relation to temperature, *Physiol. Res.*, **1**: 247–288.

LEITCH, I., 1916: Some experiments on the influence of temperature on the rate of growth in *Pisum sativum*, *Ann. Bot.*, **30**: 25–46.

LEONIAN, L. H., 1936: Effect of auxins from some green algae upon *Phytophthora cactorum*, *Bot. Gaz.*, **97**: 854–859.

LEVINE, M., 1934: A preliminary report on plants treated with the carcinogenic agents of animals, *Bull. Torrey Bot. Club*, **61**: 103–118.

———, 1936: The response of plants to localized application of various chemical agents, *Bull. Torrey Bot. Club*, **63**: 177–200.

LEWITH, S., 1890: Über die Ursache der Widerstandsfähigkeit der Sporen gegen hohe Temperaturen. Ein Beitrag zur Theorie der Disinfection, *Arch. Path. Pharm.*, **26**: 341–354.

LIPMAN, C. B., and G. N. LEWIS, 1934: Tolerance of liquid-air temperatures by seeds of higher plants for sixty days, *Plant Physiol.*, **9**: 392–394.

LIVINGSTON, B. E., and W. B. MACK, 1931: Temperature relations of CO_2 production by young wheat seedlings, as influenced by oxygen pressure, time and presence of ethylene, abst., *Am. J. Bot.*, **18**: 889–890.

——— and F. W. HAASIS, 1933: Relation of time and maintained temperature to germination percentage for a lot of rice seed, *Am. J. Bot.*, **20**: 596–615.

———, 1934: Environments, *Sci.*, **80**: 569–576.

LLOYD, F. E., and T. C. BARNES, 1932: Changes in the cells of *Spirogyra* associated with the presence of water polymers, *Proc. Nat. Acad. Sci.*, **18**: 422–427.
LODEWICK, J. E., 1930: Effect of certain climatic factors on the diameter growth of longleaf pine in western Florida, *J. Agr. Res.*, **41**: 349–363.
LOEB, J., 1915: Rules and mechanism of inhibition and correlation in the regeneration of *Bryophyllum calycinum*, *Bot. Gaz.*, **60**: 249–277.
———, 1916: Further experiments on correlation of growth in *Bryophyllum calycinum*, *Bot. Gaz.*, **62**: 293–302.
———, 1916: On the association and possible identity of root forming and geotropic substances or hormones in *Bryophyllum calycinum*, *Sci.*, **44**: 210–211.
———, 1916: "The Organism as a Whole," G. P. Putnam's Sons, New York, 379 pp.
———, 1917: A quantitative method of ascertaining the mechanism of growth and of inhibition of growth of dormant buds, *Sci.*, **45**: 436–439.
———, 1917: Influence of the leaf upon root formation and geotropic curvature in the stem of *Bryophyllum calycinum* and the possibility of a hormone theory of these processes, *Bot. Gaz.*, **63**: 25–51.
———, 1917: The chemical basis of axial polarity in regeneration, *Sci.*, **46**: 547–551.
———, 1918: Chemical basis of correlation. I, Production of equal masses of shoots by equal masses of sister leaves in *Bryophyllum calycinum*, *Bot. Gaz.*, **65**: 150–174.
———, 1924: "Regeneration from a physico-chemical viewpoint," McGraw-Hill Book Company, Inc., New York, 143 pp.
LONG, E. R., 1915: Growth and colloid hydration in cacti, *Bot. Gaz.*, **59**: 491–497.
LOO, T., and S. LOO, 1935: Studies on the culture of isolated root-tips under sterile conditions. I, The effect of leaf extract on the growth of root-tips, *Science Repts., Contr. Lab. Plant Physiol. Nat. Cent. Univ. Nanking, China*, Ser. B, Biology, **2**: 51–79.
LOOMIS, W. E., 1923: Some relations of hardening to transplanting, *Proc. Am. Soc. Hort. Sci.*, **20**: 206–215.
———, 1927: Temperature and other factors affecting the rest period of potato tubers, *Plant Physiol.*, **2**: 287–302.
———, 1932: Growth-differentiation balance *vs.* carbohydrate-nitrogen ratio, *Proc. Am. Soc. Hort. Sci.*, **29**: 240–244.
———, 1932: The daily growth of *Zea mays*, abst., *Am. J. Bot.*, **19**: 838.
———, 1934: Daily growth of maize, *Am. J. Bot.*, **21**: 1–6.
LUDWIG, C. A., 1932: The germination of cottonseed at low temperatures, *J. Agr. Res.*, **44**: 367–380.
LUND, E. J., 1929: Continuous bioelectric currents; a mechanism for cell correlation, abst., *Am. J. Bot.*, **16**: 846–847.
———, 1929: The electric polarity of the Douglas fir and the internal distribution of its electric correlation currents, abst., *Am. J. Bot.*, **16**: 858.
———, 1929: Electric polarity in the Douglas fir, *Pub. Puget Sound Biol. Sta.*, **7**: 1–28.
———, 1929: The relative dominance of growing points in the Douglas fir, *Pub. Puget Sound Biol. Sta.*, **7**: 29–37.
——— and M. BUSH, 1930: Electric correlation potentials in the leaf of *Bryophyllum*, *Plant Physiol.*, **5**: 491–508.
———, 1931: The unequal effect of O_2 concentration on the velocity of oxidation in loci of different electric potential, and glutathione content, *Protoplasma*, **13**: 236–258.
———, 1931: External polarity potentials in the apex of the Douglas fir before and after mechanical stimulation, *Plant Physiol.*, **6**: 507–517.

———, 1931: Electric correlation between living cells in cortex and wood in the Douglas fir, *Plant Physiol.*, **6**: 631–652.
———, 1932: Control of the flux equilibrium of electrochemical processes and electric polarity in the Douglas fir by temperature, *Plant Physiol.*, **7**: 297–307.
———, 1932: Comparison of the effects of temperature on the radial and longitudinal electric polarities in wood and cortex of the Douglas fir, *Plant Physiol.*, **7**: 505–516.
LUTZ, J. M., 1935: The influence of rate of thawing on freezing injury of apples, potatoes and onions, *Proc. Am. Soc. Hort. Sci.*, **33**: 227–233.
LYON, T. L., 1905: Tillering of corn, *Neb. Sta. Bull.*, **91**: 16–24.
LYSENKO, T. D., 1932: On the problem of Iarovization of corn, millet, Sudan grass, sorghum, and soy bean, *Bull. Jarovizacii*, No. 2 and 3.
———, 1932: Fundamental results of research on vernalization of agricultural plants, *Bull. Jarovizacii*, No. 4, 1–57. Quoted by Maximow, 1934.
MACDOUGAL, D. T., 1903: The influence of light and darkness upon growth and development, *Mem. N. Y. Bot. Gard.*, **2**: 1–319.
———, 1916: The mechanism and conditions of growth, *Mem. N. Y. Bot. Gard.*, **6**: 5–26.
——— and H. A. SPOEHR, 1917: Growth and imbibition, *Proc. Am. Phil. Soc.*, **56**: 289–352.
———, 1917: Rate and course of growth of succulents, *Carnegie Inst. Wash. Year Book*, **16**: 83–85.
———, 1917: Growth of wheat (*Triticum*) and corn (*Zea*), *Carnegie Inst., Wash. Year Book*, **16**: 85–87.
———, 1918: Development of conceptions included in growth, *Carnegie Inst. Wash. Year Book*, **17**: 55–57.
———, 1918: The growth mechanism of protoplasm, *Carnegie Inst. Wash. Year Book*, **17**: 57–58.
———, 1918: The nature and course of growth in higher plants, *Carnegie Inst. Wash. Year Book*, **17**: 58–59.
———, 1918: Temperature and the hydration and growth of colloids and of cell masses, *Carnegie Inst. Wash. Year Book*, **17**: 68–70.
———, 1919: Growth in organisms, *Sci.*, **49**: 599–605.
———, 1919: Hydration and growth, *Proc. Am. Phil. Soc.*, **58**: 346–372.
———, 1920: Hydration and growth, *Carnegie Inst. Wash. Pub.* 297.
———, 1920: The physical factors in the growth of the tomato, *Bull. Torrey Bot. Club*, **47**: 261–269.
———, 1921: Growth of trees, *Sci.*, **54**: 338.
———, 1921: Growth in trees, *Carnegie Inst. Wash. Pub.* 307.
———, 1921: The distinctive agencies in the growth of the cell, *Proc. Soc. Exp. Biol. Med.*, **19**: 103–110.
——— and F. SHREVE, 1924: Growth in trees and massive organs of plants, *Carnegie Inst. Wash. Pub.* 350.
———, 1925: Accretion and distention in plant cells, *Am. Nat.*, **59**: 336–345.
MACHT, D. I., 1926: Concerning the influence of polarized light on the growth of seedlings, *J. Gen. Physiol.*, **10**: 41–52.
———, 1928: Effect of polarized light on plants, *Am. J. Bot.*, **15**: 621.
———, 1930: A study of twenty-three octyl alcohols on growth of *Lupinus* seedlings, *Am. J. Bot.*, **17**: 572–578.
———, 1931: Effect of some inorganic and organic mercurials on growth of *Lupinus albus*, *Am. J. Bot.*, **18**: 598–602.
——— and J. D. MEYER, 1933: Effect of eighteen normal aliphatic alcohols on growth of *Lupinus albus*, *Am. J. Bot.*, **20**: 145–149.

MACK, W. B., and B. E. LIVINGSTON, 1933: Relation of oxygen pressure and temperature to the influence of ethylene on carbon dioxide production and on shoot elongation in very young wheat seedlings, *Bot. Gaz.*, **94**: 625–687.

MACLEOD, F. L., M. R. ARMSTRONG, M. E. HEAP, and L. A. TOLBERT, 1935: The vitamin A content of five varieties of sweet potatoes, *J. Agr. Res.*, **50**: 181–187.

MACMILLAN, C., 1891: On the growth-periodicity of the potato tuber, *Am. Nat.*, **25**: 462–469.

MARSH, G., 1928: Relation between continuous bioelectric currents and cell respiration. IV, The origin of electric polarity in the onion root, *J. Exp. Zool.*, **51**: 309–325.

MARTIN, J. F., 1932: The cold resistance of Pacific Coast spring wheats at various stages of growth as determined by artificial refrigeration, *J. Am. Soc. Agron.*, **24**: 871–880.

MARTIN, J. H., 1927: Comparative studies of winter hardiness in wheat, *J. Agr. Res.*, **35**: 493–537.

———, 1933: The practical application of iarovization, *J. Am. Soc. Agron.*, **26**: 251.

———, 1933: Iarovization in field practice, *U. S. Plant Ind., Div. Cereal Crops Mimeographed Sheets.*

MARTINI, M. L., H. V. HARLAN, and M. N. POPE, 1930: Some growth curves of barley kernels, *Plant Physiol.*, **5**: 263–272.

MASON, S. C., 1925: The minimum temperature for growth of the date palm and the absence of a resting period, *J. Agr. Res.*, **31**: 401–415.

———, 1925: Partial thermostasy of the growth center of the date palm, *J. Agr. Res.*, **31**: 415–455.

———, 1925: The inhibitive effect of direct sunlight on the growth of the date palm, *J. Agr. Res.*, **31**: 455–469.

MASON, T. G., 1922: Growth and correlation in sea island cotton, *West Indian Bull.*, **19**: 214–238.

———, 1922: Growth and abscission in sea island cotton, *Ann. Bot.*, **36**: 457–484.

MASURE, M. P., 1932: Effect of ultra-violet radiation on growth and respiration of pea seeds with notes on statistics, *Bot. Gaz.*, **93**: 21–41.

MATZKE, E. B., 1936: The effect of street lights in delaying leaf-fall in certain trees, *Am. J. Bot.*, **23**: 446–452.

MAXIMOW, N. A., 1912: Chemische Schutzmittel der Pflanzen gegen Erfrieren, *Ber. bot. Ges.*, **30**: 52–65, 293–305, 504–516.

———, 1914: Experimentelle und kritische Untersuchungen über das Gefrieren und Erfrieren der Pflanzen, *Jahrb. wiss. Bot.*, **53**: 327–420.

——— and A. POJARKOVA, 1925: Über die physiologische Natur der Unterschiede zwischen Sommer- und Wintergetreide, *Jahrb. wiss. Bot.*, **64**: 702–730.

———, 1929: Internal factors of frost and drought resistance in plants, *Protoplasma*, **7**: 259–291.

——— and M. A. KROTKINA, 1930: Investigation on the after effect of low temperature on the length of vegetative period. Quoted by Maximow, 1934.

———, 1934: The theoretical significance of vernalization, *Imp. Bur. Gen. Aberystwyth* (Wales) *Herbage Pub. Ser. Bull.* 16.

———, 1935: "Vernalization and phasic development of plants," *Imp. Bur. Plant Genetics, School of Agr.*, Cambridge, Eng., 151 pp.

MCCALL, A. G., 1923: The influence of acidity itself on plant growth without regard to other factors, *J. Am. Soc. Agron.*, **15**: 290–297.

MCCLELLAND, C. K., 1928: Study of suckers in corn, *Miss. Sta. Bull.*, **231**: 11.

MCCLELLAND, T. B., 1928: Studies of the photoperiodism of some economic plants, *J. Agr. Res.*, **37**: 603–628.

McCollum, J. P., 1934: Vegetative and reproductive responses associated with fruit development in the cucumber, *N. Y. Sta.* (*Cornell*) *Mem.* 163.

McComb, A. L., 1934: The relation between acorn weight and the development of one year chestnut oak seedlings, *J. Forestry*, **32**: 479–484.

McKay, J. W., and T. H. Goodspeed, 1930: The effects of X-radiation on cotton, *Sci.*, **71**: 644.

McKee, R., 1935: Vernalization experiments with forage crops, *U. S. Dept. Agr. Circ.*, 377.

McKinney, H. H., and W. J. Sando, 1930: The behavior of winter wheat in artificial environments, *Sci.*, **71**: 668–670.

——— and ———, 1933: Russian methods for accelerating sexual reproduction in wheat. Further information regarding "Iarovization," *J. Heredity*, **24**: 165–166.

——— and ———, 1933: Earliness and seasonal growth habit in wheat as influenced by temperature and photoperiodism, *J. Heredity*, **24**: 169–179.

———, *et al.*, 1934: Field experiments with vernalized wheat, *U. S. Dept. Agr. Circ.* 325.

——— and W. J. Sando, 1935: Earliness of sexual reproduction in wheat as influenced by temperature and light in relation to certain phases, *J. Agr. Res.*, **51**: 621–641.

McNair, J. B., 1934: The evolutionary status of plant families in relation to some chemical properties, *Am. J. Bot.*, **21**: 427–452.

McRostie, G. P., 1924: Longevity of alfalfa and clover seeds, *Sci. Agr.*, **4**: 236–238.

Megee, C. R., 1935: A search for factors determining winter hardiness in alfalfa, *J. Am. Soc. Agron.*, **27**: 685–698.

Mehrlich, F. P., 1931: Factors affecting growth from the foliar meristems of *Bryophyllum calycinum*, *Bot. Gaz.*, **92**: 113–141.

Meier, F. E., 1934: Lethal response of the alga *Chlorella vulgaris* to ultra-violet rays, *Smithsonian Misc. Coll.*, **92**: No. 3.

———, 1936: Growth of a green alga in isolated wave length regions, *Smithsonian Misc. Coll.* **94**: No. 17.

———, 1936: Lethal effect of short wave lengths of the ultraviolet on the alga *Chlorella vulgaris*, *Smithsonian Misc. Coll.* **95**: No. 2.

Meyer, B. S., 1932: Further studies on cold resistance in evergreens with special reference to the possible rôle of bound water, *Bot. Gaz.*, **94**: 297–321.

Meyer, W. C., 1930: Dormancy and growth studies of the American lotus, *Nelumbo lutea*, *Plant Physiol.*, **5**: 225–234.

Michener, H. D., 1935: Effects of ethylene on plant growth hormone, *Sci.*, **82**: 551–552.

Miller, E. C., 1925: Unpub. data, *Kans. Sta.*

Miller, E. V., and C. Brooks, 1932: Effect of carbon dioxide content of storage atmosphere on carbohydrate transformation in certain fruits and vegetables, *J. Agr. Res.*, **45**: 449–460.

Miller, L. P., 1933: Effect of sulphur compounds in breaking the dormancy of potato tubers and in inducing changes in the enzyme activities of the treated tubers, *Contr. Boyce Thompson Inst.*, **5**: 29–82.

Mitchell, H. L., 1934: Pot culture tests of forest soil fertility, *Black Rock Forest Bull.* 5.

Molisch, H., 1897: "Über das Gefrieren und Erfrieren der Pflanzen," Jena.

Montgomery, E. G., 1909: The economic value of tillers, *Neb. Sta. Bull.*, **112**: 31–34.

Moore, C. N., and C. P. Haskins, 1932: Note on premature flowering in grapefruit from X-rayed seeds, *Sci.*, **76**: 167–168.

——— and ———, 1933: Physiological variations in certain crop plants following seed exposure to high voltage X-rays, *Bot. Gaz.*, **94**: 801–807.

MORINAGA, T., 1926: Effect of alternating temperatures upon the germination of seed, *Am. J. Bot.*, **13**: 141–159.

———, 1926: The favorable effect of reduced oxygen supply upon the germination of certain seeds, *Am. J. Bot.*, **13**: 159–166.

MORROW, K. S., R. B. DUSTMAN, and H. O. HENDERSON, 1931: Changes in the chemical composition of mangels and rutabagas during storage, *J. Agr. Res.*, **43**: 919–930.

MÜLLER-THURGAU, H., 1880, 1886: Über das Gefrieren und Erfrieren der Pflanzen, *Landw. Jahrb.*, **9**: 133–189 (1880); **15**: 453–610 (1886).

MURNEEK, A. E., 1925: Correlation and cyclic growth in plants, *Bot. Gaz.*, **79**: 329–333.

———, 1925: The effect of fruit on vegetative growth in plants, *Proc. Am. Soc. Hort. Sci.*, **21**: 274–276.

———, 1926: Is fruiting of the apple an exhaustive process? *Proc. Am. Soc. Hort. Sci.*, **22**: 196–200.

———, 1926: Physiology of reproduction in horticultural plants. I, Reproduction and metabolic efficiency in the tomato, *Mo. Sta. Res. Bull.* 90.

———, 1926: Effects of correlation between vegetative and reproductive functions in the tomato (*Lycopersicum esculentum*), *Plant Physiol.*, **1**: 3–56.

———, 1932: Growth and development as influenced by fruit and seed formation, *Plant Physiol.*, **7**: 79–90.

———, 1933: The nature of shedding of immature apples, *Mo. Res. Bull.* 201.

MUSSO, J. O., 1933: A new method of using electricity in growing plants, *Fortschr. Landw.*, **8**: 272–274; *Chem. Abst.*, **27**: 5776.

NAVEZ, A. E., and T. W. ROBINSON, 1932: Geotropic curvature of *Avena* coleoptiles, *J. Gen. Physiol.*, **16**: 133–145.

———, 1933: Growth promoting substance and illumination, *Proc. Nat. Acad. Sci.*, **19**: 636–637.

———, 1933: "Growth promoting substance" and elongation of roots, *J. Gen. Physiol.*, **16**: 733–739.

NEBEL, B. R., 1930: Xenia and metaxenia in apples, *N. Y. Sta. (Geneva) Tech. Bull.* 170.

———, 1934: Metaxenia and xenia in apples. IV, *Gartenbauwiss.*, **9**: 45–64; *Chem. Abst.*, **29**: 7389.

NELJUBOW, D., 1901: Ueber die horizontale Nutation der Stengel von *Pisum sativum* und einigen anderen Pflanzen, *Beih. bot. Cent.*, **10**: 128–138.

———, 1911: Geotropismus in der Laboratoriumsluft, *Ber. bot. Ges.*, **29**: 97–112.

NELSON, R. C., and R. B. HARVEY, 1935: The presence in self-blanching celery of unsaturated compounds with physiological action similar to ethylene, *Sci.*, **82**: 133–134.

NEWCOMBE, F. C., 1923: Optimum temperatures for growth of some grass coleoptiles, *Papers, Mich. Acad. Sci., Arts, and Letters*, **3**: 203–210.

NEWTON, R., 1924: Colloidal properties of winter wheat plants in relation to frost resistance, *J. Agr. Sci.*, **14**: 178–191.

———, 1924: The nature and practical measurement of frost resistance in winter wheat, *Univ. Alberta Col. Agr. Res. Bull.* 1.

——— and W. R. BROWN, 1931: Frost precipitation of proteins of plant juice, *Can. J. Res.*, **5**: 87–110.

———, ———, and J. A. ANDERSON, 1931: Chemical changes in nitrogen fractions of plant juice on exposure to frost, *Can. J. Res.*, **5**: 327–332.

——— and A. G. MCCALLA, 1934: Effect of frost on wheat at progressive stages of maturity. I, Physical characteristics of the kernels, *Can. J. Res.* **10**: 414–429.

NICHOLS, G. E., 1934: The influence of exposure to winter temperatures upon seed germination in various native American plants, *Ecology*, **15**: 364–373.

NIELSEN, N., 1930: Untersuchungen über einen neuen wachstumsregulierenden Stoff: Rhizopin, *Jahrb. wiss. Bot.*, **73**: 125–191.

———, 1931: Über Wuchsstoffe der Hefe, *Biochem. Z.*, **237**: 244–246.

——— and V. HARTELIUS, 1932: The separation of growth promoting substances, *C. R. trav. lab. Carlsberg*, **19**: 1–17; *Chem. Abst.*, **26**: 4358.

NIGHTINGALE, G. T., 1922: Light in relation to the growth and chemical composition of some horticultural plants, *Proc. Am. Soc. Hort. Sci.*, **19**: 18–29.

——— and J. W. MITCHELL, 1934: Effects of humidity on metabolism in tomato and apple, *Plant Physiol.*, **9**: 217–236.

———, 1935: Effect of temperature on growth, anatomy, and metabolism of apple and peach roots, *Bot. Gaz.*, **96**: 581–639.

NIXON, R. W., 1928: The direct effect of pollen on the fruit of the date palm, *J. Agr. Res.*, **36**: 97–128.

———, 1928: Immediate influence of pollen in determining the size and time of ripening of the fruit of the date palm, *J. Heredity*, **19**: 240–254.

———, 1934: Metaxenia in dates, *Proc. Am. Soc. Hort. Sci.*, **32**: 221–226.

———, 1935: Metaxenia and interspecific pollination in *Phoenix*, *Proc. Am. Soc. Hort. Sci.*, **33**: 21–26.

NOGÜCHI, Y., 1934: Metaxenia in the Japanese persimmon. Shape and sweetness, *Japan. J. Bot.*, **7**: 61–71.

———, 1935: Modification of leaf structure by X-rays, *Plant Physiol.*, **10**: 753–762.

NOVIKOV, V. A., 1928: Cold resistance of plants. II, *J. Exp. Landw. Südosten Eur.-Russlands*, **6**: 71–100.

OAKLEY, R. A., and H. L. WESTOVER, 1921: Effect of length of day on seedlings of alfalfa varieties and the possibility of utilizing this as a practical means of identification, *J. Agr. Res.*, **21**: 599–607.

OHGA, I., 1923: Longevity of seed of *Nelumbo nucifera* (Japanese), *Bot. Mag. Tokyo*, **27**: 87–95; *Bot. Abst.*, **14**: 899.

OSMASTON, B. B., 1918: Rate of growth of bamboos, *Indian Forester*, **44**: 52–57; *Exp. Sta. Rec.*, **44**: 518–519.

PAAL, A., 1914: Über phototropische Reizleitungen, *Ber. bot. Ges.*, **32**: 499–502.

———, 1918: Über phototropische Reizleitung, *Jahrb. wiss. Bot.*, **58**: 406–458.

PACK, D. A., 1921: After-ripening and germination of *Juniperus communis*, *Bot. Gaz.*, **71**: 32–60.

———, 1921: Chemistry of after ripening, germination and seedling development of Juniper seeds, *Bot. Gaz.*, **72**: 139–151.

PALLADIN, W., 1923: "Plant Physiology," English translation, P. Blakiston's Son & Company, Philadelphia.

PAN, C. L., 1936: Length of exposure to light in relation to plant growth in rice, *J. Am. Soc. Agron.*, **28**: 58–63.

PANCHAUD, J., 1934: Action du milieu extérieur sur le métabolisme végétal, *Rev. Gén. Bot.*, **46**: 586–603; *Bot. Gaz.*, **96**: 579.

PATON, R. R., 1929: The relation of size of seedling trees to their vigor, *Ohio Sta. Bimon. Bull.*, **141**: 191–194.

PEARL, R., A. A. WINSOR, and J. R. MINER, 1928: The growth of seedlings of the canteloup *Cucumis melo* in the absence of exogenous food and light, *Proc. Nat. Acad. Sci.*, **14**: 1–4.

———, ———, and F. B. WHITE, 1928: The form of the growth curve of the canteloup (*Cucumis melo*) under field conditions, *Proc. Nat. Acad. Sci.*, **14**: 895–901.

———, T. I. EDWARDS, and J. R. MINER, 1934: The growth of *Cucumis melo* seedlings at different temperatures, *J. Gen. Physiol.*, **17**: 687–700.

———, ———, A. A. WINSOR, and C. P. WINSOR, 1934: Aeration and growth of canteloup seedlings (*Cucumis melo*), *Am. J. Bot.*, **21**: 242–250.

PEARSALL, W. H., 1923: Studies in growth. IV, Correlations in development, *Ann. Bot.*, **37**: 261–275.

———, 1925: Hydrogen-ion concentration and growth, *Sci. Prog.*, **20**: 58–67.

——— and A. M. HANBY, 1926: Growth studies. V, Factors affecting the development and form of leaves, *Ann. Bot.*, **40**: 85–103.

———, 1927: Growth studies. VI, On the relative sizes of growing plant organs, *Ann. Bot.*, **41**: 549–557.

PELTIER, G. L., and H. M. TYSDAL, 1931: Hardiness studies with 2-year-old alfalfa plants, *J. Agr. Res.*, **43**: 931–957.

——— and ———, 1932: A method for the determination of comparative hardiness in seedling alfalfas by controlled hardening and artificial freezing, *J. Agr. Res.*, **44**: 427–444.

——— and T. A. KIESSELBACH, 1934: Inducing early fruiting in winter wheat by seed activation, *J. Am. Soc. Agron.*, **26**: 503–506.

——— and ———, 1934: The comparative cold resistance of spring small grains, *J. Am. Soc. Agron.*, **26**: 681–687.

PENROSE, L. S., 1925: A note on the relation of rate of growth to structure in plants, *New Phytol.*, **24**: 294–299.

PERRY, J. T., 1932: A possible hormone-secreting region in the grass coleoptile, *Sci.*, **76**: 215–216.

PFEIFFER, N. E., 1926: Microchemical and morphological studies of effect of light on plants, *Bot. Gaz.*, **81**: 173–195.

———, 1928: Anatomical study of plants grown under glasses transmitting light of various ranges of light, *Bot. Gaz.*, **85**: 427–436.

PIETERS, A. J., 1925: Difference in internode lengths between, and effect of variation of light duration upon, seedlings of annual and biennial white sweet clover, *J. Agr. Res.*, **31**: 585–596.

PLITT, T. M., 1932: Some photoperiodic and temperature responses of the radish, *Plant Physiol.*, **7**: 337–339.

———, 1935: Temperature as a predetermining factor in the development of *Avena sativa*, *Plant Physiol.*, **10**: 269–289.

POESCH, G. H., 1931: Studies on photoperiodism of the *Chrysanthemum*, *Proc. Am. Soc. Hort. Sci.*, **28**: 389–392.

POPE, M. N., 1932: The growth curve in barley, *J. Agr. Res.*, **44**: 323–341.

POPP, H. W., 1926: A physiological study of the effect of light of various ranges of wave length on the growth of plants, *Am. J. Bot.*, **13**: 706–737.

———, 1926: Effect of light intensity on growth of soybeans and its relation to the auto-catalyst theory of growth, *Bot. Gaz.*, **82**: 306–320.

——— and F. BROWN, 1933: A review of recent work on the effect of ultra-violet radiation upon seed plants, *Bull. Torrey Bot. Club.*, **60**: 161–210.

PORTERFIELD, W. M., 1928: A study of the grand period of growth in bamboo, *Bull. Torrey Bot. Club*, **55**: 327–406.

———, 1930: Daily periodicity of growth in the bamboo, *Phyllostachys nigra*, with special reference to atmospheric conditions of temperature and moisture, *Bull. Torrey Bot. Club*, **57**: 533–557.

———, 1935: The relation of shoot roots to shoot elongation in the bamboo *Phyllostachys nigra*, *Am. J. Bot.*, **22**: 878–888.

Post, K., 1934: Production of early blooms of Chrysanthemums by the use of black cloth to reduce the length of day, *N. Y. Sta. (Cornell) Bull.* 594.

Powers, W. L., 1927: The effect of hydrogen-ion concentration on the growth of certain plants, *Soil Sci.*, **24**: 1–4.

Prankerd, T. L., 1936: Studies in the geotropism of *Pteridophyta*. VI, On rhythm in gravi-perception and reaction to gravity, *Proc. Roy. Soc.*, B, **120**: 126–141.

Prantl, K., 1873: Ueber den Einfluss des Lichts auf das Wachsthum der Blätter, *Arbeiten Bot. Inst. Würzburg*, **1**: 371–384.

Prescott, J. A., 1921: Some observations on the growth of maize in Egypt, *Sultanic Agr. Soc. Bull.* 7. Quoted by Hanna, 1925.

Priestley, J. H., and A. F. C. H. Evershed, 1922: Growth studies. I, A quantitative study of the growth of roots, *Ann. Bot.*, **36**: 225–237.

——— and W. H. Pearsall, 1922: Growth studies. II, An interpretation of some growth curves, *Ann. Bot.*, **36**: 239–249.

——— and ———, 1922: Growth studies. III, A volumeter method of measuring the growth of roots, *Ann. Bot.*, **36**: 485–488.

———, 1922: The toxic action of traces of coal gas upon plants, *Ann. App. Biol.*, **9**: 146–155.

———, 1925: Light and growth. I, The effect of brief light exposure upon etiolated plants, *New Phytol.*, **24**: 271–283.

———, 1926: Light and growth. II, On the anatomy of etiolated plants, *New Phytol.*, **25**: 145–170.

———, 1928: The meristemic tissues of the plant, *Biol. Rev.*, **3**: 1–20.

———, 1929: Cell growth and cell division in the shoot of the flowering plant, *New Phytol.*, **28**: 54–81.

———, 1930: Studies in the physiology of cambial activity. I, Contrasted types of cambial activity. II, The concept of sliding growth, *New Phytol.*, **29**: 56–73, 96–140.

———, 1930: Studies in the physiology of cambial activity. III, The seasonal activity of the cambium, *New Phytol.*, **29**: 316–354.

———, L. I. Scott, and M. E. Malins, 1933: A new method of studying cambial activity, *Proc. Leeds Phil. Soc.*, 2, **VIII**: 365–374.

Purvis, O. N., 1934: An analysis of the influence of temperature during germination on the subsequent development of certain winter cereals and its relation to the effect of length of day, *Ann. Bot.*, **48**: 919–956.

Raber, O. L., 1928: "Principles of Plant Physiology," The Macmillan Company, New York.

Raleigh, G. J., 1930: Chemical conditions in maturation, dormancy, and germination of seeds of *Gymnocladus dioica*, *Bot. Gaz.*, **89**: 273–294.

Ramaley, F., 1931: Growth of plants under continuous light, *Sci.*, **73**: 566–567.

———, 1931: Some Caryophyllaceous plants influenced in growth and structure by artificial illumination supplemental to daylight, *Bot. Gaz.*, **92**: 311–320.

———, 1934: Influence of supplemental light on blooming, *Bot. Gaz.*, **96**: 165–174.

Rea, H. E., and R. E. Karper, 1932: Propagating sorghum by cuttings, *Am. J. Bot.*, **19**: 464–476.

Redington, G., 1929: The effect of the duration of light upon the growth and development of the plant, *Biol. Rev. and Biol. Proc. Cambridge Phil. Soc.*, **4**: 180–208.

Reed, E., 1923: Hypothesis of formative stuffs as applied to *Bryophyllum calycinum*, *Bot. Gaz.*, **75**: 113–142.

Reed, H. S., 1919: Growth and variability in *Helianthus*, *Am. J. Bot.*, **6**: 252–271.

——— and F. F. Halma, 1919: On the existence of a growth inhibiting substance in the Chinese lemon, *Univ. Calif. Pub.*, **4**: 99–112.

——— and ———, 1919: The evidence for a growth-inhibiting substance in the pear tree, *Plant World*, **22**: 239–247.

——— and R. H. Holland, 1919: The growth rate of an annual plant *Helianthus*, *Proc. Nat. Acad. Sci.*, **5**: 135–144.

———, 1920: Slow and rapid growth, *Am. J. Bot.*, **7**: 327–333.

———, 1920: The dynamics of a fluctuating growth rate, *Proc. Nat. Acad. Sci.*, **6**: 397–410.

———, 1920: The nature of the growth rate, *J. Gen. Physiol.*, **2**: 545–561.

———, 1921: A method for obtaining constants for formulas of organic growth, *Proc. Nat. Acad. Sci.*, **7**: 311–316.

———, 1921: Correlation and growth in the branches of young pear trees, *J. Agr. Res.*, **21**: 849–875.

———, 1923: A note on the statistics of cyclic growth, *Proc. Nat. Acad. Sci.*, **9**: 65–67.

——— and A. R. C. Haas, 1924: The effect of hydroxyl-ion concentration on the growth of walnut roots, *Am. J. Bot.*, **11**: 78–85.

———, 1927: Growth and differentiation in plants, *Quart. Rev. Biol.*, **2**: 79–101.

———, 1929: Quantitative aspects of the problem of growth and differentiation, *Proc. Int. Cong. Plant Sci.*, **2**: 1095–1106.

———, 1932: Criteria of plant growth and their reliability, *Philippine J. Agr.*, **3**: 155–161.

Reeves, R. G., and C. C. Valle, 1932: Anatomy and microchemistry of the cotton seed, *Bot. Gaz.*, **93**: 259–277.

Regeimbal, L. O., and R. B. Harvey, 1927: The effect of ethylene on the enzymes of pineapple, *J. Am. Chem. Soc.*, **49**: 1117–1118.

Rehm, W. S., 1936: Maintained electrical polarities in region of the axillary buds in *Phaseolus multiflorus*, *Plant Physiol.*, **11**: 365–382.

Reid, M. E., 1929: Relation of composition of seed and the effects of light on growth of seedlings, *Am. J. Bot.*, **16**: 747–769.

———, 1930: Growth and nitrogen metabolism of squash seedlings. I, Variations at different seasons of the year, *Am. J. Bot.*, **17**: 272–289.

Reinhard, M. C., and K. L. Tucker, 1928: Hourly variations in effect of equal Roentgen-ray doses on growth of *Vicia faba* seedlings, *Am. J. Roentgenol.*, **19**: 71–73; *Biol. Abst.*, **4**: 356.

Reynolds, E. S., 1935: The reactions of plants to ultra-violet, *Ann. Mo. Bot. Gard.*, **22**: 759–769.

Richardson, J. E., and H. L. Mayfield, 1932: Availability of vitamins in plant tissues, *Sci.*, **76**: 498–500.

Richey, F. D., and G. E. Sprague, 1932: Some factors affecting the reversal of sex expression in the tassel of maize, *Am. Nat.*, **66**: 433–443.

Rippel, A., 1922: Die gesetzmässige Erforschung von Reaktionsgleichgewicht und Reactionsgeschwindigkeit bei den höheren Pflanzen, *J. Landw.*, **70**: 9–44; *Bot. Abst.*, **12**: 472–473.

Robbins, W. J., 1922: Effect of autolyzed yeast and peptone on growth of excised root tips in the dark, *Bot. Gaz.*, **74**: 59–79.

——— and W. E. Maneval, 1923: Further experiments on the growth of excised root tips under sterile conditions, *Bot. Gaz.*, **76**: 274–287.

———, 1926: Relation of hydrogen-ion concentration to the growth of plants, *Mo. Sta. Bull.*, **244**: 30.

———, S. Brody, A. G. Hogan, C. M. Jackson, and C. W. Greene, 1928: "Growth," Yale University Press, 182 pp.

——— and K. F. Petsch, 1932: Moisture content and high temperature in relation to the germination of corn and wheat grains, *Bot. Gaz.*, **93**: 85–92.

———, M. Bartley, and V. B. White, 1936: Growth of fragments of excised root tips, *Bot. Gaz.*, **97**: 554–579.

Robertson, D. W., and A. M. Lute, 1933: Germination of the seed of farm crops in Colorado after storage for various periods of years, *J. Agr. Res.*, **46**: 455–462.

Robertson, T. B., 1907, 1908: On the normal rate of growth of an individual and its biochemical significance, *Arch. Entwicklungsmech. Organismen*, **25**: 580–614.

———, 1923: "The Chemical Basis of Growth and Senescence," J. B. Lippincott Company, Philadelphia, 389 pp.

Robinson, B. B., 1933: Some physiological factors influencing the production of flax fiber cells, *J. Am. Soc. Agron.*, **25**: 312–327.

Rodriguez, A. G., 1932: Influence of smoke and ethylene on the fruiting of the pineapple, *J. Puerto Rico Dept. Agr.*, **16**: 5–18.

Rosa, J. T., 1921: Investigation on the hardening process in vegetable plants, *Mo. Sta. Res. Bull.* 48.

———, 1925: Ripening of tomatoes with ethylene, *Proc. Am. Soc. Hort. Sci.*, **22**: 315–322.

———, 1928: Relation of tuber maturity and of storage factors to potato dormancy, *Hilgardia*, **3**: 99–124.

———, 1928: Effects of chemical treatments on dormant potato tubers, *Hilgardia*, **3**: 125–142.

———, 1928: Changes in composition during ripening and storage of melons, *Hilgardia*, **3**: 421–443.

Rose, D. H., 1915: A study of delayed germination in economic seeds, *Bot. Gaz.*, **59**: 425–444.

Rose, R. C., 1919: After-ripening and germination of seeds of *Tilia*, *Sambucus*, and *Rubus*, *Bot. Gaz.*, **67**: 281–309.

Rotunno, N. A., 1924: Effect of size of seed on plant production with special reference to radish, *Bot. Gaz.*, **78**: 397–413.

Rowlands, M. J., and B. Wilkinson, 1930: The vitamin B content of grass seeds in relationship to manures, *Biochem. J.*, **24**: 199–204.

Rudolfs, W., 1923: Influence of temperature and initial weight of seeds upon the growth-rate of *Phaseolus vulgaris* seedlings, *J. Agr. Res.*, **26**: 537–541.

Sachs, J., 1874: Ueber den Einfluss der Lufttemperatur und des Tageslichts auf die stündlichen und täglichen Aenderungen des Längenwachsthums der Internodien, *Arbeit. Bot. Inst. Würzburg*, **1**: 99–192.

Saeger, A., 1933: Gas injury to pure cultures of *Spirodela*, *Plant Physiol.*, **8**: 479–480.

Salmon, S. C., 1917: Why cereals winterkill, *J. Am. Soc. Agron.*, **9**: 353–380.

——— and F. L. Fleming, 1918: Relation of the density of cell sap to winter hardiness in small grains, *J. Agr. Res.*, **13**: 497–506.

———, 1933: Resistance of varieties of winter wheat and rye to low temperature in relation to winter hardiness and adaptation, *Kans. Sta. Tech. Bull.* 35.

Sampson, A. W., and E. C. McCarty, 1930: The carbohydrate metabolism of *Stipa pulchra*, *Hilgardia*, **5**: 61–100.

Savage, D. A., and L. A. Jacobson, 1935: The killing effect of heat and drought on buffalo grass and blue grama grass at Hays, Kansas, *J. Am. Soc. Agron.*, **27**: 566–583.

Sayre, J. D., 1932: Methods of determining bound water in plant tissue, *J. Agr. Res.*, **44**: 669–688.

SCHAFFNER, J. H., 1923: The influence of relative length of daylight on the reversal of sex in hemp, *Ecology*, **4**: 323–334.

———, 1927: Control of sex reversal in the tassel of Indian Corn, *Bot. Gaz.*, **84**: 440–449.

SCHAFFNIT, E., and M. LÜDTKE, 1932: The effect of cold on plant cells. II, Metabolism of cultivated plants at different temperatures and with changing nutrition, *Phytopath. Z.*, **4**: 329: *Chem. Abst.*, **27**: 2473.

——— and ———, 1932: Beiträge zur Kenntnis von Kältewirkungen auf die pflanzliche Zelle. II, Über den Stoffwechsel landwirtschaftlicher Kulturpflanzen bei verschiedenen Temperaturen und wechselnder Ernährung, *Phytopath. Z.*, **4**: 329–386.

——— and A. F. WILHELM, 1933: Beiträge zur Kenntnis von Kältewirkungen auf die pflanzliche Zelle. III, Kühlversuche mit verschieden ernährten Pflanzen und Untersuchungen über deren Stoffwechselphysiologie, *Phytopath. Z.*, **5**: 505–566.

SCHANZ, F., 1919: Wirkungen des Lichts verschiedener Wellenlänge auf die Pflanzen, *Ber. bot. Ges.*, **37**: 430–442.

SCHEIBE, A., 1932: Die Keimung des Hafers in ihrer Abhängigkeit von der physiologischen Konstitution des Saatgutes, *Pflanzenbau*, **8**: 579–649.

SCHLEY, E. O., 1913: Chemical and physical changes in geotropic stimulation and response, *Bot. Gaz.*, **56**: 480–490.

SCHMIDT, D., 1924: The effect of the weight of the seed on the growth of the plant, *N. J. Sta. Bull.* 404.

SCHRÖDER, D., 1909: Ueber den Verlauf des Welkens und die Lebenszähigkeit der Laubblätter, *diss., Göttingen*, 110 pp.

SEIFRIZ, W., 1936: Breakdown of fruit and vegetable tissue due to an electric current, *Plant Physiol.*, **11**: 195–200.

SELLSCHOP, J. P. F., and S. C. SALMON, 1928: The influence of chilling above the freezing point on certain crop plants, *J. Agr. Res.*, **37**: 315–339.

SEMMENS, E. S., 1930: Hydrolysis in the living plant by polarized light, *Bot. Gaz.*, **90**: 412–427.

SEUBERT, E., 1925: Über Wachstumsregulatoren in der Koleoptile von *Avena*, *Z. Bot.*, **17**: 49–88.

SHEAR, G. M., 1931: Studies on inanition in *Arachis* and *Phaseolus*, *Plant Physiol.*, **6**: 277–294.

SHEARD, C., and A. F. JOHNSON, 1930: The effects of infra-red, visible and ultra-violet irradiation on changes in electrical potentials and currents in plants, *Sci.*, **71**: 246–248.

SHIBUSAWA, M., and K. SHIBATA, 1927: The effect of electric discharges on the rate of growth of plants, *J. Inst. Elec. Eng. of Japan*, **473**: 1–42; *Biol. Abst.*, **4**: 2257–2258.

SHIPPY, W. B., 1930: Influence of environment on the callusing of apple cuttings and grafts, *Am. J. Bot.*, **17**: 29–327.

SHIRLEY, H. L., 1928: The influence of light intensity and quality upon the growth of plants, *Am. J. Bot.*, **15**: 621–622.

———, 1929: The influence of light intensity and light quality upon the growth of plants, *Am. J. Bot.*, **16**: 354–390.

———, 1931: The influence of light and temperature upon the utilization by young seedlings of organic reserves in the seed, *Am. J. Bot.*, **18**: 717–728.

———, 1932: Light intensity in relation to plant growth in a virgin Norway pine forest, *J. Agr. Res.*, **44**: 227–244.

SHUCK, A. L., 1934: Some factors influencing the germination of lettuce seed in laboratory practice, *N. Y. Sta. (Geneva) Tech. Bull.* 222.

———, 1935: Light as a factor influencing the dormancy of lettuce seeds, *Plant Physiol.*, **10**: 193–196.

———, 1935: A growth inhibiting substance in lettuce seeds, *Sci.*, **81**: 236.

SHULL, C. A., and H. B. LEMON, 1931: Penetration of seed coats by ultra-violet radiation, *Bot. Gaz.*, **92**: 420–429.

——— and J. W. MITCHELL, 1933: Stimulative effects of X-rays on plant growth, *Plant Physiol.*, **8**: 287–296.

SIDERIS, C. P., 1926: Behavior of pineapple plants in water cultures of varying hydrogen ion concentrations, *Exp. Sta. Assoc. Hawaiian Pineapple Canners, Univ. Hawaii Bull.* 4.

———, 1926: Review of physiological and pathological studies on the pineapple plant, *Hawaiian Pineapple Canners' Sta. Bull.* 8.

SIERP, H., and A. SEYBOLD, 1926: Untersuchungen über die Lichtempfindlichkeit der Spitze und des Stumpfes in der Koleoptile von *Avena sativa*, *Jahrb. wiss. Bot.*, **65**: 592–610.

SIFTON, H. B., 1920: Longevity of the seeds of cereals, clovers and timothy, *Am. J. Bot.*, **7**: 243–251.

SINGH, B. N., G. P. KAPOOR, and R. S. CHOUDHRI, 1936: Growth studies in relation of ultra-violet radiation, *Bot. Gaz,*, **97**: 649–665.

SINGH, T. C. N., 1932: A note on the response of grain (*Cicer arietinum*) seedlings to electricity, *New Phytol.*, **31**: 64–65.

SKOOG, F., 1934: The effect of X-rays on growth substance and plant growth, *Sci.*, **79**: 256.

——— and K. V. THIMANN, 1934: Further experiments on the inhibition of the development of lateral buds by growth hormone, *Proc. Nat. Acad. Sci.*, **20**: 480–485.

SMALL, J., 1929: Hydrogen-ion concentration in plant cell and tissues, *Protoplasma, Monographien*, II, 421 pp.

SMITH, A. J. M., and R. GANE, 1933: Influence of a gaseous product from apples on the germination of seeds, *Dept. Sci. and Ind. Res., Food Invest. Bd. Rept. (Gr. Brit.)*, **1932**: 156–158.

SMITH, O., and H. L. COCHRAN, 1935: Effect of temperature on pollen germination and tube growth in the tomato, *N. Y. Sta. (Cornell) Mem.* 175.

SNOW, R., 1925: The correlative inhibition of the growth of axillary buds, *Ann. Bot.*, **39**: 841–859.

———, 1929: The transmission of inhibition through dead stretches of stem, *Ann. Bot.*, **43**: 261–267.

———, 1929: The young leaf as the inhibiting organ, *New Phytol.*, **28**: 345–358.

———, 1931: Experiments on growth and inhibition. II, New phenomena of inhibition, *Proc. Roy. Soc.*, B, **108**: 305–317.

———, 1932: Experiments of growth and inhibition. III, Inhibition and growth promotion, *Proc. Roy. Soc.*, B, **111**: 86–105.

———, 1932: Growth-regulators in plants, *New Phytol.*, **31**: 336–354.

———, 1933: The nature of the cambial stimulus, *New Phytol.*, **32**: 288–296.

———, 1935: Activation of cambial growth by pure hormones, *New Phytol.*, **34**: 347–360.

SÖDING, H., 1926: Über den Einfluss der jungen Infloreszenz auf das Wachstum ihres Schaftes, *Jahrb. wiss. Bot.*, **65**: 611–635.

———, 1929: Weitere Untersuchungen über die Wuchshormone der Haferkoleoptile, *Jahrb. wiss. Bot.*, **71**: 184–213; *Biol. Abst.*, **5**: 1621.

———, 1931: Wachstum und Wanddehnbarkeit bei Haferkoleoptile, *Jahrb. wiss. Bot.*, **74**: 127–151.

SONAVNE, K. M., 1928: Longevity of crop seeds, *Agr. J. India*, **23**: 271–276.

SPAETH, J. N., 1932: Dormancy in *Tilia* seeds, *Sci.*, **76**: 143–144.

———, 1934: A physiological study of dormancy in *Tilia* seed, *N. Y. Sta.* (*Cornell*) *Mem.* 169.

SPRAGUE, G. F., 1934: Experiments on iarovizing corn, *J. Agr. Res.*, **48**: 1113–1120.

SPRAGUE, H. B., and M. LENZ, 1929: The effect of X-ray on potato tubers for "seeds," *Sci.*, **69**: 606.

STADLER, L. J., 1930: Some genetic effects of X-rays in plants, *J. Heredity*, **21**: 3–19.

STARKEY, E. J., 1932: Photoperiodism and chrysanthemum production, *Sci.*, **76**: 594–595.

STEINBAUER, G. P., 1932: Growth of tree seedlings in relation to light intensity and concentration of nutrient solution, *Plant Physiol.*, **7**: 742–744.

STEINBERG, R. A., 1931: An apparatus for growing plants under controlled environmental conditions, *J. Agr. Res.*, **43**: 1071–1084.

STEVENS, O. A., 1935: Germination studies of aged and injured seeds, *J. Agr. Res.*, **51**: 1093–1106.

STEWART, F. C., 1936: The relative vigor and productivity of potato plants from basal and apical sets cut from tubers in different stages of sprouting, *N. Y. Sta.* (*Geneva*) *Bull.* 658.

STILES, W., 1930: On the cause of cold death of plants, *Protoplasma*, **9**: 459–468.

———, 1935: Growth hormones, *Sci. Prog.*, **30**: 313–317.

STONE, W. E., 1933: Normal growth of potato leaves in greenhouse and field, *J. Agr. Res.*, **46**: 565–578.

STRAUSBAUGH, P. D., 1921: Dormancy and hardiness in the plum, *Bot. Gaz.*, **71**: 337–358.

SUMMERS, F., 1924: Factors governing bud formation: a chapter of plant physiology, *New Phytol.*, **23**: 20–50, 78–102, 113–131.

SUNESON, C. A., and G. L. PELTIER, 1934: Effect of stage of seedling development upon the cold resistance of winter wheats, *J. Am. Soc. Agron.*, **26**: 687–692.

SWINGLE, W. T., 1928: Metaxenia in the date palm, *J. Heredity*, **19**: 257–268.

———, 1931: Metaxenia or the direct effect of the pollen parent on the maternal fruit tissues and its application in date culture, abst., *Am. J. Bot.*, **18**: 894.

TAIT, G. M., and J. E. KNOTT, 1933: Correcting the unproductiveness of acid and alkaline muck soils for the growing of vegetable crops, *N. Y. Sta.* (*Cornell*) *Bull.* 572.

TANG, P., 1931: An experimental study of the germination of wheat seed under water as related to temperature and aeration, *Plant Physiol.*, **6**: 203–248.

TARANOVSKY, W. G., 1923: Über Veränderungen welche in Pflanzen durch das Kastrationverfahren hervorgerufen werden, *Zhur. Opytn. Agron.*, **23**: 127–164.

TETLEY, M., and J. H. PRIESTLEY, 1927: The histology of the coleoptile in relation to its phototropic response, *New Phytol.*, **26**: 171–186.

THATCHER, R. W., 1913: Enzymes of apples and their relation to the ripening process, *J. Agr. Res.*, **5**: 103–116.

THIMANN, K. V., and J. BONNER, 1932: Studies of the growth hormone of plants. II, The entry of growth substance into the plant, *Proc. Nat. Acad. Sci.*, **18**: 692–701.

——— and ———, 1933: The mechanism of the action of the growth substance of plants, *Proc. Roy. Soc.*, B, **113**: 126–149.

——— and H. E. DOLK, 1933: Conditions governing the production of the plant growth hormone by *Rhizopus* cultures, *Biol. Zent.*, **53**: 49–66.

——— and F. SKOOG, 1933: Studies on the growth hormone of plants. III, The inhibiting action of the growth substance on bud development, *Proc. Nat. Acad. Sci.*, **19**: 714–716.

———, 1934: Studies on the growth hormone of plants. VI, The distribution of the growth substance in plant tissues, *J. Gen. Physiol.*, **18**: 23–34.

——— and F. Skoog, 1934: On the inhibition of bud development and other functions of growth substance in *Vicia faba*, *Proc. Roy. Soc.*, B, **114**: 317–339.

——— and F. W. Went, 1934: On the chemical nature of the root-forming hormone, *Proc. K. Akad. Wetensch. Amsterdam*, **37**, No. 7.

———, 1935: Growth substances in plants, *Ann. Rev. Biochem.*, **4**: 545–568.

——— and J. B. Koepfli, 1935: Identity of the growth-promoting and root-forming substances of plants, *Nature*, **135**: 101–102.

Thoday, D., 1933: Some physiological aspects of differentiation, *New Phytol.*, **32**: 274–287.

Thompson, H. C., 1926: Results of sweet corn suckering experiments, *N. Y. Sta. (Cornell) Bull.*, **450**: 1–13.

———, H. S. Mills, and P. H. Wessels, 1930: Results of sweet corn suckering experiments on Long Island, *N. Y. Sta. (Cornell) Bull.*, **509**: 3–11.

Thompson, R. C., 1935: Some factors associated with dormancy of lettuce seed, *Proc. Am. Soc. Hort. Sci.*, **33**: 610–616.

Thomson, J. R., 1936: Vernalisation, *Sci. Prog.*, **30**: 644–651.

Thornton, N. C., 1930: The use of carbon dioxide for prolonging the life of cut flowers with especial reference to roses, *Am. J. Bot.*, **17**: 614–626.

———, 1931: The effect of carbon dioxide on fruits and vegetables in storage, *Contr. Boyce Thompson Inst.*, **3**: 219–244.

———, 1933: Carbon dioxide storage. V, Breaking the dormancy of potato tubers, *Contr. Boyce Thompson Inst.*, **5**: 471–481.

———, 1935: Factors influencing germination and development of dormancy in cocklebur seeds, *Contr. Boyce Thompson Inst.*, **7**: 477–496.

———, 1936: Carbon dioxide storage. IX, Germination of lettuce seeds at high temperatures in both light and darkness, *Contr. Boyce Thompson Inst.*, **8**: 25–40.

Timmons, F. L., and S. C. Salmon, 1932: The resistance of certain varieties and regional strains of alfalfa to controlled low temperatures, *J. Am. Soc. Agron.*, **24**: 642–655.

Tincker, M. A. H., 1925: The effect of length of day upon the growth and reproduction of some economic plants, *Ann. Bot.*, **39**: 721–754.

———, 1928: The effect of length of day upon the growth and chemical composition of the tissues of certain economic plants, *Ann. Bot.*, **42**: 101–140.

———, 1929: The effect of length of day on plants, *J. Roy. Hort. Soc.*, **54**: 354–378.

———, 1930: Some experiments with ultra-violet ray glasses, *J. Roy. Hort. Soc.*, **55**: 79–87.

———, 1932: Response of some common garden plants to the daily period of light, *J. Roy. Hort. Soc.*, **57**: 321–325.

———, 1935: Experiments with follicular and other hormones and plant growth, *Ann. App. Biol.*, **22**: 619–629.

———, 1935: Rose seeds: their after-ripening and germination, *J. Roy. Hort. Soc.*, **60**: 399–417.

Tjebbes, K., and J. C. Uphof, 1921: Der Einfluss des elektrischen Lichtes auf das Pflanzenwachstum, *Landw. Jahrb.*, **56**: 315–328; *Bot. Abst.*, **10**: 206.

Tottingham, W. E., R. G. Shands, and E. D. Delwiche, 1931: Test of Chibnall's method of extraction for investigating winter hardiness of plants, *Plant Physiol.*, **6**: 167–176.

Trumpf, C., 1924: Ueber den Einfluss intermittierender Belichtung auf das Etiolement der Pflanzen, *Bot. Arch.*, **5**: 381–410. Quoted by Priestley, 1926.

TSCHIRCH, A., 1921: Besitzt die Pflanze Hormone? *Viertelijahrsschr. Naturforsch. Ges. Zurich*, **66**: 201–211; *Bot. Abst.*, **12**: 465–466.

TUFTS, W. P., and C. J. HANSEN, 1933: Xenia and metaxenia in Bartlett pear, *Proc. Am. Soc. Hort. Sci.*, **30**: 134–139.

TUKEY, H. B., and K. D. BRASE, 1933: Influence of the scion and of an intermediate stem piece upon the character and development of roots of young apple trees, *N. Y. Sta. (Geneva) Bull.* 218.

TURNER, M. A., 1929: Ginning percentage and lint index of cotton in relation to the number of cotton fibers per seed, *J. Text. Ind.*, **20**: 233–273.

TYSDAL, H. M., 1933: Influence of light, temperature and soil moisture on the hardening process in alfalfa, *J. Agr. Res.*, **46**: 483–515.

UMRATH, K., 1928: Elektrische Potentiale pflanzlicher Gewebe, *Protoplasma*, **4**: 539–546.

URSPRUNG, A., and G. BLUM, 1924: Eine Methode zur Messung des Wand- und Turgordruckes der Zelle, nebst Anwendungen, *Jahrb. wiss. Bot.*, **63**: 1–110.

VACHA, G. A., and R. B. HARVEY, 1927: The use of ethylene, propylene, and similar compounds in breaking the rest period of tubers, bulbs, cuttings, and seeds, *Plant Physiol.*, **2**: 187–192.

VAN DE SANDE-BAKHUYZEN, H. L., 1926: Growth and growth formulas in plants., *Sci.*, **64**: 653–654.

———, 1926: Physiological phenomena at the time of flowering, *Proc. Soc. Exp. Biol. and Med.*, **24**: 143–145.

——— and C. L. ALSBERG, 1927: The growth curve in annual plants, *Physiol. Rev.*, **7**: 151–187.

———, 1928: Studies upon wheat grown under constant conditions. I, II, *Plant Physiol.*, **3**: 1–31.

———, E. P. GRIFFING, and C. L. ALSBERG, 1937: "Studies on Wheat Growth under Constant Conditions," *Food Res. Inst., Stanford Univ.*, Calif., 400 pp.

VAN OVERBEEK, J., 1935: The growth hormone and the dwarf type of growth in corn, *Proc. Nat. Acad. Sci.*, **21**: 292–299.

———, 1936: Different action of auxin-*a* of hetero-auxin (preliminary note), *Proc. Nat. Acad. Sci.*, **22**: 187–190.

VAN WEIJ, H. G., 1932: Der Mechanismus des Wuchsstofftransporte, *Rec. trav. bot. Néerl.*, **29**: 379–496.

VASSILIEV, I. M., and M. G. VASSILIEV, 1936: Changes in carbohydrate content of wheat plants during the process of hardening for drought resistance, *Plant Physiol.*, **11**: 115–125.

VICKERY, H. B., and G. W. PUCHER, 1931: The chemical changes that occur during the curing of tobacco leaves, *Sci.*, **73**: 397–399.

VIRTANEN, A. I., 1936: Vitamins and plants, *Nature*, **137**: 779–780.

VOGT, E., 1915: Über den Einfluss des Lichts auf das Wachstum der Koleoptile von *Avena sativa*, *Z. Bot.*, **7**: 193–270; *Bot. Gaz.*, **60**: 423.

VON EULER, H., and T. PHILIPSON, 1932: Water soluble growth factors, *Biochem. Z.*, **245**: 418–439; *Chem Abst.*, **26**: 3281.

VYSOTSKII, K. A., 1928: The electro-ionization of the cotton plant, *Proc. Turkestan Plant Breeding Sta.*, **13**: 109–110; *Biol. Abst.*, **8**: 603.

WALDO, G. F., 1930: Fruit-bud development in strawberry varieties and species, *J. Agr. Res.*, **40**: 393–409.

———, 1930: Fruit-bud formation in everbearing strawberries, *J. Agr. Res.*, **40**: 409–417.

WALDRON, L. R., 1931: Frost injury to spring wheat with a consideration of drought resistance, *J. Am. Soc. Agron.*, **23**: 625–638.

———, 1932: Differences of injury by frost to wheat plants grown comparably, *J. Am. Soc. Agron.*, **24**: 494–500.

Wallace, R. H., 1927: The production of intumescences in transparent apple by ethylene gas as affected by external and internal conditions, *Bull. Torrey Bot. Club*, **54**: 499–542.

———, 1931: Studies on the sensitivity of *Mimosa pudica.* I, The effect of certain animal anesthetics upon sleep movements, *Am. J. Bot.*, **18**: 102–111.

———, 1931: Studies on the sensitivity of *Mimosa pudica.* II, The effect of animal anesthetics and certain other compounds upon seismonic sensitivity, *Am. J. Bot.*, **18**: 215–235.

———, 1931: Studies on the sensitivity of *Mimosa pudica.* III, The effect of temperature, humidity, and certain other factors upon seismonic sensitivity, *Am. J. Bot.*, **18**: 288–307.

Waller, A. D., 1900: The electrical effects of light upon green leaves, *Proc. Roy. Soc.*, **67**: 129–137.

Waller, J. C., 1924: On the fluctuations of potential in green leaves under the influence of light, *J. Physiol.*, **58**, *Proc.* XXXVIII–XXXIX.

———, 1925: Plant electricity. I, Photoelectric currents associated with the activity of chlorophyll in plants, *Ann. Bot.*, **39**: 515–538.

———, 1929: Plant electricity. II, Towards an interpretation of the photoelectric currents of leaves, *New Phytol.*, **28**: 291–302.

Walster, H. L., 1920: Formative effect of high and low temperatures upon growth of barley: A chemical correlation, *Bot. Gaz.*, **69**: 97–126.

Walter, H., 1923: Protoplasma- und Membranquellung bei Plasmolyse, *Jahrb. wiss. Bot.*, **62**: 145–213.

———, 1929: Die osmotischen Werte und die Kälteschäden unserer wintergrünen Pflanzen während der Winterperiode 1929, *Ber. bot. Ges.*, **47**: 338–348.

Waltman, C. S., 1931: Effect of hydrogen-ion concentration on the growth of strawberries in sand and in soil, *Ky. Sta. Bull.*, **321**: 331–352.

Walton, C. F., and C. A. Fort, 1931: Mannite and dextran in the jelly of molasses from juice of frozen and deteriorated cane, *Ind. and Eng. Chem.*, **23**: 1295–1297.

Wann, F. B., and E. F. Hopkins, 1927: Further studies on growth of *Chlorella* as affected by hydrogen-ion concentration, *Bot. Gaz.*, **83**: 194–201.

Wanser, H. M., 1922: Photoperiodism of wheat: a determining factor in acclimatization, *Sci.*, **56**: 313–315.

Weaver, J. E., and W. J. Himmel, 1929: Relation between the development of root systems and shoot under long- and short-day illumination, *Plant Physiol.*, **4**: 435–459.

Weimer, J. L., 1929: Some factors involved in the winter killing of alfalfa, *J. Agr. Res.*, **39**: 263–283.

Welton, F. A., 1921: Longevity of seeds, *Ohio Sta. Mo. Bull.*, **6**: 18–25.

Went, F. A. F. C., 1933: Growth substance (auxin) in plants, *Nature*, **132**: 452.

Went, F. W., 1926: On growth accelerating substances in the coleoptile of *Avena sativa*, *Proc. K. Akad. Wetensch. Amsterdam Verslag Gewone Vergader. Afd. Natuurk*, **35**: 723–732; *Biol. Abst.*, **7**: 1293.

———, 1928: Wuchsstoff und Wachstum, *Rec. trav. bot. Néerl.*, **25**: 1–116.

———, 1929: On a substance causing root formation, *Proc. K. Akad. Wetensch. Amsterdam*, **32**: 35–39; *Exp. Sta. Rec.*, **67**: 661.

———, 1934: A test method for Rhizocaline, the root forming substance, *Proc. K. Akad. Wetensch. Amsterdam*, **37**, No. 7.

———, 1934: On the pea test method for auxin, the plant growth hormone, *Proc. K. Akad. Wetensch, Amsterdam*, **37**, 547–555; *Exp. Sta. Rec.*, **72**: 457.

———, 1935: Auxin, the plant growth-hormone, *Bot. Rev.*, **1**: 162–182.

——— and K. V. THIMANN, 1937: "Phytohormones," The Macmillan Company, New York and London, XI + 294 pp.

WESSELS, P. H., 1932: Soil-acidity studies with potatoes, cauliflower and other vegetables on Long Island, *N. Y. Sta. (Cornell) Bull.* 536.

WEST, C., G. E. BRIGGS, and F. KIDD, 1920: Methods and significant relations in a quantitative analysis of plant growth, *New Phytol.*, **19**: 200–207.

———, 1924: Effects of freezing and low temperatures upon plants, *Sci. Prog.*, **18**: 386–392.

WHITE, P. R., 1931: Plant tissue cultures. The history and present status of the problem, *Arch. Exp. Zellforsch.*, **10**: 501–518.

———, 1932: Influence of some environmental conditions on the growth of excised root tips of wheat seedlings in liquid media, *Plant Physiol.*, **7**: 613–628.

———, 1933: Plant tissue cultures. Results of preliminary experiments on the culturing of isolated stem-tips of *Stellaria media*, *Protoplasma*, **19**: 97–116.

———, 1933: The SH radical and some other sources of sulfur as affecting growth of isolated root tips of wheat seedlings, *Protoplasma*, **19**: 132–135.

———, 1933: Liquid media as substrata for the culturing of isolated root tips, *Biol. Zent.*, **53**: 359–364.

———, 1933: Concentrations of inorganic ions as related to growth of excised root tips of wheat seedlings, *Plant Physiol.*, **8**: 489–508.

———, 1934: Potentially unlimited growth of excised tomato tips in a liquid medium, *Plant Physiol.*, **9**: 585–600.

WHYTE, R. O., and P. S. HUDSON, 1933: Vernalization, *Brit. Imp. Bur. Plant Genetics*, **9**: 1–27.

WICKWIRE, G. C., L. KNEER, and W. E. BURGE, 1931: The relation of quantity of stimulating effect of insulin, glycerol, urea and dipotassium phosphate on sugar utilization by *Spirogyra*, abst., *Am. J. Bot.*, **18**: 894–895.

WIGODER, S. B., and R. PATTEN, 1929: Some effects of Röntgen rays on seedlings, *Proc. Roy. Irish Acad.*, **39B**: 146–155; *Biol. Abst.*, **6**: 704.

WILLIAMS, C. B., 1912: Corn sucker investigations, *N. C. Sta. Rept.*, **35**: 16.

——— and W. C. ETHERIDGE, 1912: Economic value of corn suckers, *N. C. Sta. Rept.*, **35**: 37–40.

WILLIAMS, M., 1923: Observations on the action of X-rays on plant cells, *Ann. Bot.*, **37**: 217–225.

———, 1925: Some observations on the action of radium on certain plant cells, *Ann. Bot.*, **39**: 547–562.

WILLIAMS, R. J., 1928: Nutrilites, *Sci.*, **67**: 607–608.

———, C. M. LYMAN, G. H. GOODYEAR, J. H. TRUESDAIL, and D. HOLADAY, 1933: "Pantothenic acid"; a growth determination of universal biological occurrence, *J. Am. Chem. Soc.*, **55**: 2912–2927.

WILSON, A. L., 1930: Hydrogen-ion concentration in relation to the growth of onions, *Proc. Am. Soc. Hort. Sci.*, **27**: 524–528.

WITHROW, R. B., and M. W. RICHMAN, 1933: Artificial radiation as a means of forcing greenhouse crops, *Ind. Sta. Bull.* 380.

———, 1934: Plant-forcing with electric lights, *Ind. Sta. Circ.* 206.

———, 1936: Light and its effects on plant growth, *Agr. Eng.*, **17**: 150–152.

——— and H. M. BENEDICT, 1936: Photoperiodic responses of certain green annuals as influenced by intensity and wave length of artificial light used to lengthen the daylight period, *Plant Physiol.*, **11**: 225–250.

WOFFENDEN, L. M., and J. H. PRIESTLEY, 1924: The toxic action of coal gas upon plants. II, The effect of coal gas upon cork and lenticel formation, *Ann. App. Biol.*, **11**: 42–54.

WOLFE, H. S., 1931: Effect of ethylene on the ripening of bananas, *Bot. Gaz.*, **92**: 337–367.

WOLFE, T. K., and M. S. KIPPS, 1926: Comparative value of scarified and of unhulled seed of biennial white sweet clover for hay production, *J. Am. Soc. Agron.*, **18**: 1127–1129.

WOODS, J. J., 1935: Availability of nutrients in raspberry plots in relation to winter injury, *Sci. Agr.*, **16**: 1–7.

WORZELLA, W. W., 1935: Inheritance of cold resistance in winter wheat with preliminary studies on the technic of artificial freezing tests, *J. Agr. Res.*, **50**: 625–635.

WRIGHT, R. C., W. T. PENTZER, T. M. WHITEMAN, and D. H. ROSE, 1931: Effect of various temperatures on the storage and ripening of tomatoes, *U. S. Dept. Agr. Tech. Bull.* 268.

——— and W. M. PEACOCK, 1934: Influence of storage temperatures on the rest period and dormancy of potatoes, *U. S. Dept. Agr. Tech. Bull.* 424.

WRIGHT, W., 1933: Radial growth of the xylem and the starch reserves of *Pinus sylvestris:* a preliminary survey, *New Phytol.*, **32**: 77–96.

WYND, F. L., and H. J. FULLER, 1931: Some effects of ultra-violet radiation upon the calcium and phosphorus content of higher plants, *Ann. Mo. Bot. Gard.*, **18**: 565–572.

ZACHAROWA, T. M., 1925: Über den Einfluss niedriger Temperaturen auf die Pflanzen, *Jahrb. wiss. Bot.*, **65**: 61–87; *Exp. Sta. Rec.*, **57**: 419.

ZIMMERMAN, P. W., and A. E. HITCHCOCK, 1929: Root formation and flowering of dahlia cuttings when subjected to different day lengths, *Bot. Gaz.*, **87**: 1–14.

——— and W. CROCKER, 1930: The effect of sulphur dioxide fumes on plants, abst., *Am. J. Bot.*, **17**: 1042.

———, ———, and A. E. HITCHCOCK, 1930: The response of plants to illuminating gas, *Proc. Am. Soc. Hort. Sci.*, **27**: 53–56.

———, A. E. HITCHCOCK, and W. CROCKER, 1931: The effect of ethylene and illuminating gas on roses, *Contr. Boyce Thompson Inst.*, **3**: 459–481.

———, W. CROCKER, and A. E. HITCHCOCK, 1932: Initiation and stimulation of roots from exposure to carbon monoxide, abst., *Am. J. Bot.*, **19**: 846.

———, ———, and ———, 1932: The effect of carbon monoxide on plants, abst., *Am. J. Bot.*, **19**: 847.

———, ———, and ———, 1933: Initiation and stimulation of roots from exposure of plants to carbon monoxide gas, *Contr. Boyce Thompson Inst.*, **5**: 1–18.

———, ———, and ———, 1933: The effect of carbon monoxide on plants, *Contr. Boyce Thompson Inst.*, **5**: 195–211.

——— and A. E. HITCHCOCK, 1933: Initiation and stimulation of adventitious roots caused by unsaturated hydrocarbon gases, *Contr. Boyce Thompson Inst.*, **5**: 351–369.

——— and W. CROCKER, 1934: Plant injury caused by vapors of mercury and compounds of mercury, *Contr. Boyce Thompson Inst.*, **6**: 167–187.

——— and ———, 1934: Toxicity of air containing sulfur dioxide gas, *Contr. Boyce Thompson Inst.*, **6**: 471–478.

——— and F. WILCOXON, 1935: Several chemical growth substances which cause initiation of roots and other responses in plants, *Contr. Boyce Thompson Inst.*, **7**: 209–229.

——— and A. E. HITCHCOCK, 1935: The response of roots to "root forming" substances, *Contr. Boyce Thompson Inst.*, **7**: 439–445.

———, ———, and F. WILCOXON, 1936: Several esters as plant hormone, *Contr. Boyce Thompson Inst.*, **8**: 103–112.

ZINKERNAGEL, H., 1932: Über den Einfluss des Leuchtgases auf Wurzeln von *Allium cepa, Ber. bot. Ges.*, **50**: 134–153; *Biol. Abst.*, **8**: 1660.

AUTHOR INDEX

(**Boldface** type indicates bibliographic reference)

A

D

G

H

I

J

K

L

M

T

U

V

Y

SUBJECT INDEX

B

C

D

F

G

M

N

O

T